CALCULUS
from Graphical, Numerical, and Symbolic Points of View
VOLUME 2

ARNOLD OSTEBEE AND PAUL ZORN
St. Olaf College

1995
PRELIMINARY
EDITION

SAUNDERS COLLEGE PUBLISHING
Harcourt Brace College Publishers

Fort Worth Philadelphia San Diego New York Orlando Austin
San Antonio Toronto Montreal London Sydney Tokyo

This material is based upon work supported by the National Science Foundation under Grant No. USE-9053363.

Any opinions, findings, and conclusions or recommendations expressed in this material are those of the authors and do not necessarily reflect the views of the National Science Foundation.

Ostebee/Zorn: CALCULUS FROM GRAPHICAL, NUMERICAL, AND SYMBOLIC POINTS OF VIEW, VOL. 2, Preliminary Edition

ISBN 0-03-010603-6

456 018 987654321

Cover credit: Steve Gottlieb © 1993

Contents

NOTE: *This is Volume 2 of the 1995 preliminary edition of* Calculus from Graphical, Numerical, and Symbolic Points of View, *by Arnold Ostebee and Paul Zorn. The two volumes, together, comprise a text for a one-year course in elementary calculus of functions of a single variable.*

This volume contains, in addition to Chapters 5–13, several sections from Chapters 3 and 4. These sections cover topics that instructors may not have treated in Calculus I (e.g., inverse trigonometric functions), or topics that, in our judgment, students can benefit from reviewing (e.g., the idea of a differential equation).

About this book: notes for instructors

This book aims to do exactly what its title suggests: treat calculus from graphical, numerical, and symbolic points of view. In this preface we elaborate briefly on what this means.

Philosophy

Several common threads run through *Toward a Lean and Lively Calculus*, *Calculus for a New Century*, discussions at many professional meetings, the report of the NCTM/MAA Joint Task Force on Curriculum for Grades 11–13, and the NCTM *Curriculum and Evaluation Standards for School Mathematics*. First, there is a consistent call for *leaner and more conceptual courses*, driven by and focused on central ideas. Second, there is a realization that courses should *reflect modern technology both in content and in pedagogy*.

A diagnosis. Many calculus courses, we believe, slight the conceptual foundations of the subject and overemphasize routine techniques—formal differentiation, antidifferentiation, convergence testing, etc. Analytic objects (integral, derivative, convergence, etc.) are represented and manipulated only algebraically (i.e., *via* symbolic manipulation of explicit elementary functions). For example, textbooks often treat limits, derivatives, and integrals—all *analytic* objects—only as *algebraic* operations on *algebraic* functions. We try to take a broader view.

Whether one views calculus as an introduction to pure mathematics or as a foundation for applications (or both!), the conclusion is the same—concepts, not techniques, are truly fundamental to the course. Whatever uses they make of the calculus, students need more than a compendium of manipulative techniques. The *sine qua non* for a useful command of the calculus is a conceptual understanding that is deep and flexible enough to accommodate diverse applications.

A prescription. Our key strategy for improving conceptual understanding is combining, comparing, and moving among graphical, numerical, and algebraic "representations" of central concepts. This strategy pervades and unifies our exposition. Bringing graphical and numerical, as well as algebraic, viewpoints to bear on calculus ideas is the philosophical foundation of our text. By representing and manipulating calculus ideas and objects graphically, numerically, and algebraically, we believe that students gain a better, deeper, and more useful understanding.

The role of technology

Combining graphical, numerical, and symbolic viewpoints in calculus can be forbiddingly time-consuming and distracting without technological assistance. With technology, these viewpoints become practically accessible; hence, we use technology freely. So should our students.

Having said this, we emphasize that we do *not* intend to "automate" or "computerize" the calculus. We regard computers and graphing calculators strictly as instructional *tools*—albeit very powerful ones—to facilitate the crucial combination of graphical, numerical, and algebraic viewpoints. Our text exploits the capabilities of these tools to help students focus on the ideas that lie at the heart of the calculus.

How much technology? What kinds?

This text freely uses and refers to numerical and graphical computations, but it is independent of any particular technology. Any of the familiar high-level products—*Mathematica, Maple, Derive*—are certainly sufficient, but hardly necessary. Many graphing calculators (e.g., the TI-81, 82, and 85, the HP-48G, etc.) and special-purpose microcomputer software packages are adequate.

Our use of technology is most conveniently described in terms of *functionalities*. The requirements for Calculus I and Calculus II differ somewhat:

Calculus I: Chapters 1–5 (the traditional content of Calculus I) draw freely on machine graphics; almost any up-to-date *form* will do. Most graphing calculators would suffice; so would almost any flexible microcomputer graphing program, such as *MasterGrapher* or *MicroCalc*.

Calculus II: To make the best use of the remaining (Calculus II—not covered in this volume) material, students will require access to a modest level of numerical computation (mainly for estimating analytic quantities: integrals, series, etc.) Many microcomputer software packages provide the necessary functionality; so do some programmable graphing calculators. In addition, access to simple symbolic operations (e.g., formal differentiation, Taylor series expansion) is desirable, but not strictly necessary.

Symbol manipulation and hand computation

Although we emphasize concepts and use technology freely, we by no means ignore symbol manipulations in general, or hand computations in particular. We take the symbolic points of view of our title fully as seriously as the graphical and the numerical. We cover, for instance, the usual techniques for formal differentiation and antidifferentiation. Why do we do so?

That machines can do calculus manipulations does not render by-hand operations obsolete. Some manipulative practice and skill builds and supports conceptual understanding. Hand computation illustrates concepts concretely, builds "symbol sense" (the algebraic counterpart of numerical intuition) and an ability to estimate, and gives students a sense of mastery. It does *not* follow, though, that harder, more baroque computational problems are necessarily better or more useful; we deemphasize them. Both research results and our own experience suggest that diverting some time and attention to concepts does little, if anything, to reduce students' hand computational facility.

Distinctive features of the text

Our text differs significantly from standard treatments. Here are some of these differences, together with some of our assumptions, goals, and strategies.

Combining graphical, numerical, and algebraic viewpoints. Throughout the text we insist that students manipulate and compare graphical, numerical, and algebraic representations of mathematical objects. In studying functions, for example, students manipulate not only elementary functions but also functions presented graphically and tabularly. In the context of formal differentiation, exercises ask students to apply the chain rule to combinations of functions presented in various ways. Graphical and numerical techniques, with error estimates, complement algebraic antidifferentiation methods. For series, routine convergence tests are emphasized less than finding—and defending—numerical limit estimates.

A mainstream course. Our main strategy—combining graphical, numerical, and algebraic viewpoints to clarify concepts and make them concrete—aims *explicitly* at mainstream students, who especially need such help. We regard a more conceptual calculus as also more applicable. To *use* calculus ideas and techniques, students must know what they are doing and why, not merely how. Thus, our text is appropriate for a general audience: mathematics majors, science and engineering majors, and non-science majors.

Concepts vs. rigor. Proving theorems in full generality is less valuable, we think, than helping students understand concretely what theorems say, why they're reasonable, and why they matter. Too often, fully rigorous proofs address questions that students are unprepared to ask.

Still, we believe that introducing calculus students to the idea of proof—and to some especially important classical proofs—is essential. We prove major results, but *emphasize* only those that we believe contribute significantly to understanding calculus concepts. In examples and problems, too, we pay attention to developing analytic skills and synthesizing mathematical ideas.

More varied exercises. However clear its exposition, a textbook's problems generate most of students' mathematical activity and occupy most of their time. Through the problems we assign, we tell students concretely what we think they should know and do.

Routine drills and challenging theoretical problems are standard in texts; ours includes some of both. More distinctive, though, are problems that fall *between* these poles:

- Problems that combine and compare algebraic, graphical, and numerical viewpoints and techniques.

- Problems that require "translation" among various representations and interpretations of calculus ideas (e.g., to interpret derivative information in terms of either slope or rate of change).

- Problems that use calculus as a *language* for interpreting and solving problems. Students are asked to translate problems into mathematical terms, solve these problems using the tools of calculus, and reinterpret mathematical results in the context of the original problem.

Strategies for better problem solving. Emphasizing problem solving is nowadays *de rigeur* in calculus textbooks; ours is no exception. What, concretely, does such an emphasis entail in content and strategy?

Conceptual understanding, we believe, is the weakest link in students' ability to solve nontrivial problems. Thus, *the goal of better problem solving skills is implicit in the goal of deeper conceptual understanding*. To that end:

- We use numerical and graphical methods both to improve students' understanding of concepts and to enlarge their kit of tools to solve problems. Students, we find, are surprisingly quick to master and apply elementary numerical and graphical methods, even error estimation.

- When technology is available, students are freed, indeed forced, to *analyze* the structure of a problem and *plan* a solution strategy. We emphasize general problem-solving strategies (reduction to more tractable subproblems, estimation, search for patterns, etc.) explicitly wherever we can.

- We provide a greater *qualitative* variety of exercises including problems that are posed more generally, problems that call for more synthesis, and problems that rely on a larger set of solution techniques. In this richer environment, we hope students will come to see mathematics as an open-ended, creative activity, not a rigid collection of recipes.

Advice from you

We appreciate hearing instructors' comments, suggestions, and advice on this preliminary edition. Many of the suggestions received from users of earlier versions of this text have been incorporated in this version. Our physical and e-mail addresses are below.

Arnold Ostebee and Paul Zorn
Department of Mathematics
St. Olaf College
1520 St. Olaf Avenue
Northfield, Minnesota 55057-1098

e-mail: ostebee@stolaf.edu zorn@stolaf.edu

October, 1994

How to use this book: notes for students

All authors want their books to be *used*: read, studied, thought about, puzzled over, reread, underlined, disputed, understood, and, ultimately, enjoyed. So do we.

That might go without saying for *some* books—beach novels, user manuals, field guides, etc.—but it may need repeating for a calculus textbook. We know as teachers (and remember as students) that mathematics textbooks are too often read *backwards*: faced with Exercise 231(b) on page 1638, we've all shuffled backwards through the pages in search of something similar. (Very often, moreover, our searches were rewarded.)

A calculus textbook isn't a novel. It's a peculiar hybrid of encyclopedia, dictionary, atlas, anthology, daily newspaper, shop manual, *and* novel—not exactly light reading, but essential reading nevertheless. Ideally, a calculus book should be read in *all* directions: left to right, top to bottom, back to front, and even front to back. That's a tall order. Here are some suggestions for coping with it.

Read the narrative. Each section's narrative is designed to be read from beginning to end. The examples, in particular, are supposed to illustrate ideas and make them concrete—not just serve as templates for homework exercises.

Read the examples. Examples are, if anything, more important than theorems, remarks, and other "talk." We use examples both to show already-familiar calculus ideas "in action," and to set the stage for new ideas.

Read the pictures. We're serious about the "graphical points of view" mentioned in our title. The pictures in this book are not "illustrations" or "decorations." Pictures like this one–

Graph of $y = x \sin(1/x)$ **near** $x = 0$

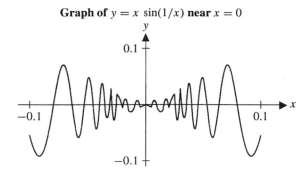

are everywhere, even in the middle of sentences. That's intentional: graphs are an important part of the language of calculus. An ability to think "pictorially"—as well as symbolically and numerically—about mathematical ideas may be the most important benefit calculus can offer.

Often they're put in margin notes, like this one.

Read with a calculator and pencil. This book is full of requests◄◄ to check a calculation, sketch a graph, or "convince yourself" that something makes sense. Take these "requests" seriously. Mastering mathematical ideas takes more than reading; it takes doing, drawing, and thinking.

Read the language. Mathematics is not a "natural language" like English or French, but it has its own vocabulary and usage rules. Calculus, especially, relies on careful use of technical language. Words like **rate, amount, concave, stationary point,** and **root** have precise, agreed-upon mathematical meanings. Understanding such words goes a long way toward understanding the mathematics they convey; misunderstanding the words leads inevitably to confusion. Whenever in doubt, consult the index.

Read the appendices. The human appendix generally lies unnoticed—unless trouble starts, when it's taken out and thrown away. Don't treat *our* appendices that way. Though perhaps slightly enlarged, they're full of healthy matter: reviews of precalculus topics, help with "story problems," proofs of various kinds, even a graphical "atlas" of functions. Used as directed the appendices will help appreciably in digesting the material.

Read the instructors' preface (if you like). Get a jump on your teacher.

In short: *read the book.*

A last note

Why study calculus at all? There are plenty of good practical and "educational" reasons: because it's good for applications; because higher mathematics requires it; because it's good mental training; because other majors require it; because jobs require it. Here's another reason to study calculus: because calculus is among our species' deepest, richest, farthest-reaching, and most beautiful intellectual achievements. We hope this book will help you see it in that spirit.

A last request

Last, a request. We sincerely appreciate—and take very seriously—students' opinions, suggestions, and advice on this book. We invite you to offer your advice, either through your teacher or by writing us directly. Our addresses appear below.

Arnold Ostebee and Paul Zorn
Department of Mathematics
St. Olaf College
1520 St. Olaf Avenue
Northfield, Minnesota 55057-1098

October, 1994

Chapter 5

The Integral

5.1 Areas and integrals

The **tangent line problem** and the **area problem** are the two big geometric problems of calculus.➤ As we've seen, the idea of *derivative*, together with the rules for computing derivatives, solves the tangent line problem. For an impressive variety of functions, even complicated ones, it's by now a routine matter to calculate the slope of the tangent line at any point on the graph.

These problems have been big for many centuries: the Greeks worked hard, and ingeniously, at both. Had they had our algebraic advantages, they would have gone farther than they did.

The area problem and the integral

The general area problem asks how to measure the area of a plane region. For special regions—rectangles, triangles, squares, circles, trapezoids, etc.—well known➤ formulas do the job.

And long *known—the Greeks knew them.*

Area problems are more challenging—and far more interesting—for regions bounded by less familiar curves, such as graphs of functions. The most important such area problem, phrased in calculus language, is to measure an area like that shown shaded in the "generic" picture below:

What's the shaded area?

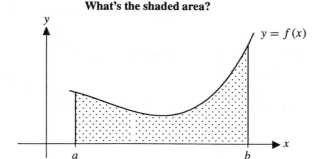

Stated carefully, the problem reads like this:

> *Find the area of the region bounded above by the graph of f, below by the x-axis, on the left by the vertical line $x = a$, and on the right by the vertical line $x = b$.*

That's a mouthful—we'll usually say just "the area under the graph of f from a to b."➤

The shorthand phrase is less tedious, but also less precise. If in doubt use the full form.

Signed area

In this chapter it will often be convenient to consider **signed area**, as opposed to area in the everyday sense. The adjective "signed" means that *area below the x-axis counts as negative.* (In the picture above all the shaded area is above the *x*-axis, so the question of sign doesn't arise.) We'll give more details below.

A compact notation for signed area is even more convenient:

Definition: (The integral as signed area.) Let f be a function defined for $a \le x \le b$. Either of the equivalent expressions

$$\int_a^b f \quad \text{or} \quad \int_a^b f(x)\, dx$$

denotes the signed area bounded by $x = a$, $x = b$, $y = f(x)$, and the *x*-axis.

Note:

In words. For either $\int_a^b f(x)\,dx$ or $\int_a^b f$, read "the **integral** of f from a to b." The function f is the **integrand**.

Which is better? Both $\int_a^b f$ and $\int_a^b f(x)\,dx$ denote the same area. The first form looks simpler; for now, we'll usually choose it. The other ("dx") form has advantages that will sometimes matter later on.◄

Differences between the two forms are akin to those between the dy/dx and f' notations for derivative.

How to find it? We've defined the symbolic expression $\int_a^b f$ to *stand for* area. So far, that's all we've done. The problem of *calculating* areas remains.

■ **Example 1.**   Several areas are shown below, labeled as integrals. Use familiar area formulas to evaluate each integral.

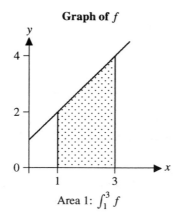

Area 1: $\int_1^3 f$

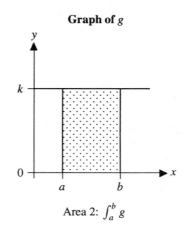

Area 2: $\int_a^b g$

Graph of h

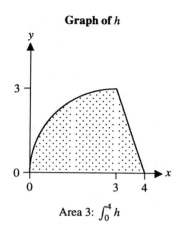

Area 3: $\int_0^4 h$

Solution: All three areas are easy to find. Area 1 is a trapezoid, with base 2 and vertical sides 2 and 4—hence the area➡ is 6. Area 2 is a rectangle; by the usual formula,➡ its area measures $(b - a) \cdot k$. Area 3 is a quarter circle plus a triangle; a close look at these figures reveals a total area of $9\pi/4 + 3/2$, or about 8.569. In integral notation:

$$\int_1^3 f = 6; \qquad \int_a^b g = k(b - a); \qquad \int_0^4 h = \frac{9\pi}{4} + \frac{3}{2}. \qquad \square$$

A trapezoid with base b and heights h_1 and h_2 has area $b(h_1 + h_2)/2$. Can you prove this?

base × height

Signed area: positive and negative contributions

Integrals measure *signed area*. In calculating integrals, therefore, area *above* the graph of f and *beneath* the x-axis counts as negative. Thus if an *integrand* f takes negative values in an interval $[a, b]$, then so may the *integral* $\int_a^b f$.

■ **Example 2.**   Let $f(x) = 1 - x^2$. Find (or estimate) values for the integrals $I_1 = \int_0^2 f$ and $I_2 = \int_{-2}^2 f$.

Solution: The areas in question look like this:

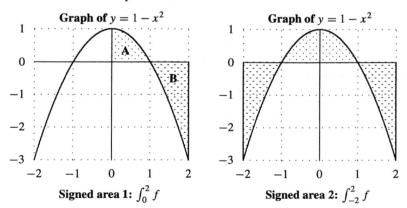

Signed area 1: $\int_0^2 f$ Signed area 2: $\int_{-2}^2 f$

Computing I_1 and I_2 *exactly* is, for now, difficult.➡ *Estimating* the integrals, on the other hand, is easy. Eyeballing general sizes➡ suggests that A has area around 2/3; similarly, B has area around 3/2. From these guesses, we'd estimate (counting B as negative!)

$$I_1 = \int_0^2 f \approx \frac{2}{3} - \frac{3}{2} = -\frac{5}{6}.$$

Soon it will be easy.

Note that each square has area 1.

It is in the ballpark; the exact answer turns out to be −2/3.

Convince yourself; how is even-ness of f involved?

How close is this estimate? We can't tell—yet—for sure, but it seems in the ball-park.◂ In any event, symmetry shows◂ that

$$I_2 = 2 \cdot I_1 \approx -\frac{5}{3}.$$

□

■ **Example 3.** Let $g(x) = x^3$. Find or estimate $\int_{-1}^{1} g(x)\,dx$ and $\int_{0}^{1} g(x)\,dx$.

Solution: Here's the graph:

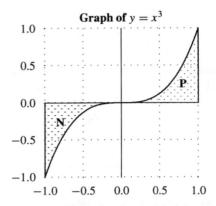

Graph of $y = x^3$

Or, equivalently, because g is odd.

Because of the graph's symmetry◂ the positive and negative areas P and N cancel; hence

$$\int_{-1}^{1} g(x)\,dx = 0.$$

(*Any* odd integrand over *any* interval that's symmetric about $x = 0$ behaves the same way.)

Symmetry is no help with $\int_0^1 g$. A close look suggests that P's area is about that of one small square, or 0.25. In other words,

$$\int_{0}^{1} g(x)\,dx \approx 0.25.$$

And why!

(We'll see soon that◂ this estimate is actually exact.)

□

Properties of the integral

Viewing the integral as area makes many of the integral's important properties simple and natural. For example, this picture—

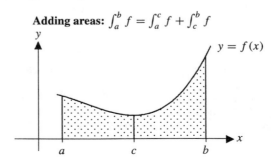

Adding areas: $\int_a^b f = \int_a^c f + \int_c^b f$

But often useful.

is a convincing illustration of the following simple◂ property of the integral:

If $a < c < b$, then $\displaystyle\int_a^b f = \int_a^c f + \int_c^b f$

The next theorem collects several other properties of the integral, similar to the one above.➤

Theorem 1. **(New integrals from old.)** Let f and g be continuous functions on $[a, b]$; let k denote a real constant. Then

1. $\displaystyle\int_a^b (f(x) \pm g(x))\, dx = \int_a^b f(x)\, dx \pm \int_a^b g(x)\, dx.$

2. $\displaystyle\int_a^b kf(x)\, dx = k \int_a^b f(x)\, dx.$

3. If $f(x) \le g(x)$ for all x in $[a, b]$, then

$$\int_a^b f(x)\, dx \le \int_a^b g(x)\, dx.$$

Note:

Like the derivative. The first two properties above are akin to properties of the derivative: like the derivative, the integral behaves well with respect to sums and constant multiples. The following picture illustrates the **constant multiple rule** for integrals:

Constant multiple rule for integrals: why $\int_a^b 2f = 2\int_a^b f$

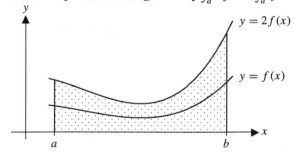

A useful case. The third property is often used in the case where either f or g is a constant. Here's one useful version:

Fact: **(Bounding an integral.)** Suppose that for some numbers m and M, the inequality $m \le f(x) \le M$ holds for all x in $[a, b]$. Then

$$m \cdot (b - a) = \int_a^b m\, dx \le \int_a^b f(x)\, dx \le \int_a^b M\, dx = M \cdot (b - a).$$

A picture best shows what's going on:➤

Bounding areas: why $m(b-a) \leq \int_a^b f \leq M(b-a)$

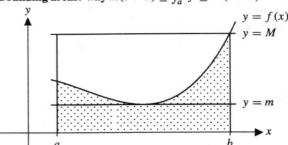

As the picture illustrates, the value of the integral (i.e., the area of the shaded region) lies between the areas of the smaller and the larger rectangles.

It is a fact—a remarkable one.

■ **Example 4.**   Using the fact[*] that $\int_0^\pi \sin x \, dx = 2$, find $\int_0^{\pi/2} \sin x \, dx$ and $\int_0^\pi (3 \sin x + 2 \cos x) \, dx$.

Draw them to convince yourself!

Solution: Symmetry of the sine and cosine graphs[*] shows:

$$\int_0^{\pi/2} \sin x \, dx = \frac{1}{2} \int_0^\pi \sin x \, dx = 1; \qquad \int_0^\pi \cos x \, dx = 0.$$

Therefore, by the sum and constant multiple rules,

$$\int_0^\pi (3 \sin x + 2 \cos x) \, dx = 3 \int_0^\pi \sin x \, dx + 2 \int_0^\pi \cos x \, dx = 6 + 0 = 6.$$

Does it?

A look at the graph should[*] make the general size of the answer plausible:

Why $\int_0^\pi (3 \sin x + 2 \cos x) \, dx = 6$

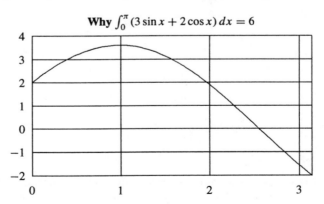

The "net area" *does* seem to be about 6.   □

■ **Example 5.**   For any positive constant b,

$$\int_0^b 1 \, dx = b \qquad \text{and} \qquad \int_0^b x \, dx = \frac{b^2}{2}.$$

Explain why. Then use the theorem to find $\int_0^b (Cx + D) \, dx$, for any constants C and D.

Solution: Graphs➤➤ of the functions $f(x) = 1$ and $g(x) = x$ over the interval   *Draw them!*
$[0, b]$ show that the first two integrals have the claimed values. To finish, we use the
sum and constant multiple rules for integrals:

$$\int_0^b (Cx + D)\, dx = \int_0^b (Cx)\, dx + \int_0^b D\, dx$$

$$= C \int_0^b x\, dx + \int_0^b D\, dx$$

$$= C \frac{b^2}{2} + Db. \qquad \square$$

Average value and the integral

In the picture below, the rectangle is chosen to have the same area as the shaded
region:

Equal areas: Average value of a function

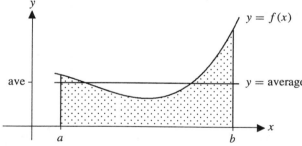

The *height* of that rectangle is, in a natural way, the **average value** of the function f
over the interval $[a, b]$. The following definition restates in analytic language what
the picture shows geometrically:

> **Definition:**   Let f be defined on an interval $[a, b]$. The quantity
>
> $$\frac{\int_a^b f(x)\, dx}{b - a}$$
>
> is the **average value of f over $[a, b]$.**

The ideas of integral and average value are very close cousins; understanding either
helps with the other. The picture above summarizes this two-way relationship: either
the integral defines the average value (as the *height* of a certain rectangle) or the
average value defines the integral (as the *area* of the same rectangle).

■ **Example 6.** The graphs below show speed functions, $s_A t$ and s_B, for two cars A and B.

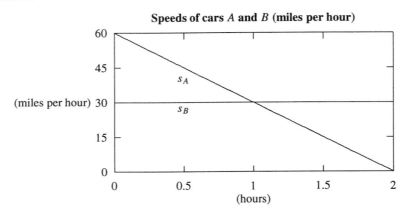

Speeds of cars A and B (miles per hour)

What's the average value of each function over the interval $[0, 2]$? What do the results mean in "car talk?"

Convince yourself.

Solution: Looking at areas shows readily that

$$\int_0^2 s_A(t)\,dt = 60 = \int_0^2 s_B(t)\,dt.$$

Thus, using the definition above, *both* speed functions have average value 30 over the interval $[0, 2]$.

The answers make excellent sense automotively. The area under each graph represents the total distance—60 miles—that each car covers in the two-hour interval. Dividing the distance by the time elapsed gives an average speed of 30 mph for each car.   □

Interpreting the integral

So far we've approached the integral $\int_a^b f$ only in geometric terms. Nevertheless, the integral has many *other* natural and important interpretations. We'll see several in this course. Here we mention just one.

Speed and distance

If a function $f(t)$ represents the *speed* of a moving object at time t, then $\int_a^b f(t)\,dt$ represents the *distance traveled* by the object over the interval $a \le t \le b$. Why?

Draw a graph to convince yourself.   If the speed function $f(t)$ happens to be a *constant*, say k, then

$$\int_a^b f(t)\,dt = k \cdot (b - a) = \text{speed} \cdot \text{time} = \text{distance},$$

as claimed. Remarkably enough, the same result holds even if $f(t)$ *isn't* constant. The previous example should make this plausible; in the next sections, we'll see exactly why it's true.

Velocity and net distance

Speed, by definition, is always positive. Velocity can be *either* positive or negative, depending on direction. As a result, integrating a velocity function $v(t)$ gives slightly different information from that above. If $v(t)$ tells the *velocity* of an object moving along a line at time t, then $\int_a^b f(t)\,dt$ represents the *net distance traveled* over the interval $a \le t \le b$. In particular, net distance, like velocity, can take either sign.

Integrating from right to left: a technicality

The English description of $\int_a^b f(x)\,dx$—"the integral of f *from* $x = a$ *to* $x = b$"— suggests an implicit *direction* of movement: x *starts* at a and *ends* at b.**➤**

The variable name x isn't sacred—t is another popular choice.

It's *almost* always natural to think of x as moving *from left to right*. Up to now, we've done just that: in every integral $\int_a^b f$ we've treated so far, we've assumed or stated that $a \le b$. For example, we've discussed

$$\int_0^\pi \sin x\,dx, \quad \text{but not} \quad \int_\pi^0 \sin x\,dx \quad \text{or} \quad \int_\pi^0 (3\sin x + 2\cos x)\,dx.$$

Sometimes**➤** the latter sort of "right-to-left" integrals *do* arise. The following convention, based on signs, handles all the possibilities with a minimum of fuss and bother:

We'll see some in the next section.

(5.1.1)
$$\int_a^b f(x)\,dx = -\int_b^a f(x)\,dx.$$

The convention says, for instance, that since $\int_0^1 x^3\,dx = 0.25$,**➤**

See Example 3.

$$\int_1^0 x^3\,dx = -\int_0^1 x^3\,dx = -0.25.$$

Why does Equation 5.1.1 make sense? The speed-distance context**➤** offers one answer:

We saw it just above.

> If $f(t)$ *represents speed at time t, and $a < b$, then $\int_a^b f(t)\,dt$ gives the (positive) distance traveled from time a to time b.*

It's physically reasonable, therefore, to consider as *negative* the distance covered over the *reverse* interval.**➤**

Even if Equation 5.1.1 seems arbitrary now, it will prove useful, even natural, in later work.

Integrals and areas: the story so far and a look ahead

In this section we defined the integral $\int_a^b f$ as a certain area. Then we observed some properties and uses of the integral, arguing mainly from geometric intuition.

What's missing, so far, is any really practical or effective method of *measuring* areas. In the simplest cases, as we saw, elementary area formulas (for squares, rectangles, etc.) are some help. For all *but* the simplest functions, however, such formulas are useless. We address the problem of *calculating* integrals, in two different ways, in the next three sections.

In Sections 5.2 and 5.3 we do so by relating the new idea of the integral to the older (and better-understood) idea of the derivative. This key connection between derivative and integral, known as the **fundamental theorem of calculus**, is among the most important results of mathematics.

In Section 5.4 we give *another* definition of the integral, this time as a limit of approximating sums. Logically, the situation resembles that for the derivative, which we defined informally as the slope of a line, but formally as a limit of *differences*.

(Must area exist?) There seems little question that areas under simple graphs—lines, parabolas, sinusoids, etc.—exist. Things are less clear for "not-nice" (e.g., discontinuous) functions, whose graphs may be extremely ragged. Consider, for instance, this graph:

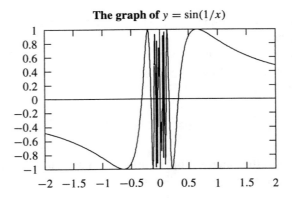

The graph of $y = \sin(1/x)$

Can such a nasty curve sensibly be said to bound area?

The answer turns out (for this curve, anyway) to be yes. Defining area (let alone computing it) in such unpromising circumstances, however, requires special care. In Section 5.4 we'll redefine the integral, using limits, in order to handle cases like the one shown. In practice, there's no need to worry unduly; the vast majority of standard calculus functions behave nicely enough to accord well with our everyday intuition about area.

Exercises

1. Let g be the function shown graphically below. When asked to estimate $\int_1^2 g(x)\,dx$, a group of calculus students submitted the following answers: -4, 4, 45, and 450. Only one of these responses is reasonable; the others are "obviously" incorrect. Which is the reasonable one? Why?

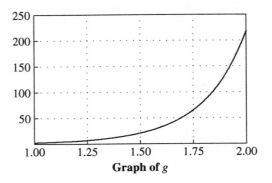

Graph of g

2. Let f be the function shown graphically below.

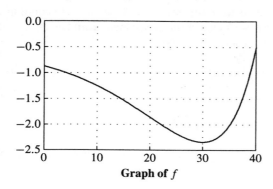

Graph of f

(a) Which of the following is the best estimate of $\int_0^{40} f(x)\,dx$: -65, -30, 0, 35, 60. Justify your answer.

(b) Estimate the average value of f over the interval $[10, 30]$.

3. Four students disagree on the value of the integral $\int_0^{\pi/2} \cos^8 x\,dx$. Jack argues for $\pi \approx 3.14$, Joan for $35\pi/256 \approx 0.43$, Ed for $2\pi/9 - 1 \approx -0.30$, and Lesley for $\pi/4 \approx 0.79$. Use the graph below to determine who is right. (One *is* right!) Explain how you know that the other values are incorrect.

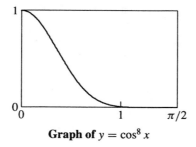

Graph of $y = \cos^8 x$

4. The graph of a function h is shown below.

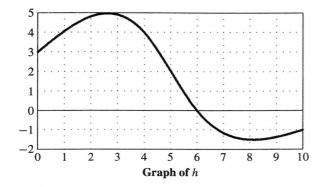

Graph of h

List, from smallest to largest: (i) $h'(5)$; (ii) the average value of h over the interval $[0, 10]$; (iii) the average rate of change of h over the interval $[0, 10]$; (iv) $\int_0^{10} h(x)\,dx$; (v) $\int_0^5 h(x)\,dx$; (vi) $\int_6^{10} h(x)\,dx$.

5. Estimate each of the quantities (i)–(vi) listed in the previous problem.

6. Let h be the function shown graphically in Exercise 4 and suppose that $h(t)$ represents the *eastward* velocity (in meters/second) of an object moving along an east-west axis at time t seconds. Estimate each of the following quantities (use appropriate units!) and interpret in the language of velocity, distance, etc.

(a) $\int_0^{10} h(t)\, dt$

(b) $\frac{1}{10} \int_0^{10} h(t)\, dt$

(c) $\frac{1}{10}(h(10) - h(0))$

(d) $|h(8)|$

(e) $\int_0^{10} |h(t)|\, dt$

(f) $\frac{1}{10} \int_0^{10} |h(t)|\, dt$

7. Suppose that a car travels on an east-west road with eastward velocity $v(t) = 60 - 20t$ mph at time t hours.

(a) Evaluate $\int_0^4 v(t)\, dt$. Interpret the answer in car talk.

(b) Find the average value of $v(t)$ on the interval $[0, 4]$. Show your result graphically.

(c) Let $s(t)$ be the car's speed at time t. Evaluate $\int_0^4 s(t)\, dt$ and interpret the answer in car talk.

(d) What is the car's average speed between $t = 0$ and $t = 4$?

8. Sketch the graph of a continuous function f such that $\int_{-2}^3 f(t)\, dt = -10$ and

(a) $\int_{-2}^3 |f(t)|\, dt = 10$.

(b) $\int_{-2}^3 |f(t)|\, dt \neq 10$.

9. The graph of a function f is shown below.

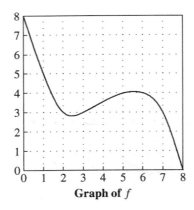

Graph of f

(a) Which of the following is the best estimate of $\int_1^6 f(x)\, dx$: $-24, 9, 20, 38$? Justify your answer.

(b) Find positive integers A and B such that $A \leq \int_3^7 f(x)\, dx \leq B$. Explain how you know that the values of A and B you chose have the desired properties.

(c) $\int_6^8 f(x)\, dx \approx 4$. Does this approximation overestimate or underestimate the exact value of the integral? Justify your answer.

(d) Estimate the average value of f over the interval $[0, 2]$.

10. For each function below, use a graph of f to evaluate $\int_0^2 f(x)\, dx$, $\int_1^4 f(x)\, dx$, $\int_{-5}^{-1} f(x)\, dx$, and $\int_{-2}^3 f(x)\, dx$.

(a) $f(x) = 1$ (c) $f(x) = -a$

(b) $f(x) = a$

11. For each function below, use a graph of f to evaluate $\int_0^2 f(x)\,dx$, $\int_1^4 f(x)\,dx$, $\int_{-5}^{-1} f(x)\,dx$, and $\int_{-2}^3 f(x)\,dx$.

(a) $f(x) = 3x$ (c) $f(x) = 5 - 2x$

(b) $f(x) = 2x + 5$

12. The graph of a function f (shown below) consists of two straight lines and two quarter-circles.

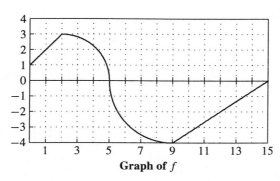

Graph of f

Evaluate each of the following integrals.

(a) $\int_0^2 f(x)\,dx$.   (f) $\int_0^{15} f(x)\,dx$.

(b) $\int_2^5 f(x)\,dx$.   (g) $\int_0^{15} |f(x)|\,dx$.

(c) $\int_0^5 f(x)\,dx$.   (h) $\int_{15}^9 f(x)\,dx$.

(d) $\int_5^9 f(x)\,dx$.   (i) $\int_{12}^{15} f(x)\,dx$.

(e) $\int_5^5 f(x)\,dx$.   (j) $\int_9^{12} f(x)\,dx$.

13. Let f be the function shown graphically in the previous exercise.

(a) Show that $\int_2^4 f(x)\,dx > 5$.   (c) Is $\int_0^6 f(x)\,dx < 10$?

(b) Show that $\int_6^9 f(x)\,dx < -9$.   (d) Is $\int_2^7 f(x)\,dx$ positive or negative?

14. (a) Evaluate $\int_{-3}^3 (x + 2)\,dx$.   (c) Evaluate $\int_{-3}^3 (|x| + 2)\,dx$.

    (b) Evaluate $\int_{-3}^3 |x + 2|\,dx$.   (d) Evaluate $\int_{-3}^3 (2 - |x|)\,dx$.

15. Evaluate $\int_0^2 f(x)\,dx$, where $f(x) = \begin{cases} 1 + x, & \text{if } 0 \le x \le 1 \\ 2 - x, & \text{if } 1 < x \le 2. \end{cases}$

16. Evaluate $\int_0^1 \sqrt{1 - (x - 1)^2}\,dx$.

17. Evaluate $\int_1^3 \left(6 - \sqrt{4 - (x - 3)^2}\,dx \right)\,dx$.

18. Evaluate $\int_0^3 f(x)\,dx$, where $f(x) = \begin{cases} 1 - \sqrt{1 - x^2}, & \text{if } 0 \leq x \leq 1 \\ 1 + \sqrt{4 - (x-3)^2}, & \text{if } 1 < x \leq 3. \end{cases}$

19. Explain why $\int_1^2 x^3\,dx = \int_3^4 (x-2)^3\,dx = \int_{-3}^{-2} (x+4)^3\,dx$.

20. Suppose that $\int_{-2}^5 f(x)\,dx = 18$, $\int_{-2}^5 g(x)\,dx = 5$, and $\int_{-2}^5 h(x)\,dx = -11$. Evaluate each of the following integrals.

    (a) $\int_{-2}^5 \big(f(x) + g(x)\big)\,dx$

    (b) $\int_{-2}^5 \big(f(x) + h(x)\big)\,dx$

    (c) $\int_{-2}^5 \big(f(x) + g(x) - h(x)\big)\,dx$

    (d) $\int_{-2}^5 3f(x)\,dx$

    (e) $\int_{-2}^5 -4g(x)\,dx$

    (f) $\int_5^{-2} f(x)\,dx$

    (g) $\int_{-2}^5 \big(h(x) + 1\big)\,dx$

    (h) $\int_0^7 g(x-2)\,dx$

21. Suppose that (i) $\int_0^2 f(x)\,dx = 2$; (ii) $\int_1^2 f(x)\,dx = -1$; (iii) $\int_2^4 f(x)\,dx = 7$.

    (a) Evaluate $\int_1^4 f(x)\,dx$.

    (b) Evaluate $\int_0^4 3f(x)\,dx$.

    (c) Evaluate $\int_0^1 f(x)\,dx$.

    (d) Evaluate $\int_0^1 f(x+1)\,dx$.

    (e) Evaluate $\int_0^1 \big(f(x) + 1\big)\,dx$.

    (f) Evaluate $\int_2^4 f(x-2)\,dx$.

    (g) Evaluate $\int_2^4 \big(f(x) - 2\big)\,dx$.

    (h) Explain why f must be negative somewhere in the interval $[1, 2]$.

    (i) Explain why $f(x) \geq 3$ for some values of x in the interval $[2, 4]$.

    (j) Draw the graph of a function f with properties (i)–(iii).

22. Suppose that f is a function such that $\int_0^3 f(x)\,dx = -1$.

    (a) Suppose that f is an *even* function. Explain why $\int_{-3}^3 f(x)\,dx = -2$.

    (b) If f is an *odd* function, what is the value of $\int_{-3}^3 f(x)\,dx$? Explain.

23. Evaluate each of the following integrals using graphical arguments.

    (a) $\int_0^\pi \cos x\,dx$

    (b) $\int_{\pi/2}^{3\pi/2} \sin x\,dx$

    (c) $\int_{-2}^2 \big(7x^5 + 3\big)\,dx$

    (d) $\int_{-1}^1 \big(4x^3 - 2x\big)\,dx$

24. Evaluate each of the following integrals using graphical arguments and the fact that $\int_0^\pi \sin x\,dx = 2$.

    (a) $\int_0^{2\pi} \sin x\,dx$

    (b) $\int_0^{2\pi} |\sin x|\,dx$

    (c) $\int_0^\pi (\sin x + 1)\,dx$

    (d) $\int_0^{\pi/2} \sin x\,dx$

    (e) $\int_{-\pi/2}^{\pi/2} \cos x\,dx$

    (f) $\int_0^{\pi/2} (\cos x + x)\,dx$

    (g) $\int_0^{100\pi} \sin x\,dx$

    (h) $\int_0^{100\pi} |\sin x|\,dx$

    (i) $\int_0^{100\pi} \cos x\,dx$

25. Explain why $\displaystyle\int_1^3 \frac{1-x}{x^2}\,dx < \int_1^2 \frac{1-x}{x^2}\,dx$.

26. Is $\int_0^2 \left(2x^3 - x^2\right) dx$ greater than $\int_0^2 \left(3x^2 + x\right) dx$? Justify your answer.

27. Suppose that f is continuous and nonconstant on $[a, b]$. Show that if m and M are constants such that $m \le f(x) \le M$ when $x \in [a, b]$, then $m(b - a) < \int_a^b f(x)\, dt < M(b - a)$.

28.   (a) Explain why $\int_0^{2\pi} (1 + \cos x)\, dx \ge 0$.

     (b) Show that $\int_0^{2\pi} (1 + \cos x)\, dx > 0$.   [HINT: Consider the intervals $[0, \pi]$ and $[\pi, 2\pi]$ separately.]

29. Show that $0 \le \int_0^\pi x \sin x\, dx \le \pi^2/2$.

30. Show that $\pi/6 \le \int_{\pi/6}^{\pi/2} \sin x\, dx \le \pi/3$.

31. Show that $-\pi/3 \le \int_{2\pi/3}^{\pi} \cos x\, dx \le -\pi/6$.

32.   (a) Prove that $2x/\pi \le \sin x \le x$ when $0 \le x \le 1$.

     (b) Use part (a) to show that $\pi/16 \le \int_0^{\pi/4} \sin x\, dx \le \pi^2/32$.

33.   (a) Use the racetrack principle to show that $1 + x \le e^x \le 1 + 3x$ when $0 \le x \le 1$.

     (b) Use part (a) to find upper and lower bounds on $\int_0^1 e^x\, dx$.

34.   (a) Show that $\pi(1 + 1/e) \le \int_0^{2\pi} e^{\sin x}\, dx \le \pi(1 + e)$.

     (b) Use the result in part (a) to find upper and lower bounds for $\int_0^{50\pi} e^{\sin x}\, dx$.

35.   (a) Show that $\pi/6 \le \int_0^{\pi/2} \cos x\, dx \le 5\pi/12$.   [HINT: $1/2 \le \cos x \le 1$ when $0 \le x \le \pi/3$.]

     (b) Using ideas similar to those in part (a), find upper and lower bounds on the value of $\int_0^{\sqrt{\pi/2}} \cos \left(x^2\right) dx$.

36.   (a) Suppose that $f(x) \ge 0$ when $0 \le x \le 1$. Show that $\int_0^1 x^k f(x)\, dx \le \int_0^1 f(x)\, dx$ for all integers $k \ge 0$.

     (b) Does the result in part (a) remain valid if the assumption "$f(x) \ge 0$ when $0 \le x \le 1$" is not made? Explain.

37. Show that $\int_0^{\pi/4} \cos(2x)\, dx \ne \int_0^{\pi/2} \cos x\, dx$. Which integral is larger? Justify your answer.

38. Show that $\int_0^{\sqrt{\pi}} \sin \left(x^2\right) dx \ne \int_0^\pi \sin x\, dx$. Which integral is larger? Justify your answer.

39. Show that $\int_0^{\pi/2} \sqrt{1 + \cos(2x)}\, dx = \sqrt{2} \int_0^{\pi/2} \cos x\, dx$.

40. Show that $0 \le \int_0^{\sqrt{\pi/2}} \sin \left(x^2\right) dx \le \sqrt{\pi/2} - 1 + \frac{1}{2} \sin 1 \approx 0.674$.

41. Suppose that f is an odd function and continuous everywhere. Explain why the average value of f over any interval $[-a, a]$ is zero.

42. Suppose that f is an even function and continuous everywhere. Explain why the average value of f over the interval $[-a, a]$ is equal to the average value of f over the interval $[0, a]$.

43. Suppose that f is a continuous function and that $-2 \leq f(x) \leq 5$ when $1 \leq x \leq 3$. Show that the average value of f lies in the interval $[-2, 5]$.

44. Suppose that f is a continuous function. If the average value of f over the interval $[0, 1]$ is 2 and the average value of f over the interval $[1, 3]$ is 3, what is the average value of f over the interval $[0, 3]$?

45. Suppose that f is a continuous function. If the average value of f over the interval $[-3, 1]$ is 2 and the average value of f over the interval $[-3, 7]$ is 5, what is the average value of f over the interval $[1, 7]$?

46. Show that the expression $\int_a^b \left(f(x) - c \right)^2 dx$ assumes its minimum value when c is the average value of f on the interval $[a, b]$.

47. Suppose that f is continuous on $[0, 1]$. Explain why $\int_0^1 f(x)\, dx = \int_0^1 f(1 - x)\, dx$.

48. Suppose that f is continuous on $[a, b]$. Show that $\left| \int_a^b f(x)\, dx \right| \leq \int_a^b |f(x)|\, dx$. [HINT: $-|f(x)| \leq f(x) \leq |f(x)|$.]

49. Suppose that f and g are continuous on $[a, b]$ and $\int_a^b f(x)\, dx \leq \int_a^b g(x)\, dx$.

    (a) Must $f(x) \leq g(x)$ for every x in $[a, b]$? If so, explain why. If not, give a counterexample.

    (b) Must there be a number c such that $a \leq c \leq b$ and $f(c) \leq g(c)$? If so, explain why. If not, give a counterexample.

50. Archimedes (ca. 250 B.C.E.) proved that the area under a parabolic arch is $2bh/3$ where b is the width of the base of the arch and h is the height. [NOTE: Archimedes actually proved a more general theorem about the area of any region cut off from a parabola by a line.]

    (a) Use Archimedes' result to show that $\int_{-a}^a x^2\, dx = 2a^3/3$. [HINT: Draw the curve $y = x^2$ and the line $y = a^2$.]

    (b) Use part (a) and the fact that x^2 is an even function to show that $\int_0^a x^2\, dx = a^3/3$. [HINT: $\int_{-a}^a x^2\, dx = \int_{-a}^0 x^2\, dx + \int_0^a x^2\, dx$.]

    (c) Use part (b) to show that $\int_a^b x^2\, dx = (b^3 - a^3)/3$. [HINT: $\int_a^b x^2\, dx = \int_0^b x^2\, dx - \int_0^a x^2\, dx$.]

51. Let $f(x) = 2x + 3$.

    (a) Sketch a graph of f in the viewing window $[1, 2] \times [5, 9]$.

    (b) Evaluate $\int_1^2 f(x)\, dx$. Shade the region of the part (a) graph represented by this integral.

    (c) Show that $f^{-1}(x) = (x - 3)/2$.

    (d) Evaluate $\int_5^9 f^{-1}(x)\, dx$ and shade the region of the part (a) graph represented by this integral.

    (e) Verify that $\int_1^2 f(x)\, dx = 2f(2) - f(1) - \int_{f(1)}^{f(2)} f^{-1}(x)\, dx$.

52. Suppose that f is a continuous function and invertible on $[a, b]$. Then it can be shown that $\int_a^b f(x)\, dx = bf(b) - af(a) - \int_{f(a)}^{f(b)} f^{-1}(x)\, dx$.

(a) Give a graphical proof of this result when $0 < a < b$, and $0 < f(a) < f(b)$.   [HINT: Draw a sketch of f in the viewing window $[0, b] \times [0, f(b)]$.]

(b) Show that $\int_1^e \ln x \, dx = e - \int_0^1 e^x \, dx$.

(c) Show that $\int_a^b \sqrt{x} \, dx = b^{3/2} - a^{3/2} - \int_{\sqrt{a}}^{\sqrt{b}} x^2 \, dx$.

53. Use results stated in Exercises 50 and 52 to show that $\int_0^a \sqrt{x} \, dx = 2a^{3/2}/3$.

5.2   The area function

Positive, negative, or zero!

In the previous section we defined the integral $\int_a^b f$ to be the *signed area* of the region from $x = a$ to $x = b$, bounded above or below by the graph of f. For given f, a, and b, this area is a certain fixed *number*.◄

In this section we define and study a *function* based on the signed area bounded by the graph of f. This **area function**, which we'll denote A_f, is "built" from f, using the integral.

We've played the new-function-from-old game before. In the most important case, we built the *derivative* function f' from an "original" function f, and then studied what each tells about the other. Here we repeat the process, this time with f and A_f. The relationship between f and A_f will prove remarkably similar to that between f' and f.

Defining the area function A_f

First the formal definition:

> **Definition:** Let f be a function and a any point of its domain. For any input x, the **area function** A_f is defined by the rule
> $$A_f(x) = \int_a^x f(t)\, dt = \text{the signed area defined by } f, \text{ from } a \text{ to } x.$$

A "generic" picture illustrates the idea:

The area function: $A_f(x) =$ the shaded area

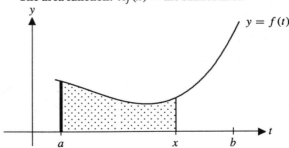

The idea is best understood through pictorial examples. First, three brief remarks:

Variable names: why so many? Why did we use another variable, t, in the defining expression $A_f(x) = \int_a^x f(t)\, dt$? Because x was already taken; we used it to denote the right-hand boundary of the defining region. To avoid using x in two different ways, we introduced another variable.

The domain of A_f. For which inputs x does the rule $A_f(x) = \int_a^x f$ make sense?

E.g., discontinuities.

We discussed it in the previous section.

        Barring bad behavior◄ of f, A_f has the same domain as f itself. In particular, $A_f(x)$ makes sense even if x is to the *left* of a; in this case we use the sign convention◄ $\int_a^x f = -\int_x^a f$.

The choice of a. The role of a is to fix one edge (the *left* edge, in the picture above) of the region whose area gives $A_f(x)$—the other edge varies freely.

        As the definition says, *any* a in the domain of f is a legal choice. We'll see below that different choices of a give different (but only slightly different) functions A_f.

■ **Example 1.** Let $f(x) = 3$ and $a = 0$. Describe the area function $A_f(x) = \int_0^x f(t)\,dt$ for positive inputs x. Does A_f have a simple formula?

Solution: Because $f(x) = 3$, $A_f(x) = 3x$. The first picture below shows why;➤ in the second, graphs of f and A_f appear together.➤

A rectangle with height 3 and base x has area $3x$.

For now we ignore negative inputs x; we'll return to them later.

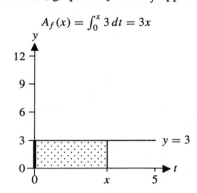

$$A_f(x) = \int_0^x 3\,dt = 3x$$

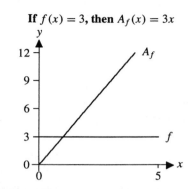

If $f(x) = 3$, then $A_f(x) = 3x$

Notice too that $A_f(x) = 3x$ is an *antiderivative* of $f(x) = 3$. This result is no accident, as we'll see.➤   □

We'll return to this important fact soon.

■ **Example 2.** As before, let $f(x) = 3$ and $a = 0$. Discuss the area function $A_f(x) = \int_0^x f(t)\,dt$ for *negative* values of x. Does the formula $A_f(x) = 3x$ still hold?

Solution: Yes—$A_f(x) = 3x$ for *all* values of x. The key step is a careful look at *signs*. Here are the pictures:

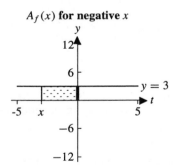

$A_f(x)$ for negative x

If $f(x) = 3$, then $A_f(x) = 3x$

To make sense of them, recall➤ how we handle "right to left" integrals: Since $x < 0$,

From the previous section.

$$\int_0^x f = -\int_x^0 f = -\,(\text{shaded area}).$$

Since $x < 0$ and the rectangle has height 3, $3x$ is, just as claimed, *minus* the shaded area.

Combining this example with the last one leads to this conclusion:

If $f(x) = 3$ and $a = 0$, then $A_f(x) = \int_0^x f = 3x$ for *all* x.   □

A happy moral. What $\int_a^x f$ means geometrically is clear if $x \geq a$.➤ If $x < a$, things look trickier; the previous example was all about this possibility.

See the first picture in this section if you've forgotten.

Working the previous example took some careful analysis of plus and minus signs. After the dust settled, though, the happiest possible outcome was revealed:

The same formula for $A_f(x)$ works for *all* x.

This is evidence that our signed area conventions are not just convenient, but "right."

Thus the cases $x < a$ and $x \geq a$, though apparently different, lead to the same result.◂

The same principle holds for any function f defined by a reasonable formula. Since it does, we'll seldom worry about (or even mention) the issue of whether $x < a$ or $x \geq a$.

■ **Example 3.**   Consider the linear function $f(x) = x$ and the area function $A_f(x) = \int_0^x t\, dt$. What's A_f now?

Solution: As the left graph below shows, the region defining $A_f(x)$ for a positive input x is a *triangle*, with base x and height x.◂ The formula for A_f, therefore, is

We'll skip the case of negative x, for reasons outlined above.

$$A_f(x) = \frac{\text{base} \times \text{height}}{2} = \frac{x^2}{2}.$$

Graphs of $f(x)$ and $A_f(x)$ appear together at the right.

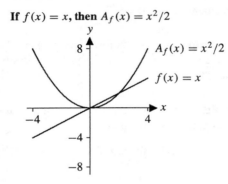

$A_f(x) = \int_0^x t\, dt = x^2/2$ **If** $f(x) = x$, **then** $A_f(x) = x^2/2$

This is still no accident. Stay tuned.

Again, A_f turns out to be an antiderivative of f.◂   □

■ **Example 4.**   (**Another** f, **same** a.) Consider the new linear function, $f(x) = 2 - x$. Find the new area function $A_f(x) = \int_0^x f$. Is A_f again an antiderivative for f?

Solution: The usual area picture applies; area below the x-axis counts as negative.◂

Notice the different shadings above and below the t-axis.

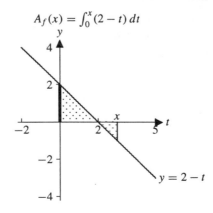

$A_f(x) = \int_0^x (2 - t)\, dt$ **If** $f(x) = 2 - x$, **then** $A_f(x) = 2x - x^2/2$

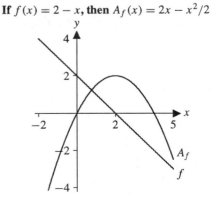

Or trapezoids, if x < 2.

With a little care, one can use triangles◂ to find the net shaded area. On the left-hand graph above the upper triangle has area 2; the lower triangle has area $(x-2)^2/2$. Thus the net area, for given x, is⬧

Check the algebra.

$$A_f(x) = 2 - \frac{(x-2)^2}{2} = 2x - \frac{x^2}{2}.$$

(Note: The picture shows a value of x greater than 2. If $0 < x < 2$, the picture is different but the resulting formula is the same.)⬧ The graphs of both f and A_f appear at the right, above.

The exercises pursue this point further.

Another⬧ approach to the problem is to use additive properties of the integral⬧ and ideas from earlier examples:

Perhaps easier.

cf. Section 5.1.

$$A_f(x) = \int_0^x (2-t)\,dt = \int_0^x 2\,dt - \int_0^x t\,dt = 2x - \frac{x^2}{2}.$$

Either way the point is the same:

> *In all examples so far, the area function A_f is an antiderivative of f.*

$\square$

■ **Example 5.** (Same f, another a.) Consider the same linear function, $f(x) = 2 - x$, this time setting $a = 1$, so that $A_f(x) = \int_1^x f$. What's the area function now?

Solution: The only difference from the previous example concerns the lower endpoint of integration. Thus the new function A_f differs from the old one only by the *area of the trapezoidal region from $x = 0$ to $x = 1$.*⬧ In symbols:

This area is 3/2—convince yourself.

$$
\begin{aligned}
A_f(x) &= \int_1^x f(t)\,dt = \int_0^x f(t)\,dt - \int_0^1 f(t)\,dt \\
&= 2x - \frac{x^2}{2} - \frac{3}{2}.
\end{aligned}
$$

The graphs support all this:

$A_f(x) = \int_1^x (2-t)\,dt$

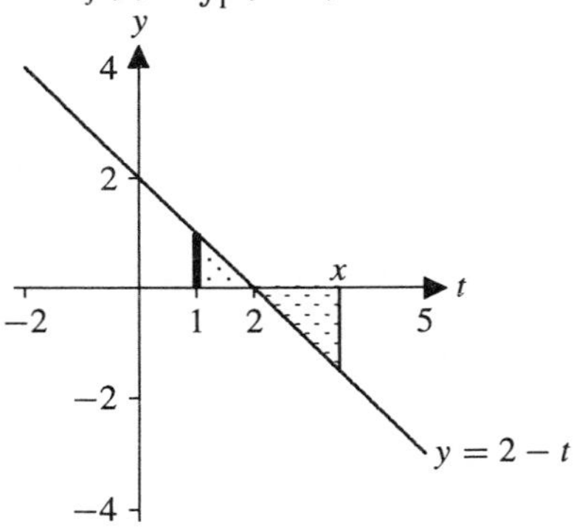

If $f(x) = x - 2$ and $a = 1$ then $A_f(x) = 2x - x^2/2 - 3/2$

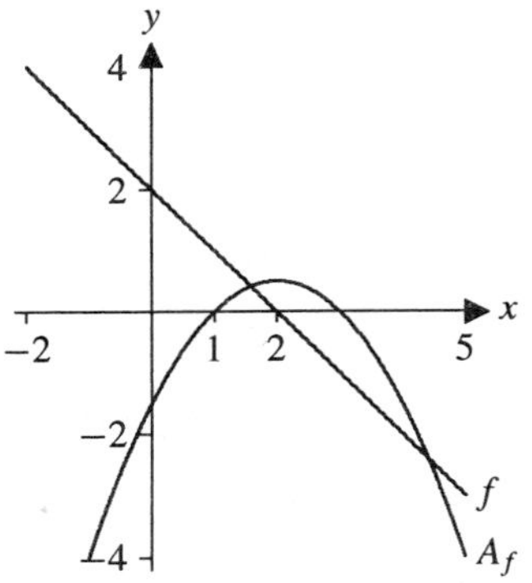

Yet again, A_f is an antiderivative of f;⬧ this time, the one for which $A_f(1) = 0$. $\square$

Two antiderivatives of the same f can differ only by a constant, as they do here.

In each example above we've used an elementary area formula to find an explicit algebraic expression for the area function. In the next example, no such simple formula is available.

■ **Example 6.** Let $f(x) = 1/x$ and $A_f(x) = \int_1^x f(t)\,dt$. Discuss the area function A_f. Does it look familiar?

Solution: The graph below left shows $f(x)$; the shaded area represents $A_f(3)$:

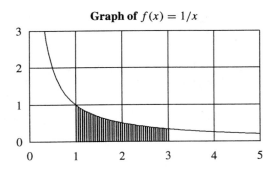

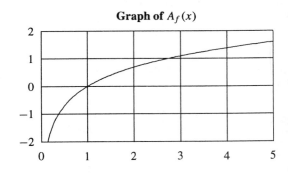

Note that $A_f(x) < 0$ for $x < 1$. Do you see why?

Values of A_f can be estimated, using the grid. The shaded area, for example, seems to be a little more than one unit. By plotting such estimates we can produce, at least approximately, a graph of A_f, like the one at upper right.◄◄

The graph of A_f should look familiar—in fact, it shows the natural logarithm function. We'll explain this striking result in the next section.   □

Properties of A_f: what the examples show

In every example above, direct calculation showed the area function A_f to be an antiderivative of f. This was no accident. Quite the contrary:

> **Theorem 2. (The fundamental theorem of calculus, informal version.)** For *any* well-behaved function f and any "base point" a, A_f is an antiderivative of f.

One of the most important in mathematics.

We'll discuss and prove this important theorem◄◄ in the next section. In the rest of this section, we amass evidence for it.

To do so, we'll list various properties of A_f. Each follows directly from the definition, and each bolsters the claim that $A'_f = f$. Added to the concrete examples above, they build a strong case for the fundamental theorem of calculus.

Properties of A_f

Let f be a continuous function, a a point of its domain, and $A_f(x) = \int_a^x f(t)\,dt$. Then:

- $A_f(a) = 0$.

- Where f is positive, A_f is increasing.

- Where f is negative, A_f is decreasing.

- Where f is zero, A_f has a stationary point.

- Where f is increasing, A_f is concave up.

- Where f is decreasing, A_f is concave down.

Use, e.g., the graphs in Example 5 to convince yourself.

Examples—in graphical form, especially—are the best evidence for each claim.◄◄ Each claim restates or sharpens the general principle of the fundamental theorem.

Exercises

1. Let $f(x) = x$ and $A_f(x) = \int_0^x t\,dt$. It was shown in the text that $A_f(x) = x^2/2$ when $x > 0$. Show that this formula is correct even when $x < 0$.

2. Let $F(x) = \int_0^x f(t)\,dt$, $G(x) = \int_1^x f(t)\,dt$, and $H(x) = \int_{-2}^x f(t)\,dt$.

   (a) Suppose that $f(x) = 1$. Find symbolic expressions for the functions F, G, and H. Are F, G, and H antiderivatives of f?

   (b) Suppose that $f(x) = 3x$. Find symbolic expressions for the functions F, G, and H. Are F, G, and H antiderivatives of f?

   (c) Suppose that $f(x) = 3x+1$. Find symbolic expressions for the functions F, G, and H. Are F, G, and H antiderivatives of f? [HINT: Use your answers to parts (a) and (b).]

3. Let $F(x) = \int_0^x f(t)\,dt$, $G(x) = \int_2^x f(t)\,dt$, and $H(x) = \int_{-1}^x f(t)\,dt$.

   (a) Suppose that $f(x) = a$, a positive constant. Find symbolic expressions for the functions F, G, and H. Are F, G, and H antiderivatives of f?

   (b) Suppose that $f(x) = bx$ where b is a positive constant. Find symbolic expressions for the functions F, G, and H. Are F, G, and H antiderivatives of f?

   (c) Suppose that $f(x) = a + bx$ where a and b are positive constants. Find symbolic expressions for the functions F, G, and H. Are F, G, and H antiderivatives of f?

4. Let a be a constant, $f(x) = 2 - x$, and $A_f(x) = \int_a^x f(t)\,dt$.

   (a) Evaluate $A_f(a)$.

   (b) Find a symbolic expression for $A_f(x)$.
   [HINT: $\int_a^x (2 - t)\,dt = \int_0^x (2 - t)\,dt - \int_0^a (2 - t)\,dt$.]

   (c) Show that A_f is an antiderivative of f.

   (d) On the same axes, plot f and A_f for $a = 0$, $a = 2$, and $a = -1$.

5. Let $f(x) = \sin x$ and $A_f(x) = \int_0^x f(x)\,dx$.

   (a) It can be shown that $A_f(\pi/2) = 1$. Use this fact to evaluate $A_f(\pi)$, $A_f(3\pi/2)$, $A_f(2\pi)$, $A_f(-\pi/2)$, $A_f(-\pi)$, $A_f(-3\pi/2)$, and $A_f(-2\pi)$.

   (b) Explain why A_f is a periodic function. [HINT: Use the fact that f is 2π-periodic function.]

   (c) Show that $0 \le A_f(x) \le 2$.

   (d) Sketch a graph of f and $A_f(x)$ on the same axes.

6. Use the fact that $\int_0^{\pi/2} \cos x\,dx = 1$ to show that $-1 \le \int_0^x \cos t\,dt \le 1$.

7. Suppose that f is a continuous function and that $\int_0^x f(t)\,dt = \sin\left(x^2\right)$.

   (a) Show that $\int_{\sqrt{\pi/2}}^x f(t)\,dt = \sin\left(x^2\right) - 1$.

   (b) Find an expression for $\int_{-\sqrt{3\pi/2}}^x f(t)\,dt$.

8. Let $F(x) = \int_a^x f(t)\,dt$ and $G(x) = \int_b^x f(t)\,dt$, where a and b are constants, and f is a continuous function. Show that $G(x) = F(x) + C$, where C is a constant.

9. Suppose that F is an antiderivative of a differentiable function f.

   (a) If F is increasing on $[a, b]$, what is true about f?

   (b) If f is negative on $[a, b]$, what is true about F?

   (c) If f' is positive on $[a, b]$, what is true about F?

   (d) If F is concave down on $[a, b]$, what is true about f'?

   (e) Suppose that G is another antiderivative of f. What relationship exists between F and G?

10. Let $f(x) = 2 - |x|$ and $A_f(x) = \int_0^x f(t)\,dt$.

   (a) Find a symbolic expression for A_f.

   (b) Plot f and A_f on the same axes.

   (c) Where is A_f increasing? What is true about f on on each of these intervals?

   (d) Where is A_f decreasing? What is true about f on each of these intervals?

   (e) Where does A_f have local extrema? What is true about f at each of these points?

   (f) Where is A_f concave up? What is true about f on each of these intervals?

   (g) Where is A_f concave down? What is true about f on each of these intervals?

   (h) Does A_f have any inflection points? If so, where are they located? What is true about f at each of these points?

   (i) How would your answers to parts (c)–(h) change if the definition of A_f were changed to $A_f(x) = \int_3^x f(t)\,dt$?

11. Let $A_f(x) = \int_0^x f(t)\,dt$ where f is the function graphed below.

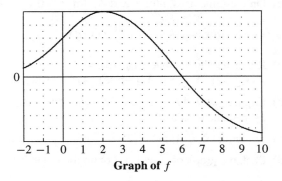

Graph of f

   (a) Which is larger: $A_f(1)$ or $A_f(5)$? Justify your answer.

   (b) Which is larger: $A_f(7)$ or $A_f(10)$? Justify your answer.

   (c) Which is larger: $A_f(-2)$ or $A_f(-1)$? Justify your answer.

   (d) Where is A_f increasing?

   (e) Explain why A_f has a local extremum at $x = 6$. Is this a local maximum or a local minimum?

(f) Let $F(x) = \int_{-2}^{x} f(t)\,dt$. Explain why $A_f(x) = F(x) + C$ where C is a negative constant.

(g) Rank the five numbers 0, $A_f(-1) - A_f(-2)$, $A_f(0) - A_f(-1)$, $A_f(1) - A_f(0)$, and $A_f(2) - A_f(1)$ in increasing order.

(h) The numbers $A_f(4) - A_f(3)$, $A_f(5) - A_f(4)$, $A_f(6) - A_f(5)$, $A_f(7) - A_f(6)$, and $A_f(8) - A_f(7)$ can each be interpreted as the slope of a secant line of A_f. What do these values suggest about the concavity of A_f on the interval $[3, 8]$?

12. Let $A_f(x) = \int_{c}^{x} f(t)\,dt$, where c is a constant and f is a continuous function.

(a) Prove that if f is positive on $[a, b]$, then A_f is increasing on $[a, b]$. [HINT: Let y and z be numbers such that $a \le y < z \le b$. Compare $A_f(y)$ and $A_f(z)$.]

(b) Use part (a) to show that if f is negative on $[a, b]$, then A_f is decreasing on $[a, b]$. [HINT: The function g defined by $g(x) = -f(x)$ is positive wherever f is negative.]

(c) Use parts (a) and (b) to show that A_f has a local extremum wherever f changes sign.

13. Let $A_f(x) = \int_{c}^{x} f(t)\,dt$, where c is a constant and f is a continuous function.

(a) Prove that if f is increasing on $[a, b]$, then A_f is concave up on $[a, b]$. [HINT: A function f is concave up on $[a, b]$ if and only if $\big(f(x) - f(a)\big)/(x - a) < \big(f(b) - f(x)\big)/(b - x)$ for every $x \in (a, b)$.]

(b) Use part (a) to show that if f is decreasing on $[a, b]$, then A_f is concave down on $[a, b]$.

(c) Suppose that f has a local extremum at a. Explain why A_f has an inflection point at a.

14. Let $G(x) = \int_{-3}^{x} f(t)\,dt$ and $H(x) = \int_{2}^{x} f(t)\,dt$ where f is the function graphed below. [NOTE: The graph of f is made up of straight lines and a semicircle.]

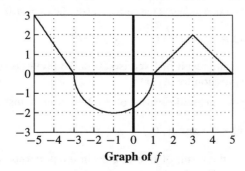

Graph of f

(a) How are the values of $G(x)$ and $H(x)$ related? Give a geometric explanation of this relationship.

(b) On which subintervals of $[-5, 5]$, if any, is H increasing?

(c) Explain why G has a local minimum at $x = 1$.

(d) Where in the interval $[-5, -2]$ does G achieve its minimum value? Its maximum value? What are these values?

(e) Where in the interval $[-5, -5]$ does H achieve its minimum value? Its maximum value? What are these values?

(f) On which subintervals of $[-5, 5]$, if any, is G concave down?

(g) Where does H have points of inflection?

15. Let $F(x) = \int_a^x f(t)\,dt$ and $G(x) = \int_b^x g(t)\,dt$, where f and g are continuous functions. Suppose that $F(c) = G(c)$ and $f(x) \leq g(x)$ when $x \geq c$. Show that $F(x) \leq G(x)$ when $x \geq c$.

16. (a) Use the figure below to show that $\int_0^x \sqrt{1 - t^2}\,dt = \frac{1}{2}x\sqrt{1 - x^2} + \frac{1}{2}\arcsin x$ when $-1 \leq x \leq 1$.

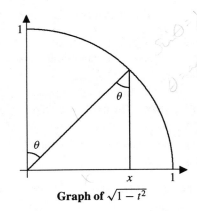

Graph of $\sqrt{1 - t^2}$

(b) Let $f(x) = \sqrt{1 - x^2}$ and let $A_f(x) = \int_0^x f(t)\,dt$. Show that A_f is an antiderivative of f.

17. Let $f(x) = \sqrt{1 - x^2}$ and let $A_f(x) = \int_{-1/2}^x f(t)\,dt$.

(a) Use the result stated in part (a) of the previous exercise to find a symbolic expression for A_f.

(b) Is A_f an antiderivative of f? Justify your answer.

18. Let $a > 0$ be a constant. Use a sketch similar to that given in Exercise 16 to find an expression for $\int_0^x \sqrt{a^2 - t^2}\,dt$ when $-a \leq x \leq a$. [HINT: Check that your expression produces the correct value when $x = a$.]

19. Let $f(x) = x^2$ and $A_f(x) = \int_0^x f(t)\,dt$.

(a) Use the result of Exercise 50 in the previous section to find a symbolic expression for $A_f(x)$.

(b) Is A_f an antiderivative of f? Justify your answer.

20. Let $f(x) = x^2$ and $A_f(x) = \int_{-3}^x f(t)\,dt$.

(a) Use the result of Exercise 50 in the previous section to find a symbolic expression for $A_f(x)$.

(b) Is A_f an antiderivative of f? Justify your answer.

5.3   The fundamental theorem of calculus

What the theorem says

In the last section we defined and studied the *area function* A_f, built from an original function f and a "base point" a. All evidence so far points irresistibly to this conclusion:

A_f *is an* antiderivative *of* f.

The formal statement of this fact is the most important theorem of our subject. Here it is, complete with mathematical fine print:

Theorem 3. (The fundamental theorem of calculus, formal version.) Let f be a continuous function, defined on an open interval I containing a. The function A_f with rule

$$A_f(x) = \int_a^x f(t)\,dt$$

is defined for every x in I, and $\dfrac{d}{dx}\big(A_f(x)\big) = f(x)$.

We'll say much more about the FTC,➤ and provide a proof at the end of this section. First, some remarks.

Our shorthand.

Why continuous? Requiring f to be *continuous* assures that the integral $\int_a^x f$ *exists*. If f is *discontinuous* (e.g., blows up somewhere) A_f may have problems, too.

Why is I open? Working on an *open interval I* is a technical convenience. Doing so avoids troubles that might occur if gaps and endpoints were permitted.

In other symbols. The last equation of the FTC is sometimes written as follows:

$$\frac{d}{dx}\left(\int_a^x f(t)\,dt\right) = f(x).$$

This form avoids the symbol A_f, but requires more careful unpacking.➤

Which form is clearer?

■ **Example 1.** Let $f(x) = 2x\sin(x^2)$ and let $A_f(x) = \int_0^x f(t)\,dt$. Find a symbolic formula for A_f; interpret results graphically.

Solution: For the simple functions of the last section, we used elementary area formulas➤ to find explicit formulas for A_f. Here, simple formulas aren't enough. What, for instance, is $A_f(2)$, shown shaded below?

For rectangles, triangles, trapezoids, etc.

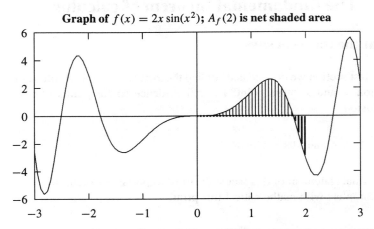

Graph of $f(x) = 2x \sin(x^2)$; $A_f(2)$ is net shaded area

The FTC solves our problem. It says that A_f is *some* antiderivative of f; all we need do is find the right one.

Notice first that for any constant C, $F(x) = -\cos(x^2) + C$ is an antiderivative of $f(x) = 2x \sin(x^2)$.*** *Every* antiderivative of f—including A_f—has this form.*** Therefore for *some* constant C,

Differentiate to see for yourself.

See the fact that follows this example.

$$A_f(x) = -\cos(x^2) + C.$$

Why? Because $\int_0^0 f = 0$.

To choose the "right" value of C, we use the fact that $A_f(0) = 0$.*** In other words,

$$A_f(0) = 0 = -\cos 0 + C \implies C = 1.$$

We've found our formula: $A_f(x) = -\cos(x^2) + 1$.

The graphs below agree; the A_f graph seems, as it should, to describe the growth of signed area based at 0.

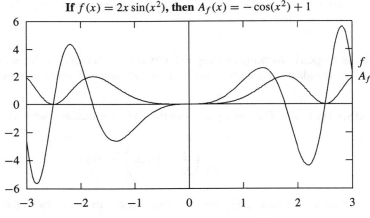

If $f(x) = 2x \sin(x^2)$, then $A_f(x) = -\cos(x^2) + 1$

Is this result reasonable, given the general size of units?

Notice, finally, that $A_f(2) = -\cos(4) + 1 \approx 1.65364$.***   □

Finding *all* antiderivatives

In the previous example we claimed that any two antiderivatives of the function $f(x) = 2x \sin(x^2)$ may differ only by an additive constant. This important principle deserves specific mention:

> **Fact:** Suppose that $F'(x) = G'(x)$ for all x in an interval I. Then for some constant C, $F(x) = G(x) + C$ for all x in I.

The fact means, in practice, that finding *one* antiderivative for a function f on an interval I is (almost) tantamount to finding *all* antiderivatives: if $F(x)$ is *any* antiderivative, every other antiderivative is of the form $F(x) + C$.

This fact, although most useful for *integrals*, is really a property of *derivatives*; it follows from the mean value theorem.

Continuous functions have antiderivatives. The fundamental theorem guarantees (among other things) that every continuous function f on I *has* an antiderivative on I—namely, the area function A_f.

For ordinary, well-behaved functions of elementary calculus, this is no great surprise. For naughtier functions, like the one shown below, things are murkier.

Does $f(x) = x\sin(1/x)$ have an antiderivative?

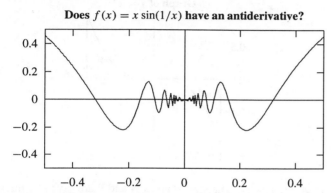

The fundamental theorem says that, naughty or not, f *has* an antiderivative. (*Finding* one in symbolic form may be difficult or impossible—the FTC offers no help there.)

Why the FTC is fundamental

The FTC deserves its high-sounding name for both theoretical and practical reasons. It's fundamental *theoretically* because it connects the two main concepts of the calculus: the derivative and the integral. In geometric terms, the FTC relates the tangent line problem to the area problem: each is a sort of inverse of the other.

The FTC's *practical* consequences are at least as important; implicitly or explicitly, we'll use them again and again. Most important of all, the FTC leads to a practical method➤ of *calculating* specific integrals.

We don't have one yet. So far, we've relied entirely on elementary area formulas.

Using the FTC

Before proving the FTC, let's see it in action.

As a first use, we'll prove another theorem, closely related to the FTC above, but handier in computations:

Theorem 4. (Fundamental theorem, second version.) Let f be continuous on $[a, b]$, and let F be *any* antiderivative of f. Then

$$\int_a^b f(x)\,dx = F(b) - F(a).$$

Proof: This result follows readily from the original FTC. If F is any antiderivative of f, then $F(x)$ can differ from $A_f(x)$—another antiderivative of f, according to the FTC—only by some constant C. In other words, $F(x) = A_f(x) + C$. But then

$$F(b) - F(a) \;=\; \big(A_f(b) + C\big) - \big(A_f(a) + C\big) = A_f(b) - A_f(a)$$

$$=\; \int_a^b f - \int_a^a f = \int_a^b f,$$

just as claimed.   □

Calculating areas

■ **Example 2.**   In Section 5.1 we claimed—but didn't show—that the regions N and P shown shaded have signed areas -0.25 and 0.25, respectively. Show it now.

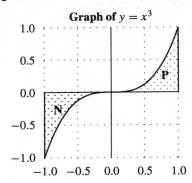

Graph of $y = x^3$

Solution: Let $f(x) = x^3$. The signed areas N and P are given by integrals:

$$\text{signed area N} = \int_{-1}^{0} f(x)\,dx; \quad \text{signed area P} = \int_{0}^{1} f(x)\,dx.$$

Because $F(x) = x^4/4$ is an antiderivative of $f(x)$, the previous theorem applies:

$$\int_{-1}^{0} f(x)\,dx = F(0) - F(-1) = -\frac{1}{4}; \quad \int_{0}^{1} f(x)\,dx = F(1) - F(0) = \frac{1}{4},$$

as we hoped.   □

Bracket notation.   The **bracket notation** offers a convenient shorthand for calculating integrals. For any function F,

$$F(x)]_a^b \qquad \text{means} \qquad F(b) - F(a).$$

Watch for brackets in the next example.

■ **Example 3.**   Evaluate $\int_0^1 x^2\,dx$ and $\int_0^1 x^{10}\,dx$. Interpret each integral as an area.

Solution: The previous theorem says that it's enough to find *any* antiderivatives, plug in endpoints, and subtract. We'll do just that:

$$\int_0^1 x^2\,dx \;=\; \frac{x^3}{3}\bigg]_0^1 = \frac{1}{3};$$

$$\int_0^1 x^{10}\,dx \;=\; \frac{x^{11}}{11}\bigg]_0^1 = \frac{1}{11}.$$

Each integral measures an area:

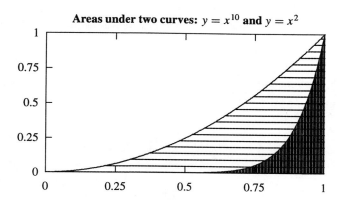

Areas under two curves: $y = x^{10}$ **and** $y = x^2$

The light- and dark-shaded areas represent, respectively, the integrals $\int_0^1 x^2 \, dx = 1/3$ and $\int_0^1 x^{10} \, dx = 1/11$. The picture shows that the answers are numerically reasonable.

■ **Example 4.**   Calculate the area shown shaded below:

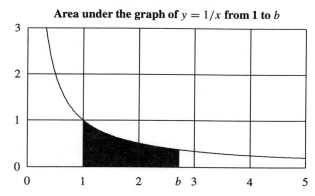

Area under the graph of $y = 1/x$ **from 1 to** b

Which value of b makes the area 1?

Solution: The shaded area is the value of the integral $\int_1^b 1/x \, dx$.   Since the function $\ln x$ is an antiderivative for $1/x$, the FTC says:

$$\int_1^b \frac{1}{x} \, dx = \ln x]_1^b = \ln b - \ln 1 = \ln b.$$

If $b = e$—as it is in the picture—then the area is $\ln e = 1$.   □

From rate to amount

Restated slightly, the second version of the fundamental theorem reads like this:

Fact: Let f be a well-behaved function on $[a, b]$, with derivative f'. Then

$$\int_a^b f'(t) \, dt = f(b) - f(a).$$

In words:

Integrating f' (the rate *function) over $[a, b]$ gives the change in f (the* amount *function) over the same interval.*

This fact has important practical uses—with it, we can deduce *amount* information from *rate* information.

■ **Example 5.** Demand for electric power varies more or less predictably over the course of a day. Drawing on experience, engineers in one mythical small town (a bedroom suburb, perhaps) use the formula

$$r(t) = 4 + \sin(0.263t + 4.7) + \cos(0.526t + 9.4)$$

to model the *rate* of power consumption, in megawatts, t hours after midnight. A graph of r appears below. Is the model reasonable? What does the shaded area represent?

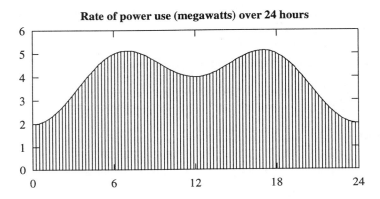

Rate of power use (megawatts) over 24 hours

Solution: The graph looks plausible (though certainly oversimplified.) Residential power demand peaks in the morning, declines during working hours, and rises again in the evening. The formula for r is complicated, but the ingredients—periodic functions and various constants—look right.[◀◀]

Remember how additive and multiplicative constants affect periodic functions?

The shaded area is the *integral* $\int_0^{24} r(t)\, dt$. To understand it in the present context, let $a(t)$ be the total *amount* of power consumed up to time t, starting at midnight.[◀◀] Then $a'(t) = r(t)$; by the fact above,

By definition, therefore, $a = 0$ when $t = 0$.

$$\int_0^{24} r(t)\, dt = a(24) - a(0) = a(24) = \text{total consumption, all day .}$$

With the "classic" FTC (and a little help from above in finding an antiderivative) we can *calculate* this amount. It's easy to *check*[◀◀] that

But harder to guess!

$$4t - \frac{\cos(0.263t + 4.7)}{0.263} + \frac{\sin(0.526t + 9.4)}{0.526}$$

is an antiderivative for $r(t)$. The rest is straight calculation:

$$\int_0^{24} r(t)\, dt = \left. 4t - \frac{\cos(0.263t + 4.7)}{0.263} + \frac{\sin(0.526t + 9.4)}{0.526} \right]_0^{24}$$

$$\approx 95.781 \text{ megawatt hours.} \qquad \qquad \square$$

When *not* to use the FTC

The FTC, useful➤ as it is, won't solve *all* our integral problems. Evaluating integrals by antidifferentiation means, first, *finding a usable antiderivative.*

Beautiful, too.

For a surprising number of integrands, even apparently simple ones, finding an antiderivative formula is hard, or even impossible. None of the following functions, for instance, has an elementary antiderivative:

$$\sin(x^2); \qquad \frac{x}{\ln x}; \qquad 3 + \sin x + 0.3\arcsin(\sin 7x)).$$

Nevertheless, all have perfectly sensible integrals over finite intervals. The picture below shows what $\int_0^{10} f$ means for the last function above:

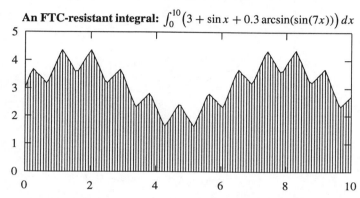

An FTC-resistant integral: $\int_0^{10}\big(3 + \sin x + 0.3\arcsin(\sin(7x))\big)\,dx$

Antiderivative or no antiderivative, the integral certainly exists.➤ Soon we'll study techniques➤ for *estimating* integrals like the one above.

Can you estimate it? We get about 30.

Ones that don't *involve antiderivatives.*

A useful notation for antiderivatives—and a caution

The **indefinite integral notation** for antiderivatives, as in

$$\int x^2\,dx = \frac{x^3}{3} + C \quad \text{and} \quad \int \cos x\,dx = \sin x + C,$$

can be a convenient shorthand. It reflects, too, the fundamental connection between antiderivatives and integrals. A little care is necessary, however:

The whole family. $\int f(x)\,dx$ denotes a *family* of functions, one for each value of C.

Similar signs, different meanings. The **definite integral** $\int_a^b f(x)\,dx$ and the **indefinite integral** $\int f(x)\,dx$ have very different meanings: the first is a *number*, the second a family of *functions*.

Proving the FTC

The FTC is clearly *useful*. By now, we hope, it's *plausible*. But why is it *true*?

The idea of the proof

Recall the setup: f is a continuous function, defined on an interval I containing $x = a$, and

$$A_f(x) = \int_a^x f(t)\,dt.$$

The theorem claims that for x in I,

$$A'_f(x) = \lim_{h \to 0} \frac{A_f(x + h) - A_f(x)}{h} = f(x).$$

To show this, we'll work directly with the difference quotient.

Notice first that $A_f(x + h) - A_f(x) = \int_a^{x+h} f - \int_a^x f = \int_x^{x+h} f$. The pictures below shows why:

$A_f(x + h) - A_f(x) =$ **the shaded area**

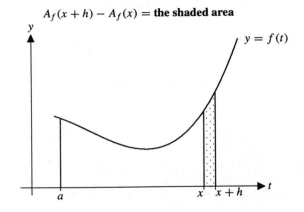

$A_f(x + h) - A_f(x)$—**a closer look**

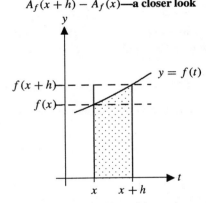

The picture just above shows that for small positive values of h,

$$A_f(x + h) - A_f(x) = \int_x^{x+h} f \approx f(x) \cdot h.$$

(The last quantity is the area of the shorter *rectangle*.) Thus for h near 0,

$$\frac{A_f(x + h) - A_f(x)}{h} \approx \frac{f(x) \cdot h}{h} = f(x).$$

Therefore, as $h \to 0$, the difference quotient tends to $f(x)$, as desired.   □

More on the proof: fine print

A rigorous proof can be constructed from the informal argument above. The missing ingredient is a precise, quantitative approach to the approximation $\int_x^{x+h} f \approx f(x) \cdot h$. It helps if f happens to be *increasing* near x—as it is in the figure above. In that

Draw a picture!   case, ◄◄

$$f(x) \cdot h \le A_f(x+h) - A_f(x) = \text{shaded area} \le f(x+h) \cdot h,$$

so

$$f(x) \le \frac{A_f(x+h) - A_f(x)}{h} \le f(x+h).$$

As $h \to 0$, the quantities on both left and right tend to $f(x)$. So, therefore, must the difference quotient.

(A similar idea works if f is not increasing, but some technical details are different.) □

Exercises

1. Let $F(x) = \int_0^x f(t)\, dt$ where f is the function whose graph is shown below.

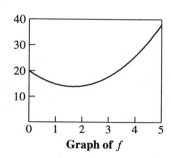

Graph of f

(a) Which of the following is the graph of F? Justify your answer.

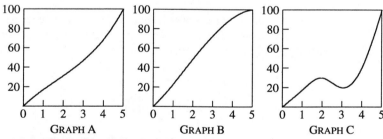

GRAPH A   GRAPH B   GRAPH C

(b) Suppose that $g(0) = 2$ and $g'(x) = f(x)$ for all x. Explain how the graphs of F and g are related.

2. Let $F(x) = \int_0^x f(t)\, dt$ where f is the function graphed below. (The graph of f is made up of straight lines and a semicircle.)

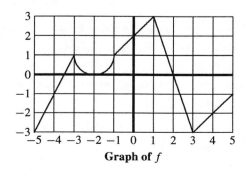

Graph of f

(a) Evaluate $F(-1)$, $F(0)$, and $F(2)$.

(b) Identify all the critical points of F in the interval $[-5, 5]$.

(c) Identify all the inflection points of F in the interval $[-5, 5]$.

(d) What is the average value of f on the interval $[-5, 5]$?

(e) Let $G(x) = \int_0^x F(t)\, dt$. On which subintervals of $[-5, 5]$, if any, is G concave upward?

3. Let $F(x) = \int_0^x f(t)\, dt$ where $f(t)$ is the function shown below:

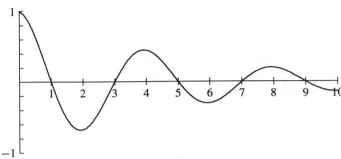

Graph of f

(a) Does $F(x)$ have any local maxima within the interval $[0, 10]$? If so, where are they located?

(b) At what value of x does $F(x)$ attain its minimum value on the interval $[0, 10]$?

(c) On which subinterval(s) of $[0, 10]$, if any, is the graph of $F(x)$ concave up? Justify your answer.

4. Let $G(x) = \int_0^x g'(t)\, dt$ where g is a differentiable function defined on the interval $[0, 10]$ such that the graph of g passes through the point $(5, 1)$ and the derivative of g is the function sketched below.

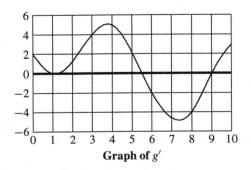

Graph of g'

(a) How are the graphs of g and G related? Explain.

(b) How many solutions of the equation $g(x) = 0$ exist in the interval $[0, 10]$? Explain.

(c) How many solutions of the equation $G(x) = 0$ exist in the interval $[0, 10]$? Explain.

5. Evaluate each of the following definite integrals using the Fundamental Theorem of Calculus.

(a) $\int_1^4 \left(x + x^{3/2}\right) dx$

(e) $\int_1^b x^n \, dx \quad [n \neq -1]$

(b) $\int_0^\pi \cos x \, dx$

(f) $\int_2^{2.001} \frac{x^5}{1000} \, dx$

(c) $\int_{-2}^5 \frac{dx}{x+3}$

(g) $\int_0^{0.001} \frac{\cos x}{1000} \, dx$

(d) $\int_0^b x^2 \, dx$

(h) $\int_0^{\sqrt{\pi}} x \sin \left(x^2\right) dx$

6. Suppose that $\int_0^x f(t) \, dt = 3x^2 + e^x - \cos x$. Find $f(2)$.

7. Find an equation of the line tangent to the graph of $F(x) = \int_1^x \sqrt[3]{t^2 + 7} \, dt$ at $x = 1$.

8. Let f be a function with the following properties:

   (i) $f'(x) = ax^2 + bx$

   (ii) $f'(1) = 6$

   (iii) $f''(1) = 18$

   (iv) $\int_1^2 f(x) \, dx = 18$

Find an algebraic expression for $f(x)$.

9. Let $f(x) = \begin{cases} -2, & \text{if } x < 0 \\ 1, & \text{if } x \geq 0 \end{cases}$ and let $F(x) = \int_{-2}^x f(t) \, dt$.

   (a) Evaluate $\int_{-1}^1 f(x) \, dx$.

   (b) Sketch a graph of $F(x)$ and $f(x)$ on the same axes for $x \in [-2, 2]$.

   (c) Does $F'(1)$ exist? If so, evaluate it using the definition of the derivative. If not, explain why it doesn't exist.

   (d) Does $F'(-1)$ exist? If so, evaluate it using the definition of the derivative. If not, explain why it doesn't exist.

   (e) Explain why $F'(0)$ doesn't exist. Why doesn't this contradict the Fundamental Theorem of Calculus?

10. A company is planning to phase out a product because demand for it is declining. Demand for the product is currently 800 units/month and dropping by 10 units/month each month. To maintain customer and employee relations, the company has announced that it will continue to produce the product for one more year. At the present time, the company is producing the product at a rate of 900 units/month and 1680 units of the product are in inventory.

   (a) Give expressions for the demand and production rates as functions of time t (take $t = 0$ as the present).

   (b) Let $D(t)$ be the demand rate for the product. Explain why $\int_0^t D(s) \, ds$ is the total demand for the product in t months' time.

   (c) Give an expression for the inventory at time t. [HINT: The amount of the product in inventory is the difference between supply and demand at the end of t months.]

(d) The company would like to reduce production at a constant rate of R units/month, with R chosen so as to reduce its inventory to zero at the end of 12 months. What should R be to reduce the inventory to zero at $t = 12$?

11. The function $\ln x$ is sometimes *defined* for $x > 0$ by integration:

$$\ln x = \int_1^x \frac{dt}{t}.$$

Assume $a > 0$ and $x > 0$. Derive the identity $\ln(ax) = \ln a + \ln x$ by carrying out and justifying each of the following steps:

(a) Show that $\left(\int_1^{ax} \frac{dt}{t} \right)' = \frac{1}{x}$. [HINT: Use the chain rule.]

(b) Explain why the result in part (a) implies that that $\ln(ax) = \ln x + C$.

(c) By choosing an appropriate x, show that the value of the constant C in part (b) is $\ln a$.

5.4   Approximating sums: the integral as a limit

The integral $\int_a^b f$, as defined so far, is the *signed*▸ area of the region bounded by the graph of f from $x = a$ to $x = b$. In the last section we saw, in the fundamental theorem of calculus, the crucial inverse relationship between integrals and derivatives. According to the FTC, we can readily calculate any integral for which we can find an antiderivative.

Remember what "signed" means? It's important.

Despite their close relationship, integration and antidifferentiation are *not* the same thing. For many integrals (even "nice" ones such as $\int_0^1 \sin(x^2)\,dx$) antidifferentiation is useless, because no usable antiderivative formula exists. Nevertheless, such integrals clearly exist, as graphs show.

In this section we take a new approach to the integral,▸ defining it, this time, as a certain limit of "approximating sums." The limit definition offers new and useful perspectives on the integral and what it means. Equally important, it lends itself well▸ to the problem of *estimating* integrals numerically.

The approach is new; the integral is the same!

Better than the FTC.

This section resembles Section 2.3, where we first met the limit definition of the *derivative*. There, as here, the general problem was to translate an intuitively reasonable geometric concept▸ into precise analytic language. For both derivative and integral, the idea of limit proves crucial; it links *approximations* to *exact values*.

There, the slope of the tangent line; here, the area under a graph.

Estimating integrals by approximating sums

The formal definition of integral as a limit involves some formidable-seeming terminology and notation. The underlying ideas, however, are natural and straightforward—especially when thought of graphically.

To back up this claim, we'll approach the definition *via* a leisurely example, introducing terms and ideas as we go.

■ **Example 1.** Part of the graph of $f(x) = x^3 - 3x^2 + 8$ appears below; the shaded area is the integral $\int_{0.5}^{3.5} f(x)\,dx$.▸ Use the FTC to evaluate the integral *exactly*. Then use approximating sums to *estimate* the same integral.

First of all, estimate the area roughly, by eye alone. Write your answer in the margin. Don't cheat.

What's the area?

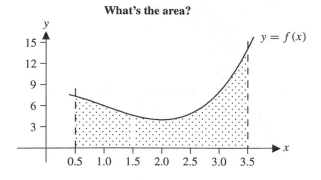
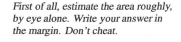

Solution: With the FTC, an exact answer is easy to find by antidifferentiation:▸

Check the steps; they're routine.

$$\int_{0.5}^{3.5} \left(x^3 - 3x^2 + 8\right) dx = \left(\frac{x^4}{4} - x^3 + 8x\right)\Big]_{0.5}^{3.5} = \frac{75}{4} = 18.75.$$

How might we *estimate* this area? The following figures suggest four plausible strategies:

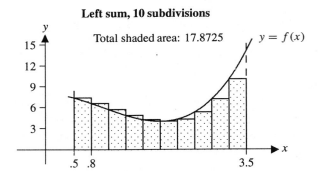

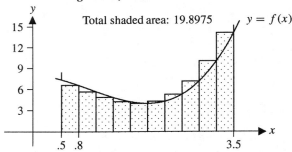

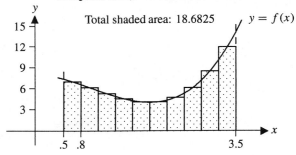

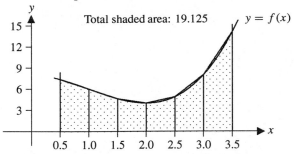

In each case we *approximated* the desired area by a sum of simpler areas—rectangles or trapezoids. (For trapezoids we used 6, not 10, subdivisions in order to show more clearly the difference between the integral and the trapezoid estimate.) Adding up the areasof these simpler figures gives, in each case, a natural estimate to our desired area. ➤ □

Study the pictures; which estimate looks best? Which is best?

Calculating approximate areas

How, exactly, did we calculate each area estimate given above?

A close look at each picture gives the answer. Consider first L_{10}, the **left approximating sum with 10 subdivisions**. The shaded area consists of ten rectangles; each has *base* 0.3. ➤ The rectangles' *heights* vary—the height of each is the value of the function f at the *left* edge of the base. ➤

I.e., (3.5 − 0.5)/10.

Hence the name left sum.

The *second rectangle*, for example, has the x-interval [0.8, 1.1] for its base; its height is $f(0.8)$, i.e., the value of f at the *left* endpoint. Since $f(x) = x^3 - 3x^2 + 8$, $f(0.8) = 6.592$. Therefore for the second rectangle,

$$\text{area} = 0.3 \cdot 6.592 = 1.9776.$$

The *total* area of all ten "left" rectangles, therefore, is the sum

$$
\begin{aligned}
L_{10} &= f(0.5) \cdot 0.3 + f(0.8) \cdot 0.3 + f(1.1) \cdot 0.3 + \cdots + f(3.2) \cdot 0.3 \\
&= 7.375 \cdot 0.3 + 6.592 \cdot 0.3 + 5.701 \cdot 0.3 + \cdots + 7.159 \cdot 0.3 + 10.05 \cdot 0.3 \\
&= 17.8725.
\end{aligned}
$$

The **right approximating sum with 10 subdivisions**, R_{10}, differs only slightly—the height of each rectangle is the value of f at the *right* endpoint of the base interval. ➤ Hence the total area of all ten "right" rectangles is the sum

Think this sentence through carefully. Do you see how the picture says the same thing?

$$
\begin{aligned}
R_{10} &= f(0.8) \cdot 0.3 + f(1.1) \cdot 0.3 + f(1.4) \cdot 0.3 + \cdots + f(3.5) \cdot 0.3 \\
&= 6.592 \cdot 0.3 + 5.701 \cdot 0.3 + \cdots + 10.05 \cdot 0.3 + 14.13 \cdot 0.3 \\
&= 19.8975.
\end{aligned}
$$

For M_{10}, the **midpoint approximating sum with 10 subdivisions**, heights are evaluated at the *midpoint* of each subinterval. Therefore

$$
\begin{aligned}
M_{10} &= f(0.65) \cdot 0.3 + f(0.95) \cdot 0.3 + f(1.35) \cdot 0.3 + \cdots + f(3.35) \cdot 0.3 \\
&= 7.007 \cdot 0.3 + 6.150 \cdot 0.3 + \cdots + 8.465 \cdot 0.3 + 11.93 \cdot 0.3 \\
&= 18.6825.
\end{aligned}
$$

Using a **trapezoid approximating sum** looks, geometrically, like a good idea. *Calculating* it, moreover, is almost as easy as for the "rectangular" sums above. A key idea ➤ makes things simpler: the area of a trapezoid is the average of the areas of the two rectangles determined by its shorter and longer sides. From this fact it follows:

A picture makes this clear; draw one.

> The trapezoid approximation with n subdivisions is the average *of the corresponding* left *and* right *approximations.*

For *our* trapezoid sum (we'll call it T_6): ➤

Numerical details left to you.

$$
\begin{aligned}
T_6 &= \frac{1}{2}(L_6 + R_6) \\
&= \frac{1}{2}(17.4375 + 20.8125) = 19.125.
\end{aligned}
$$

Sigma notation; partitions

Sums with many terms ➤ are tedious to write. **Sigma notation** (*aka* $\sum$-notation) is more efficient than brute force—even better, it shows clearly the similarities and the differences among summands.

Those above, for instance.

■ **Example 2.** Discuss and evaluate each expression:

$$\sum_{k=1}^{4} k; \quad \sum_{j=1}^{4} j; \quad \sum_{i=1}^{10} 3i.$$

First it was k, then j; i through n are all popular choices.

Solution: By definition, $\sum_{k=1}^{4} k = 1 + 2 + 3 + 4$, or simply 10. Although typographically slightly different, the second sum means *exactly* the same thing: $\sum_{j=1}^{4} j = 1 + 2 + 3 + 4 = 10$. The moral: *The name of the **index variable**⁴ doesn't matter.*

The third expression is more complicated, but the ideas are the same:

$$\sum_{i=1}^{10} 3i = 3 \cdot 1 + 3 \cdot 2 + 3 \cdot 3 + \cdots + 3 \cdot 9 + 3 \cdot 10 = 3 \cdot (1 + 2 + \cdots + 10)$$

$$= 3 \cdot 55 = 165.$$

(Notice that the common factor can be pulled outside the sum: $\sum_{i=1}^{10} 3i = 3 \sum_{i=1}^{10} i$. This is the Σ-version of the distributive law.) □

■ **Example 3.** Use sigma notation to rewrite the left, right, and midpoint approximating sums (L_{10}, R_{10}, and M_{10}) we calculated above.

Literally!

Solution: All three approximating sums are based⁴ on subdividing the interval $[0.5, 3.5]$ into 10 equal subintervals, each of length $\Delta x = 0.3$. Here are their endpoints, listed in order:

$$0.5 < 0.8 < 1.1 < 1.4 < \cdots < 3.2 < 3.5.$$

(This ordered set of 11 endpoints is called a **partition** of the x-interval $[0.5, 3.5]$.) For convenience, let's *name* these endpoints—in the same order—$x_0, x_1, x_2, \ldots, x_{10}$.

Now sigma notation is convenient:

$$L_{10} = \sum_{i=0}^{9} f(x_i) \Delta x = f(0.5) \cdot 0.3 + f(0.8) \cdot 0.3 + \cdots + f(3.2) \cdot 0.3;$$

$$R_{10} = \sum_{i=1}^{10} f(x_i) \Delta x = f(0.8) \cdot 0.3 + f(1.1) \cdot 0.3 + \cdots + f(3.5) \cdot 0.3;$$

$$M_{10} = \sum_{i=0}^{9} f\left(\frac{x_i + x_{i+1}}{2}\right) \Delta x = f(0.65) \cdot 0.3 + f(0.95) \cdot 0.3 + \cdots + f(3.35) \cdot 0.3.$$

Why does the R_{10} sum run from $i = 1$ to $i = 10$?, not from $i = 0$ to $i = 9$?

It is a fact. Do you see why?

The $\sum$ form makes the approximating sums appear simple and compact.⁴

If we'd rather see numbers than symbols, we can always use the fact⁴ that $x_i = 0.5 + 0.3i$ to rewrite our sums. Here are two of them:

$$L_{10} = \sum_{i=0}^{9} f(x_i) \Delta x = 0.3 \cdot \sum_{i=0}^{9} f(0.5 + 0.3i);$$

$$M_{10} = \sum_{i=0}^{9} f((x_i + x_{i+1})/2) \Delta x = 0.3 \cdot \sum_{i=0}^{9} f(0.65 + 0.3i).$$ □

Getting better all the time

It stands to reason that an approximating sum—of any type—should do *better* at approximating an integral as the *number* of subdivisions increases. The following pictures (still pertaining to $\int_{0.5}^{3.5} f(x)\,dx$) agree:

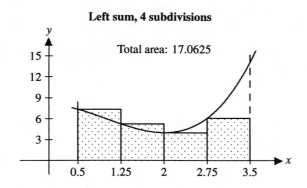

Left sum, 4 subdivisions

Total area: 17.0625

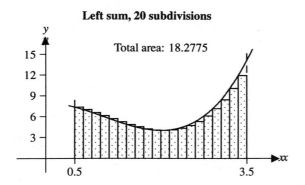

Left sum, 20 subdivisions

Total area: 18.2775

Neither left approximation appears perfect, but the error committed with 20 subdivisions seems much less.

Numerical values of various sums➺ show the same thing:

All computed by machine, of course.

	Approximating sums for the integral $\int_{0.5}^{3.5} f(x)\,dx$			
n	left sum	right sum	midpoint sum	trapezoid sum
2	17.0625	27.1875	17.0625	22.125
4	17.0625	22.125	18.3281	19.5938
8	17.6953	20.2266	18.6445	18.9609
16	18.1699	19.4355	18.7236	18.8027
32	18.4468	19.0796	18.7434	18.7632
64	18.5951	18.9115	18.7484	18.7533
128	18.6717	18.8299	18.7496	18.7508
256	18.7107	18.7898	18.7499	18.7502

Riemann sums and the limit definition of integral

Riemann sums and their ingredients

Left, right, and midpoint approximating sums are all special types of **Riemann sums.** Roughly speaking, *any* "rectangular" approximating sum is a Riemann sum. Riemann's idea of an approximating sum to an integral $\int_a^b f(x)\,dx$ is very liberal. He allows rectangles with *unequal* bases; their heights can be determined anywhere— even randomly—within their respective subintervals.

Here, for instance, is the graphical version of a somewhat irregular Riemann sum for our by-now familiar integral:

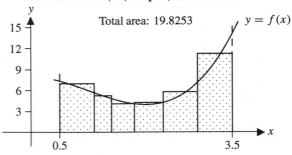

A Riemann sum, 6 (unequal) subdivisions

Not a *good* approximation perhaps, but a legitimate one.

We won't require such generality, but here, for the record, is the formal definition of Riemann sum. The necessary ingredients are a function f and an interval $I = [a, b]$.

Definition: Let I be partitioned into n subintervals by any $n + 1$ points

$$a = x_0 < x_1 < x_2 < \cdots < x_{n-1} < x_n = b;$$

let $\Delta x_i = x_i - x_{i-1}$ denote the width of the i-th subinterval. Within each subinterval $[x_{i-1}, x_i]$, choose any point c_i. The sum

$$\sum_{i=1}^{n} f(c_i)\Delta x_i = f(c_1)\Delta x_1 + f(c_2)\Delta x_2 + \cdots + f(c_n)\Delta x_n$$

is a **Riemann sum with n subdivisions** for f on $[a, b]$.

Our left, right, and midpoint sums fit this definition with room to spare—each is built from a **regular partition** of $[a, b]$ (i.e., one with subintervals of equal length) and some simple, consistent, scheme for choosing the **evaluation points** c_i.

(A *trapezoid* approximating sum, on the other hand, *isn't* quite a Riemann sum, since its approximating figures aren't rectangles. Nevertheless, trapezoidal sums do an excellent job of approximating the integral.)

The integral defined as a limit

Graphical intuition suggests that all of our various approximating sums—L_n, R_n, M_n, T_n—should "converge" toward the true signed area $\int_{0.5}^{3.5} f(x)\,dx$ as n increases. Numerical evidence argues irresistibly for a value somewhere near 18.75.

The limit definition of integral makes these ideas precise:

> **Definition:** Let the function f be defined on the interval $I = [a, b]$. The **integral of f over I**, denoted $\int_a^b f(x)\,dx$, is the number to which all Riemann sums S_n tend as n tends to infinity and as the widths of all subdivisions tend to zero. In symbols:
>
> $$\int_a^b f(x)\,dx = \lim_{n \to \infty} S_n = \lim_{n \to \infty} \sum_{i=1}^{n} f(c_i)\,\Delta x_i\,,$$
>
> if the limit exists.

Note:

When is a function integrable? A function for which the limit in the definition exists is called **integrable** on $[a, b]$. In practice, being integrable on $[a, b]$ is easy—almost every calculus-style function that's defined and *bounded* (i.e., doesn't blow up in absolute value) on $[a, b]$ is integrable. The surprise, rather, is how hard it is to find a *non*-integrable function.➤➤

Here's one: $f(x) = 1$ if x is rational; $f(x) = 0$ if x is irrational.

Well-behaved functions. The limit in the definition, taken at face value, is a slippery customer. Understanding every ramification of permitting arbitrary partitions and sampling points, for example, can be tricky. Luckily, the matter is much simpler for the well-behaved functions (e.g., continuous functions) we typically meet in calculus. For such functions, almost any respectable sort of approximating sum does what we'd expect—approaches the true value of the integral as n tends to infinity.

Notes on notation

The limit definition helps explain the otherwise mysterious dx in the notation $\int_a^b f(x)\,dx$. Consider, e.g., the case of a *right* approximating sum with n *equal* subdivisions, each of length $(b - a)/n = \Delta x$. Then, by definition,

$$\int_a^b f(x)\,dx = \lim_{n \to \infty} \sum_{i=1}^{n} f(x_i)\,\Delta x.$$

Now the left side looks much like the right; "dx" on the left corresponds, in a natural way, to "Δx" on the right.➤➤

We've seen this before:
$$dy/dx = \lim_{\Delta x \to 0} \Delta y / \Delta x.$$

Approximating sums and computers

Approximating sums offer simple, effective, and accurate approximations to almost any integral. Simple for a computer, anyway—calculating approximating sums *by hand* is impractical except for simple integrands and few subdivisions. Calculating approximating sums is the sort of bureaucratic task computers do best–simple, repetitive, and tedious.

Exercises ────────────────────

1. Let g be the function graphed below. Estimate the value of $\int_{-5}^{5} g(x)\,dx$ by evaluating left, right, and midpoint sums each with 5 equal subintervals, then draw sketches that illustrate these sums geometrically.

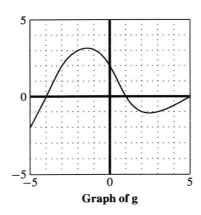

Graph of g

2. Let f be the function graphed below.

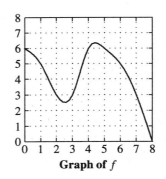

Graph of f

(a) Is $22 < \int_1^7 f(x)\,dx$? Justify your answer.

(b) Estimate the value of $\int_1^7 f(x)\,dx$ using a right sum with three equal subintervals.

(c) Estimate the value of $\int_1^7 f(x)\,dx$ using a left sum with three equal subintervals.

(d) Estimate the value of $\int_0^8 f(x)\,dx$ using a right sum with four equal subintervals.

(e) Estimate the value of $\int_0^8 f(x)\,dx$ using a left sum with four equal subintervals.

3. Let $I = \int_0^5 \sqrt[3]{2x}\,dx$. Use sigma notation to write the left, right, and midpoint approximations to I. (Assume that N = 10 equal subintervals are used.)

4. Let $I = \int_0^5 \sqrt{3x}\,dx$. Use sigma notation to write the left, right, and midpoint approximations to I. (Assume that N equal subintervals are used.)

5. Find a definite integral that is approximated by the midpoint sum

$$\frac{2}{10} \sum_{k=1}^{40} \cos\left(\frac{2k-1}{10}\right).$$

What is the value of this integral?   [HINT: Write out the first few terms of the sum.]

6. Find a definite integral that is approximated by the right sum $\dfrac{2}{100} \displaystyle\sum_{k=1}^{100} \sin\left(\dfrac{2k}{100}\right)$. What is the value of this integral?

7. At time t (measured in hours), $0 \le t \le 24$, a firm uses electricity at the rate of $E(t)$ kilowatts. The power company's rate schedule indicates that the cost per kilowatt-hour at time t is $c(t)$ dollars. Assume that both E and c are continuous functions.

   (a) Set up an N-term left sum that approximates the cost of the electricity consumed in a 24-hour period.

   (b) What definite integral is approximated by the sum you found in part (a)?

8. The table below contains values of a continuous function f at several values of x.

x	1	2	3	4	5	6
$f(x)$	0.14	0.21	0.28	0.36	0.44	0.54

   (a) Estimate $\int_2^5 f(x)\,dx$ using a left sum with 3 equal subintervals. (Be sure to show your work.)

   (b) Do the same using the trapezoid rule.

   (c) Draw a sketch that illustrates the computations you did in parts (a) and (b).

   (d) Show that $f(2) \cdot 2 + f(4) \cdot 3$ is a Riemann sum approximation to $\int_1^6 f(x)\,dx$. Justify your answer.   [HINT: Draw a picture.]

9. The rate at which the world's oil is being consumed is continuously increasing. Suppose that the rate (measured in billions of barrels per year) is given by the function $r(t)$, where t is measured in years and $t = 0$ is January 1, 1990.

   (a) Write a definite integral that represents the total quantity of oil used between the start of 1990 and the start of 1995.

   (b) Suppose that $r(t) = 32e^{0.05t}$. Find an approximate value for the definite integral from part (a) using a right sum with $n = 5$ equal subintervals.

   (c) Interpret each of the five terms in the sum from part (b) in terms of oil consumption.

   (d) Evaluate the definite integral from part (a).

10. Evaluate $\displaystyle \lim_{n \to \infty} \frac{1}{n} \sum_{j=1}^{n} \left(\frac{j}{n}\right)^3$ by expressing it as a definite integral and then evaluating this integral using the Fundamental Theorem of Calculus.

11. Evaluate $\displaystyle \lim_{n \to \infty} \frac{2}{n} \sum_{j=1}^{n} \left(\frac{2j}{n}\right)^3$ by expressing it as a definite integral and then evaluating this integral using the Fundamental Theorem of Calculus.

12. (a) Write the sum $\frac{5}{2}(2.3)^2 + \frac{5}{6}(2.8)^2 + \frac{5}{12}(3.3)^2 + \frac{5}{20}(3.8)^2$ using sigma notation.
    [HINT: $2 = 1 \cdot 2$, $6 = 2 \cdot 3$, $12 = 3 \cdot 4$, and $20 = 4 \cdot 5$.]

    (b) Is the sum in part (a) a Riemann sum approximation to $\displaystyle \int_0^4 x^2\,dx$? Explain.

13. For each of the following definite integrals, draw a sketch illustrating the approximations L_4, R_4, M_4, and T_4. Then compute each approximation and compare it with the exact value of the integral computed using the FTC. How could you have used the sketches to predict the results of these comparisons?

(a) $\displaystyle\int_{-2}^{2} x^2\, dx$ (c) $\displaystyle\int_{-2}^{2} (x+1)\, dx$

(b) $\displaystyle\int_{-2}^{2} x^3\, dx$ (d) $\displaystyle\int_{-\pi/2}^{\pi/2} \sin x\, dx$

14. The table below gives speedometer readings at various times over a one-hour interval.

Speed readings over one hour							
time (min)	0	10	20	30	40	50	60
speed (mph)	42	38	36	57	0	55	51

(a) Plot a plausible speed graph for the one-hour period.

(b) Estimate the total distance traveled using

    (i) A trapezoid approximating sum, 6 subdivisions

    (ii) A left approximating sum, 6 subdivisions

    (iii) A midpoint approximating sum, 3 subdivisions

  Which answer is most convincing? Why?

(c) Plot a plausible distance graph for the one-hour period.

5.5 Approximating sums: interpretations and applications

In the previous section we defined the integral as a limit of approximating sums. In symbols:

$$\int_a^b f(x)\,dx = \lim_{n\to\infty} \sum_{i=1}^{n} f(c_i)\,\Delta x_i.$$

The limit itself involves some technical subtleties, but the basic idea of sums (left sums, right sums, midpoint sums, etc.) approximating an integral is natural and straightforward—especially if thought of graphically.

In this section we look more closely at approximating sums. In the process we'll both see the integral itself in another light and apply the idea of approximating sums to several problems of measurement.

Different views of approximating sums

Approximating sums are **discrete** (i.e., stepwise) approximations to integrals of **continuously** varying functions. This important relationship can be interpreted in several useful ways. We discuss three.

Approximating sums and simpler areas

The simplest and most natural view of approximating sums is geometric; geometrically, approximating sums represent simpler versions (polygons, usually) of "curvy" regions.➥ The shaded region below, for example, might be called a **polygonal approximation** to the area from a to b under the graph of f.

Most of the pictures in this section and the last reflect this point of view.

A polygonal approximation to a curved area

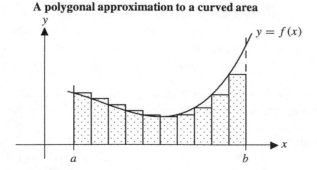

The total *area* of the polygonal region is (in the case shown) the value of a *left* sum; it's evidently a reasonable approximation to $\int_a^b f$, the "true" area under the graph.

Approximating sums and simpler functions

Another view of approximating sums for an integral $\int_a^b f$ concerns the function f. From this point of view an approximating sum amounts to *replacing the original function f with a new, simpler function*, one that's linear on each subinterval➥ of the partition in question. In the graph below, for example, the function f of the previous

I.e., a "piecewise linear" function.

picture was "replaced" with a 6-step linear approximation.

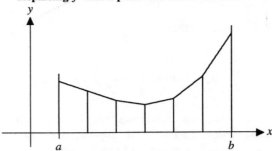

Replacing f with a piecewise-linear function

The area underneath is the value of the trapezoid approximating sum T_6.

Weighted sums

Left, right, midpoint, or any other kind

A Riemann sum◂ for $\int_a^b f$ has the form

$$\sum_{i=1}^{n} f(c_i)\,\Delta x_i = f(c_1)\Delta x_1 + f(c_2)\Delta x_2 + \cdots + f(c_n)\Delta x_n.$$

An ordinary average is another type of weighted sum.

In other words, an approximating sum is a **weighted sum**◂ of values of f at various points c_i (the **sampling points**) in the interval $[a, b]$. The exact choice of sampling points, and the weights attached, depend on the type of approximating sum chosen.

■ **Example 1.** Let $I = \int_0^1 \sin x\, dx$. Let R_4 and M_4 be the midpoint and right approximating sums for I, each with 4 subdivisions. What are the sampling points? What are the weights? What are the weighted sums?

Convince yourself that the forms are correct.

Solution: Calculating M_4 and R_4 is routine:◂

$$
\begin{aligned}
R_4 &= \sin(0.25)\cdot 0.25 + \sin(0.5)\cdot 0.25 + \\
    &\qquad \sin(0.75)\cdot 0.25 + \sin(1)\cdot 0.25 \approx 0.56248; \\
M_4 &= \sin(0.125)\cdot 0.25 + \sin(0.375)\cdot 0.25 + \\
    &\qquad \sin(0.625)\cdot 0.25 + \sin(0.875)\cdot 0.25 \approx 0.460897.
\end{aligned}
$$

The sampling points for R_4 are 0.25, 0.5, 0.75, and 1—the *right* endpoints of four equal subintervals. For M_4, the sampling points are 0.125, 0.375, 0.625, and 0.875, the midpoints of the same subintervals.

Unequal subintervals would give unequal weights.

The weights in an approximating sum are the coefficients of the sample values $f(c_i)$. In the sums above, every sample value has weight $1/4$, the length of the subinterval it comes from.◂ This result seems just; there's no obvious reason to favor one value over another. The weighted sums are the numerical values of R_4 and M_4, shown above.   □

Applications of approximating sums

Approximating sums are good for more than formally defining the integral. Indeed, compared to symbolic integrals, approximating sums are often simpler conceptually, handier in applications, and much, much easier to calculate.◂ For integrals that involve complicated formulas, or no formulas at all,◂ approximating sums may be the *only* practical resort.

For machines, especially.

Such integrals are common in practice.

Beyond formulas—integration with tabular data

Integration is important in the real world. Alas, real-world functions seldom come with handy formulas attached. The FTC, for all its other glories, is not much good without symbolic formulas to act on. Approximating sums are the answer.

■ **Example 2.** If $v(t)$ is the speed (in mph) of a car at time t (in hours), then $\int_a^b v(t)\, dt =$ miles traveled from $t = a$ to $t = b$. To measure distance traveled from $t = 0$ to $t = 1$, "all" we need is the integral $\int_0^1 v(t)\, dt$.

Real cars don't adhere to explicit speed formulas. What *can* be observed, practically speaking, are numerical speed data, like those below. (Notice the irregular time intervals; real data sets often have gaps.)

Speed readings over one hour										
time (min)	0	5	10	15	20	30	35	40	45	55	60
speed (mph)	35	38	36	57	71	55	51	23	10	27	35

How far did the car travel in one hour?

Solution: Without more information (e.g., about speeds *between* the times measured), we'll never know exactly. All we can do is use approximating sums to estimate $\int_0^1 v(t)\, dt$ as best we can.

Our discrete speed data are mathematically compatible with many possible speed functions. The simplest possibility, probably, is drawn by "connecting the dots" (first figure below):➤

We converted time from minutes to hours.

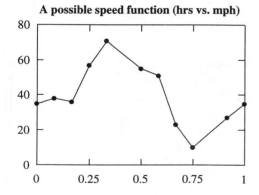

A possible speed function (hrs vs. mph)

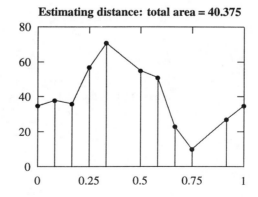

Estimating distance: total area = 40.375

The total bounded area (above right) is easily computed using ten trapezoids;➤ we got 40.375—a reasonable estimate for distance traveled over the hour.

We'll spare you the arithmetic.

The speed function plotted above isn't the only possibility. *Any* curve through the known data points makes mathematical sense.➤ The important point is that the trapezoid estimate depends *only* on the 11 observed data—not on the particular curve drawn through them. The picture below illustrates this idea: it shows a *new* speed function,➤ but the *same* trapezoid estimate (40.375 miles) for distance.

Some curves are physically more likely than others.

Better or worse than the old? What do you think?

New speed function (hrs vs mph) , same trapezoid area

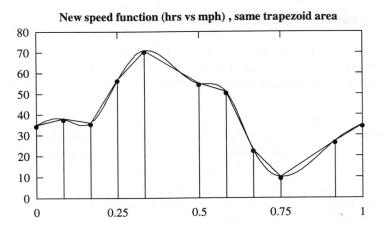

Areas in the plane; measuring by slicing

We first defined the integral in terms of area bounded by the graph of one function. Not surprisingly, integrals can also be used to measure areas of more general plane areas, e.g., areas bounded by two or more graphs. Approximating sums help show how and why.

■ **Example 3.** Using an integral, measure the area shown below. (It's bounded by the graphs of $y = \sin x$ and $y = \cos x$.)

Area between two curves: $y = \sin x$ **and** $y = \cos x$

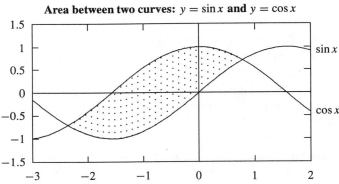

Solution: The picture itself suggests an approximate solution. We'll *slice* the area into vertical strips, approximate each strip by a *rectangle*, and add up the rectangles' areas. Here's the picture for 10 rectangles:◄

The first rectangle has height 0.

Approximating area with 10 "left-rule" rectangles

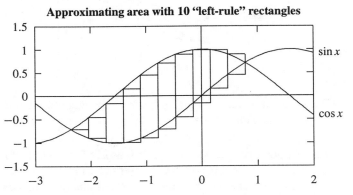

Look carefully:

- The curves intersect at $x = -3\pi/4$ and $x = \pi/4$.➤ (Because $\cos x - \sin x \geq$ 0 over the interval $[-3\pi/4, \pi/4]$, the integral gives the true shaded area desired—even though part of the area lies below the x-axis.)

  Do you see why?

- The height of each rectangle is determined at its *left* edge $x = x_i$; the height itself is $\cos(x_i) - \sin(x_i)$.

- The total area of all ten rectangles is L_{10}, the *left sum* with 10 equal subdivisions for the function $\cos x - \sin x$ on the interval $[-3\pi/4, \pi/4]$. (It isn't necessary to solve our problem, but by machine, calculating L_{10} is easy; we get 2.80513.)

Geometric intuition says that as $n \to \infty$, the area L_n tends to the exact area we want. By the limit definition of integral, L_n *also* tends to the integral in question. We conclude:

> *Our desired area is the integral* $\displaystyle\int_{-3\pi/4}^{\pi/4} (\cos x - \sin x)\,dx.$

Finding an exact answer is now easy, thanks to the FTC:➤

But check the details.

$$\begin{aligned} \text{area} &= \int_{-3\pi/4}^{\pi/4} (\cos x - \sin x)\,dx \\ &= \sin x + \cos x \Big]_{-3\pi/4}^{\pi/4} \\ &= 2\sqrt{2} \approx 2.82843. \end{aligned}$$

□

A more general fact➤ applies:

Draw a general picture to illustrate.

> **Fact:** Let f and g be any continuous functions. Let R be the region bounded above by g, below by f, and on the left and right by $x = a$ and $x = b$. Then R has area $\int_a^b (g - f)$.

Two basic area formulas

The simplest plane regions, from our point of view, are bounded *above* by the graph of one function, below by the x-axis, and on the left and right by vertical lines, like this:

Area at its simplest

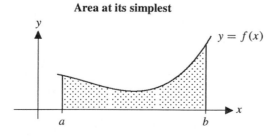

In this simplest case, the integral $\int_a^b f(x)\,dx$ measures the shaded area.

   More interesting, but slightly more complicated, regions may be bounded by various types of curves—not necessarily all graphs of functions. Two common types appear below:➤

Some plane regions are of neither type shown. However, many regions, even quite complicated ones, are combinations of regions of the two types below.

Region 1: top and bottom curves

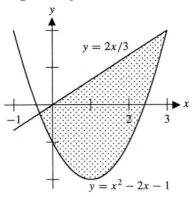

Region 2: right and left curves

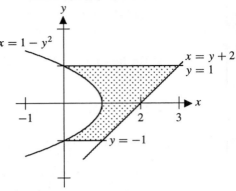

The following rules apply to finding areas like those above:

> **Fact:** Let f and g be continuous functions.
>
> **Integrating in x.** Let R be the region bounded *above* by $y = g(x)$, *below* by $y = f(x)$, on the *left* by $x = a$, and on the *right* by $x = b$. Then R has area
>
> $$\int_a^b \left(g(x) - f(x)\right) dx.$$
>
> **Integrating in y.** Let R be the region bounded on the *right* by $x = g(y)$, on the *left* by $x = f(y)$, *below* by $y = c$, and *above* by $y = d$. Then R has area
>
> $$\int_c^d \left(g(y) - f(y)\right) dy.$$

For Region 2 above, and others like it, integrating in y may be much simpler.

Using the rules

Using the rules effectively may require a combination of graphical, algebraic, and symbolic operations. We illustrate by example.

■ **Example 4.** Find the area of Region 1 above.

Solution: The graphs *appear* to intersect at $x = 3$. Indeed they do. By the quadratic formula,

Convince yourself of this!

$$\tfrac{2}{3}x = x^2 - 2x - 1 \iff x = 3 \text{ or } x = -\tfrac{1}{3}.$$

(This shows, in fact, that the two graphs intersect also at $x = -1/3$, to the left of the shaded region.)

Now, by the first rule (and a close look at the picture):

$$\text{area} = \int_0^3 \left(\tfrac{2}{3}x - \left(x^2 - 2x - 1\right)\right) dx.$$

A simple antiderivative calculation shows the result to be 6.   □

■ **Example 5.** Find the area of Region 2 above.

Solution: Integration in x seems impractical, since defining upper and lower curves is troublesome. However, a close look shows that the second form of the rule above applies, as follows:

$$\text{area} = \int_{-1}^{1} \left(y + 2 - \left(1 - y^2 \right) \right) \, dy.$$

Another simple integration➡ shows that the area is 8/3. □   *In the variable y, this time.*

Any way you slice it

Some areas can be calculated by integration either in x or in y.

■ **Example 6.** Find the area of the region R bounded by the curves $x = 0$, $y = 2$, and $y = e^x$.

Solution: Here's the picture:

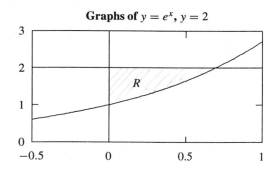

Graphs of $y = e^x$, $y = 2$

The two curves intersect where $e^x = 2$, i.e., at $x = \ln 2$. Therefore

$$\text{area of } R = \int_{0}^{\ln 2} \left(2 - e^x \right) \, dx.$$

To find the same area in terms of y, we rewrite the equation $y = e^x$ in the equivalent form $x = \ln y$;➡ the associated curve becomes our *right* boundary. Then   *Why is this legitimate?*

$$\text{area of } R = \int_{1}^{2} \ln y \, dy.$$

Either integral is easy to calculate. For the last one (recalling that $y \ln y - y$ is an antiderivative of $\ln y$, we get

$$\text{area} = y \ln y - y]_{1}^{2} = 2 \ln 2 - 1 \approx 0.3863.$$

□

Exercises

1. Let A be the region between the curves $y = x$ and $y = x^2$.

   (a) Draw a sketch of the region A.

   (b) Use 5 left rule rectangles to approximate the area of A. (Give a numerical answer.)

   (c) Draw a picture (similar to the one on p. 52) that illustrates the sum you computed in part (b). Use this picture to decide whether your answer overestimates or underestimates the area of A.

   (d) Find the area of A *exactly* by integration.

2. Find the area of the region bounded by the curves $y = \sin x$ and $y = \cos x$, and the lines $x = 1$ and $x = 3$ *exactly* (i.e., express your answer in terms of values of elementary functions rather than as a decimal number.)

Sketch the region bounded by the given curves and find the area of the region *exactly*.

3. $y = x^4$, $y = 1$

4. $y = x$, $y = x^3$

5. $y = x^2$, $y = x^3$

6. $y = x^2 - 1$, $y = x + 1$

7. $x = y^2 - 4$, $y = 2 - x$

8. $y = \sqrt{x}$, $y = 0$, $x = 4$

9. $y = \sqrt{x}$, $y = x^2$

10. $y = 9(4x + 5)^{-1}$, $y = 2 - x$

11. $y = 9\left(4x^2 + 5\right)^{-1}$, $y = 2 - x^2$

12. $y = \sin x$, $y = \cos x$, $x = 1$, $x = 3$

13. $y = 2 + \cos x$, $y = \sec^2 x$, $x = -\pi/4$, $x = \pi/6$

14. $y = e^x$, $y = 0$, $x = 0$, $x = 1$

15. $y = 2^x$, $y = 5^x$, $x = -1$, $x = 1$

5.6   Chapter summary

Chapter 5 introduces the **integral**. The derivative and the integral are the two most important concepts of calculus. Remarkably, they're closely related: the **fundamental theorem of calculus** describes how and why.

The integral as signed area.   In Section 5.1 the integral is introduced graphically, as measuring in terms of **signed area**. ("Signed" means that area below the x-axis counts negatively.) The graphical viewpoint, though insufficient for *exact* calculations, explains clearly what the integral is, and why some of its elementary properties hold.

The area function.   Given a function f and "starting point" $x = a$, we define the related **area function** A_f:

$$A_f(x) = \int_a^x f = \text{signed area bounded by } f \text{ from } a \text{ to } x.$$

Once its peculiar-looking definition is understood, the area function is easy to understand. Simple examples suggest, moreover, that the area function is an **antiderivative** for f. This idea, suitably embellished, is the simplest form of the **fundamental theorem of calculus**.

The fundamental theorem.   Put more formally, the fundamental theorem says that if A_f is defined as above, then

$$\frac{d}{dx}\left(A_f(x)\right) = f(x).$$

This result connects derivatives and integrals—the two main calculus ideas. For that reason, it's deservedly called fundamental.

The same idea, restated in another form, is called the **second version of the fundamental theorem**:

If F is *any* antiderivative of f, then

$$\int_a^b f(x)\,dx = F(b) - F(a).$$

In other words:

To find $\int_a^b f$, find any *antiderivative, plug in endpoints, and subtract.*

As one useful consequence, the theorem permits easy calculations of certain areas.

The integral as a limit of approximating sums.   Like the derivative, the integral must be defined rigorously as a limit—of approximating sums. Section 5.4 describes the procedure. Although symbolically messy, approximating sums make good graphical sense. Several varieties are introduced, compared, and calculated. (With computers, calculating approximating sums is easy; without, it's almost impossible.)

Approximating sums and applications of the integral.   Thinking of the integral as a limit of sums suggests a variety of applications. Several are presented; the main point is that the integral is capable of various interpretations.

One application of the integral concerns rates and amounts:

> *If $f(t)$ tells the* rate *at which a quantity varies at time t, then* $\int_a^b f$ *tells the* amount *by which the same quantity varies over the interval from* $t = a$ *to* $t = b$.

Another application of the integral is to calculating areas of plane regions defined by graphs of functions. Such regions come in many shapes and forms. Choosing exactly the right integral to calculate a given area can be a challenging problem. We illustrate various techniques for doing so.

Chapter 6

Finding Antiderivatives

6.1 Antiderivatives: the idea

Integrals, antiderivatives, and the FTC

This chapter is about antidifferentiation—finding, for a given function f, a new function F for which $F' = f$.

The fundamental theorem of calculus➤ (FTC) offers ample motivation for doing so. It says that if f is a continuous function and F is any antiderivative of f, then

$$\int_a^b f(x)\, dx = F(b) - F(a).$$

Finding a nice➤ antiderivative function F, in effect, translates an integral problem into something easier: plugging a and b into F. A simple example illustrates this principle—and its importance.

Here, "nice" means "easy to evaluate."

■ **Example 1.** Calculate $\displaystyle\int_0^\pi \sin x\, dx$. What does the answer mean graphically?

Solution: Any function of the form $F(x) = -\cos x + C$ is an antiderivative of $f(x) = \sin x$. For simplicity, let's use $C = 0$.➤ By the FTC,

Other values of C give the same answer.

$$\int_0^\pi \sin x\, dx = -\cos x\Big]_0^\pi = -\cos \pi + \cos 0 = 2.$$

Geometrically, the integral represents the area under one "arch" (from $x = 0$ to $x = \pi$) of the sine function. Surprisingly enough, this area measures exactly 2 square units. □

Finding antiderivatives

To use the FTC we need, first, to find an antiderivative function F. Must there *be* an antiderivative? If so, how many antiderivatives are there?

Existence. If f is continuous (we'll assume so➤ throughout this chapter) the short answer➤ is "yes": f *does* have an antiderivative. In fact, the other version of the FTC says that the *area function*

It's reasonable to do so. If f is very badly behaved, $\int_a^b f$ may not even exist.

But not the whole answer.

$$A_f(x) = \int_a^x f(t)\, dt$$

59

is an antiderivative of f.

It's nice to know in the abstract that every continuous function f has an antiderivative F. A better question for us, however, is whether f has an antiderivative function *we can use*. To be useful for evaluating integrals, an antiderivative F must have a concrete *formula*, one that accepts numerical inputs and produces numerical outputs.

Most recently in the previous example. See also Section 5.3.

See the Fact on page 28.

How many antiderivatives? As we've seen repeatedly,^{◂◂} antiderivatives are not "unique." Far from it: if $F(x)$ is an antiderivative of $f(x)$, then so is $F(x) + C$, for *any* constant C. Thus a given function f has *infinitely many* different antiderivatives.

Not *very* different, however: we showed in Section 5.3^{◂◂} that any two antiderivatives can differ *only* by an additive constant on any domain interval. In other words, finding *one* antiderivative is essentially tantamount to finding *all* antiderivatives.

Elementary functions, derivatives, and antiderivatives

Functions with "formulas" are called **elementary**; they're built from the standard function **elements**—polynomials, trigonometric and exponential functions, and their inverses.^{◂◂}

Roughly speaking, elementary functions are those a cheap scientific calculator can handle.

Every elementary function has an elementary *derivative*. (This is just a pompous way of saying that differentiating a formula gives another formula.) The situation for antiderivatives is different: an antiderivative of an elementary function may or may *not* be elementary. Even deciding whether a function *has* an elementary antiderivative—let alone finding one—can be very difficult indeed. Thus the "generic" problem of this chapter:

> *Given an* elementary *function* f, *find an* elementary *function* F *such that* $F' = f$.

Either by human or by machine. Maple's int *command doesn't always succeed, for example.*

can not *always* be solved.^{◂◂} The problem itself is sometimes called **integration in closed form**; an elementary solution (if one exists) is called a **closed-form solution**.

Antidifferentiation in closed form: art, science, or parlor trick?

The uncertainty discussed above adds an element of mystery and intrigue to our problem, elevating it somewhat above the mundane level of routine calculation. (Differentiation, by contrast, *is* routine calculation.) Finding antiderivatives can be challenging, but the rewards are proportional to the effort.

We'll attack the antiderivative problem with a modest arsenal of antidifferentiation techniques. Like differentiation rules, antidifferentiation techniques can reduce a complicated problem to one or more simpler ones. *Unlike* differentiation rules, antidifferentiation rules don't always succeed; choosing the "right" antidifferentiation rule^{◂◂} combines art, science, intuition, and (sometimes) luck.

Or rules.

Notation and terminology

We'll use several standard notations and technical terms in our tour of the antiderivative problem:

Indefinite integrals. The expression

$$\int f(x)\, dx$$

is called an **indefinite integral**; f is the **integrand** and x is the **variable of integration.** The indefinite integral is a symbolic shorthand for the antiderivative problem itself.^{◂◂} Writing

The name and notation aren't perfect; integrals and antiderivatives aren't really the same thing.

$$\int \cos x \, dx = \sin x + C$$

means, for example, that the antiderivatives of $\cos x$ are precisely the functions $\sin x + C$, where C, the **constant of integration**, may take any value.

One *caveat*: interpreted scrupulously, the indefinite integral $\int f(x)\, dx$ denotes not just one function but many—the "family" of *all* possible antiderivatives of f.➤ The distinction isn't always important in practice. Writing, say, $\int \cos x \, dx = \sin x$ seldom causes confusion.

For nice functions f, antiderivatives differ only by additive constants.

Definite integrals. The expression

$$\int_a^b f(x)\, dx,$$

with endpoints "definitely" specified, is called a **definite integral.**➤

A "definite" integral has one value, not many.

The FTC links definite and indefinite integrals. In words:

The definite integral $\int_a^b f(x)\, dx$ equals the change $F(b) - F(a)$ *from $x = a$ to $x = b$ in any indefinite integral.*

Antidifferentiation by machine. Many high-powered mathematical software attack the (hard) problem of elementary antidifferentiation. *Maple*'s int command, for example, tries to find antiderivatives in closed form. For the good reasons cited above, any antiderivative program sometimes fails—for a given problem, there may *be* no closed-form solution.

Until recent times there was no reliable method of deciding, once and for all, whether a function has an elementary antiderivative. In the late 1960's, R.H. Risch discovered an algorithm that determines *whether* a function has an elementary antiderivative and, if so, finds one. Although the Risch algorithm (and its more recent variants) can be implemented by computer, most modern software uses faster, less cumbersome methods.

Elementary antiderivatives: why bother looking?

Everything above notwithstanding, many useful functions *do* have elementary antiderivatives. We'll have great fun finding them.

But why bother? The question is a serious one. As we'll soon see,➤ numerical methods let us compute many definite integrals, to high accuracy, without fussing over antiderivatives. Here are three good reasons for our trouble:

Starting in Chapter 7.

Formulas for success. Elementary functions—those with "formulas"—are in many ways the simplest and most convenient possible functions. In applied mathematics "finding a formula" that usefully models an interesting phenomenon represents the best imaginable outcome.

As a simple but important example, consider how an object moves along a straight line, starting from rest at time $t = 0$. If the object's acceleration a is *constant* (as in "free fall"), then the velocity function v is *linear* and the distance-traveled function s is *quadratic*. More precisely:

$$\textit{If } a(t) = a, \textit{ then } v(t) = at, \textit{ and } s(t) = \frac{at^2}{2}.$$

Thus a simple *formula* for acceleration leads—in this case—to almost equally simple *formulas* for velocity and position, its first two antiderivatives.

Not all applications work so smoothly, of course. Complex combinations of forces may produce complicated acceleration functions, perhaps without elementary antiderivatives. Nevertheless, the point remains: the practical and theoretical value of *having* antiderivative formulas amply justifies some effort in *finding* them.

Predicting planetary motion. Planetary motion, the subject of **celestial mechanics**, has fascinated observers throughout history. The search for *formulas* to describe—and predict—planetary motion has gone on for many thousands of years.

Copernicus, Kepler, and Galileo, working in the 16th and early 17th centuries, began the task of describing planetary and gravitational phenomena in modern mathematical terms. Isaac Newton, in his *Principia* (published in 1687), used the tools of calculus to extend and unify earlier work. Newton showed, for instance, that his inverse-square law for gravitation (gravity varies inversely with the square of the distance) implies Kepler's first law: that the orbit of a planet is an *ellipse*, with the sun as one focus.

Exact answers. Evaluating integrals in closed form gives *exact* answers; numerical results, however accurate, are only approximate.

The difference can be important. For example, the fact that

$$\int_{-1}^{1} \frac{2}{1+x^2} \, dx \approx 3.141592614$$

is easy to show numerically. It's not very exciting, however: one number is much like another. By contrast, the fact that

$$\int_{-1}^{1} \frac{2}{1+x^2} \, dx = 2\arctan x \bigg]_{-1}^{1} = \pi,$$

which depends on antidifferentiation, is an interesting surprise—how in the world does π arise?

Integrals with parameters: many for the price of one. Many useful integrals involve **parameters**—constants whose values we don't want to specify in advance. Consider, for example, the integral

$$\int_{0}^{\pi} \sin(kx) \, dx,$$

If $k = 0$, the problem is trivial.

Think about it. Do you see why?

Check details.

for k a nonzero integer.◀ Without a specific value for k, numerical methods are helpless.◀ Antidifferentiation solves the problem for *all* values of k, at one blow:◀

$$\int_{0}^{\pi} \sin(kx) \, dx = -\frac{\cos(kx)}{k} \bigg]_{0}^{\pi} = \frac{1 - \cos(k\pi)}{k}.$$

Thus the parity of k determines the answer: it's zero if k is even, $2/k$ if k is odd.

Derivatives and antiderivatives: basic strategies

Antidifferentiation is differentiation in reverse. The expert antidifferentiator, therefore, has a ready mastery of the easier skill of differentiation and remembers this useful rule:

> *Antidifferentiation is hard; differentiation is easy. Therefore, check antiderivatives by differentiation.*

■ **Example 2.** Is $\int e^x \sin x \, dx = e^x \cos x + C$?

Solution: No. Here's why: $\left(e^x \cos x \right)' = e^x \cos x - e^x \sin x \neq e^x \sin x$. □

Caveat. Some care is necessary. When checking antiderivatives, one should realize that some functions➤➤ can be written in more than one way.

Trigonometric functions, especially.

■ **Example 3.** Is $\int \sin(2x) \, dx = \sin^2(x) + C$?

Solution: Let's check. By the chain rule,

$$\left(\sin^2(x) \right)' = 2 \sin x \cos x.$$

We hoped for $\sin(2x)$, so we're tempted to answer "no."

The *right* answer, however, is "yes." One of the many trigonometric identities says that

$$2 \sin x \cos x = \sin(2x).$$

We stand corrected. □

Basic antiderivative formulas

Differentiation techniques➤➤ start with a few *known* derivatives; combining them properly produces new derivatives. Antidifferentiation, too, requires a basis of known formulas to which others are "reduced." We'll use these:➤➤

E.g., the product and quotient rules.

Know 'em!

Basic antiderivative formulas

$$\int x^k \, dx = \frac{x^{k+1}}{k+1} + C \qquad (\text{if } k \neq -1)$$

$$\int \frac{1}{x} \, dx = \ln|x| + C$$

$$\int e^x \, dx = e^x + C$$

$$\int a^x \, dx = \frac{a^x}{\ln a} + C \qquad (\text{if } a \neq 1)$$

$$\int \sin x \, dx = -\cos x + C$$

$$\int \cos x \, dx = \sin x + C$$

$$\int \sec^2 x \, dx = \tan x + C$$

$$\int \frac{dx}{\sqrt{1-x^2}} = \arcsin x + C$$

$$\int \frac{dx}{1+x^2} = \arctan x + C$$

Note:

Why they're true. Antiderivative rules are nothing more than derivative rules "in reverse." The rule $\int \cos x \, dx = \sin x + C$, for instance, means simply that

$$(\sin x + C)' = \cos x + 0 = \cos x.$$

A subtlety: logs and absolute values. The formula $\int \dfrac{1}{x} \, dx = \ln|x| + C$ has an unexpected ingredient: the absolute value. After all, $\ln x$ itself is an antiderivative for $1/x$—why bother with an absolute value?

Positive and negative numbers.

The answer has to do with *domains*. The function $f(x) = 1/x$ is defined for all $x \neq 0$; ideally, an antiderivative should have the same domain. Unfortunately, the unadorned logarithm function $F(x) = \ln x$, accepts only *positive* inputs x. Using $G(x) = \ln|x|$ solves this problem neatly. This G has both the "right" domain⁀ and the right antiderivative. The graphs below give convincing evidence; for more formal arguments, see the exercises.

Graph of $y = 1/x$

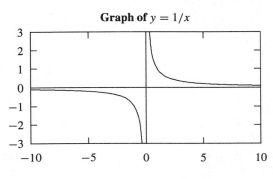

Graph of $y = \ln|x|$

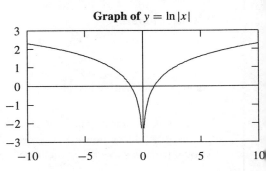

Combining antiderivatives: sums and constant multiples

Every derivative rule, read "backwards," says something about *antiderivatives*. For sums and constant multiples, the rules are especially simple.➤ If f and g are continuous functions and k is any constant, then

$$\int \left(f(x) + g(x) \right) dx = \int f(x)\, dx + \int g(x)\, dx;$$

$$\int k \cdot f(x)\, dx = k \cdot \int f(x)\, dx.$$

We'll read the chain and product rules backwards in the next two sections.

■ **Example 4.** Find $\displaystyle\int \left(3\cos x - 2e^x \right) dx$.

Solution: Use the rules and the basic formulas:

$$\int \left(3\cos x - 2e^x \right) dx = \int 3\cos x\, dx - \int 2e^x\, dx$$

$$= 3\int \cos x\, dx - 2\int e^x\, dx$$

$$= 3\sin x - 2e^x + C.$$

The answer, as usual, is easy to *check*➤ by differentiation.   □

Do so, please.

Guess-and-check

Educated guessing is a perfectly respectable strategy for antidifferentiation. Because differentiation is simple, most guesses are easily checked. Incorrect guesses, moreover, are often easy to "repair." The next two examples show what we mean.

■ **Example 5.** Find $\displaystyle\int \cos(2x)\, dx$ by guessing.

Solution: Since $\int \cos x\, dx = \sin x + C$, we might—just guessing—try $\sin(2x)$ as an antiderivative. Differentiation shows that we missed, but not by much:➤

Only by a factor of 2.

$$\left(\sin(2x) \right)' = \cos(2x) \cdot 2 \neq \cos(2x).$$

A minor alteration is all that's needed:

$$\left(\frac{\sin(2x)}{2} \right)' = \cos(2x) \implies \int \cos(2x)\, dx = \frac{\sin(2x)}{2} + C.$$ □

■ **Example 6.** Find $\displaystyle\int \frac{6}{1+4x^2}\, dx$ by creative guessing.

Solution: The integrand's form suggests an arctangent, perhaps with some constants attached, as in $F(x) = A\arctan(Bx)$. Differentiation gives➤

Check this; use the chain rule carefully.

$$F'(x) = \left(A\arctan(Bx) \right)' = \frac{A}{1+(Bx)^2} \cdot B = \frac{AB}{1+B^2 x^2}.$$

Our desired antiderivative therefore has $AB = 6$ and $B^2 = 4$. Setting $A = 3$ and $B = 2$ will do nicely; $F(x) = 3\arctan(2x)$ is a suitable antiderivative.   □

Exercises

Evaluate each of the following antiderivatives. Check your answers by differentiation.

1. $\int \left(3x^2 + 4x^{-2}\right) dx$

2. $\int \left(3e^{4x} + 2^x\right) dx$

3. $\int \left(2\sin(3x) - 4\cos(5x)\right) dx$

4. $\int 4\sec^2(3x) \, dx$

5. $\int \dfrac{dx}{1 + 4x^2}$

6. $\int \dfrac{2}{\sqrt{1 - 9x^2}} \, dx$

7. $\int \dfrac{6}{9 + x^2} \, dx$

8. $\int \dfrac{8}{\sqrt{4 - x^2}} \, dx$

9. (a) Use the trigonometric identity $\sin^2 x = \frac{1}{2}(1 - \cos(2x))$ to evaluate $\int \sin^2 x \, dx$.

   (b) Use the trigonometric identity $\cos^2 x = 1 - \sin^2 x$ and the result in part (a) to evaluate $\int \cos^2 x \, dx$.

   (c) Explain why $\displaystyle\int_0^{2\pi} \sqrt{1 - \cos(2x)} \, dx \neq \sqrt{2} \int_0^{2\pi} \sin x \, dx$.

Some antiderivatives are difficult to *find* from scratch, but may be *checked* easily by differentiation. Verify each of the following antiderivative formulas by differentiation.

10. $\int \dfrac{dx}{x^2 + a^2} = \dfrac{1}{a} \arctan\left(\dfrac{x}{a}\right) + C$

11. $\int \tan x \, dx = \ln|\sec x| + C$

    [HINT: Recall that $\left(\ln|x|\right)' = 1/x$.]

12. $\int \sec x \, dx = \ln|\sec x + \tan x| + C$

    [HINT: Some trigonometric simplification is necessary after differentiating the right side.]

13. $\int \dfrac{dx}{1 - x^2} = \dfrac{1}{2} \ln\left|\dfrac{1 + x}{1 - x}\right| + C$ [HINT: $\ln\left|\dfrac{1 + x}{1 - x}\right| = \ln|1 + x| - \ln|1 - x|$.]

14. $\int \arcsin x \, dx = x \arcsin x + \sqrt{1 - x^2} + C$

15. $\int \ln x \, dx = x \ln x - x + C$

16. $\int \arctan x \, dx = x \arctan x - \dfrac{1}{2} \ln\left(1 + x^2\right) + C$

17. $\int \dfrac{dx}{(2x + 3)^4} = -\dfrac{1}{6(2x + 3)^3} + C$

Some antiderivatives—trigonometric ones, especially—can be written in more than one way. Verify each of the following formulas using the trigonometric identities $\sin^2 x + \cos^2 x = 1$, $\sin(2x) = 2\cos x \sin x$, and $\cos(2x) = 2\cos^2 x - 1$.

18. $\displaystyle\int \cos(2x)\, dx = \frac{1}{2}\sin(2x) + C = \cos x \sin x + C$

19. $\displaystyle\int \cos^2 x \, dx = \frac{x}{2} + \frac{1}{2}\cos x \sin x + C = \frac{x}{2} + \frac{1}{4}\sin(2x) + C$

Determine whether each of the following integration formulas is true or false by differentiating the right side.

20. $\displaystyle\int e^{\sin x}\, dx = e^{-\cos x} + C$ 21. $\displaystyle\int \frac{2}{1+4x^2}\, dx = \arctan(2x) + C$

22. We've known for a long time that for *positive* x, $(\ln x)' = 1/x$. In this section we claimed (see page 64) that $\ln |x|$ is an antiderivative of $1/x$ for *all* nonzero inputs x. This problem justifies this new claim.

   (a) Do the graphs on page 64 support our claim? Explain briefly in words.

   (b) If $x > 0$, then $\ln |x| = \ln x$. So $(\ln x)' = (\ln |x|)' = 1/x$, as claimed. Now show that $(\ln |x|)' = 1/x$ even if $x < 0$.   [HINT: If $x < 0$, then $\ln |x| = \ln(-x)$.]

6.2   Antidifferentiation by substitution

The simplest and most important antidifferentiation technique is called, variously, **direct substitution**, *u*-**substitution**,[◄] or just plain **substitution**.

Antidifferentiation—by any method—is nothing more than differentiation in reverse. Substitution, in particular, can be understood as the *chain rule* running backward. We'll explain the connection more precisely later in this section. In the meantime, watch the chain rule appear again and again as we check our answers.

Substitution: how it works

For obscure reasons, the letter u is almost invariably used in this setting. We'll follow convention.

We'll use examples[◄] to see both *how* substitution works and *that* it works.

Read them all carefully; they're the heart of this section.

I.e., antidifferentiate $2x \cos\left(x^2\right)$.

■ **Example 1.**   Find $\displaystyle\int 2x \cos\left(x^2\right) dx.$[◄]

Solution: Let $u = u(x) = x^2$; then $du/dx = 2x$, so $du = 2x\,dx$. (Whoa—can du and dx be legally separated this way? The short answer is "yes," but we'll return to such questions below.) Substituting everything into the original integral gives

$$\int \cos\left(x^2\right) 2x\,dx = \int \cos u\,du.$$

The last integral is simpler than the first. The rest is routine. First we antidifferentiate, then substitute *back* for u:

$$\int \cos u\,du = \sin u + C = \sin\left(x^2\right) + C.$$

Watch the chain rule in action.

Did it work? An easy differentiation shows that it did:[◄]

$$\left(\sin\left(x^2\right) + C\right)' = \cos\left(x^2\right) \cdot 2x,$$

as we wanted.   □

Separating du and dx might seem suspicious, for example.

That was easy, quick, and possibly dirty.[◄] Notice, though, that the method—dirty or not—*did* work. It produced an answer we could easily *check*.

■ **Example 2.**   Find $\displaystyle\int \frac{x}{1+x^2}\,dx.$

Solution: Let $u(x) = 1 + x^2$; then $du/dx = 2x$ and $du = 2x\,dx$. Now we'll substitute:

$$\int \frac{x}{1+x^2}\,dx = \frac{1}{2}\int \frac{2x}{1+x^2}\,dx = \frac{1}{2}\int \frac{du}{u}.$$

The first step was for convenience—multiplying by 2 inside the integral sign "made" the desired du. (We compensated for this by dividing by 2 outside.)

In its new, simpler form, the antiderivative problem is easier:

$$\frac{1}{2}\int \frac{du}{u} = \frac{1}{2}\ln|u| + C = \frac{1}{2}\ln\left(1+x^2\right) + C.$$

Checking the answer is another chain rule exercise:

$$\left(\frac{1}{2}\ln\left(1+x^2\right) + C\right)' = \frac{1}{2}\cdot\frac{2x}{1+x^2} = \frac{x}{1+x^2}.$$ □

The idea of u-substitution: trading one integral for another

As the examples above show, a successful u-substitution transforms one indefinite integral into another, simpler than the first. The process has three steps:

Substitute. Choose➤ a function $u = u(x)$, and write $du = u'(x)\,dx$. Then substitute *both u and du* into the original integral, $\int f(x)\,dx$, to produce a new one of the form $\int g(u)\,du$.

Judiciously!

Antidifferentiate. Solve $\int g(u)\,du$ as an antidifferentiation problem in u—i.e., find a function $G(u)$ for which $G'(u) = g(u)$.

Re-substitute. Substitute *back* to eliminate u. The result, $F(x) = G(u(x))$, is an antiderivative of f; so is *any* function of the form $F(x) + C$.

Examples: a substitution sampler

Substitution is simple enough in theory. In practice, on the other hand, finding the "right" substitution is as much art as science.➤ In both realms, practice (false starts and all) makes perfect. The following examples illustrate some of the possibilities, and some useful tricks of the trade.

This makes success in substitution especially satisfying.

Choosing u: wise and foolish choices. The first step—choosing u properly—is the tricky part. Substitution works best when➤ both u and du appear conveniently in the original integrand. The new integral, moreover, should be simpler or more familiar than the old.

As in the previous examples.

■ **Example 3.** Find $\displaystyle\int \sin^3 x \cos x\,dx$.

Solution: If $u = \sin x$, then $du = \cos x\,dx$. With u and du staring at us, things look promising. The rest is routine:

$$\int \sin^3 x \, \cos x \, dx = \int u^3 \, du = \frac{u^4}{4} + C = \frac{\sin^4 x}{4} + C. \qquad \square$$

■ **Example 4.** Find $\displaystyle\int \frac{x}{1+x^4}\,dx$.

Solution: No obvious substitution suggests itself. We might try $u = 1 + x^4$, but then $du = 4x^3\,dx$, and nothing similar appears in the given integrand. A better choice, it turns out, is $u = x^2$. Then $du = 2x\,dx$, and➤

Watch the 2's carefully.

$$\int \frac{x}{1+x^4}\,dx = \frac{1}{2}\int \frac{2x}{1+x^4}\,dx = \frac{1}{2}\int \frac{du}{1+u^2}.$$

We've made progress at last; the last antiderivative is standard:

$$\frac{1}{2}\int \frac{du}{1+u^2} = \frac{1}{2}\arctan u + C = \frac{1}{2}\arctan\left(x^2\right) + C.$$

Checking the answer involves, as usual, the chain rule.➤ $\qquad \square$

Details left to you.

Traveling constants. Multiplicative constants can move freely (and legally) in and out of integrals. We exploit this freedom in the next example.➤

As we did in Example 2, above.

■ **Example 5.**   Find $\displaystyle\int \frac{3x}{5x^2+7}\,dx$.

Except for multiplicative constants.

Watch the constants carefully.

Solution: The numerator is almost [*] the derivative of the denominator; this brings $\int du/u$ to mind. If we set $u = 5x^2 + 7$, then $du = 10x\,dx$, and [*]

$$
\begin{aligned}
\int \frac{3x}{5x^2+7}\,dx &= \frac{3}{10}\int \frac{10x}{5x^2+7}\,dx \\
&= \frac{3}{10}\int \frac{du}{u} \\
&= \frac{3}{10}\ln|u| + C \\
&= \frac{3}{10}\ln\left(5x^2+7\right) + C.
\end{aligned}
$$

Differentiation shows that we're right:

$$
\left(\frac{3}{10}\ln(5x^2+7) + C\right)' = \frac{3}{10}\frac{10x}{5x^2+7} = \frac{3x}{5x^2+7}. \qquad \square
$$

Inverse substitutions: writing x and dx in terms of u and du

In examples so far we've written u and du in terms of x and dx. Sometimes it's simpler to write x and dx in terms of u and du. The next two examples illustrate this strategy.

■ **Example 6.**   Find $\displaystyle\int \frac{x}{\sqrt{2x+3}}\,dx$.

Convince yourself.

Solution: We'll start as usual. If $u = \sqrt{2x+3}$, then [*] $du = dx/\sqrt{2x+3}$. Substituting gives

$$
\int \frac{x}{\sqrt{2x+3}}\,dx = \int x\,\frac{dx}{\sqrt{2x+3}} = \int x\,du.
$$

So far so good, but what about that remaining x?

The trick is to write x in terms of u, like this:

$$
u = \sqrt{2x+3} \implies u^2 = 2x+3 \implies \frac{u^2-3}{2} = x.
$$

Now we're on our way:

$$
\begin{aligned}
\int x\,du &= \int \frac{u^2-3}{2}\,du \\
&= \frac{u^3}{6} - \frac{3u}{2} + C \\
&= \frac{(2x+3)^{3/2}}{6} - \frac{3\sqrt{2x+3}}{2} + C.
\end{aligned}
$$

For variety, let's redo the problem a bit differently, still using $u = \sqrt{2x+3}$. Notice first that

$$
\frac{u^2-3}{2} = x \implies \frac{2u}{2}\,du = dx \implies u\,du = dx.
$$

Finally, substituting for x, dx, and $\sqrt{2x+3}$ gives

$$\int \frac{x}{\sqrt{2x+3}}\, dx = \int \frac{u^2-3}{2u}\, u\, du = \int \frac{u^2-3}{2}\, du,$$

just as above.   □

■ **Example 7.**   Find $\displaystyle\int \frac{dx}{1+\sqrt{x}}$.

Solution: Let $u = \sqrt{x}$. Then, by easy calculations,

$$x = u^2 \quad \text{and} \quad dx = 2u\, du.$$

Substituting for x and dx gives

$$\int \frac{dx}{1+\sqrt{x}} = \int \frac{2u}{1+u}\, du.$$

The last integral is certainly simpler than the first. *Another* substitution makes it simpler still. If $v = 1 + u$, then $dv = du$ and $u = v - 1$, so➤   *Check the algebra.*

$$\int \frac{2u}{1+u}\, du = \int \frac{2(v-1)}{v}\, dv = 2\int\left(1 - \frac{1}{v}\right)\, dv = 2(v - \ln|v|) + C.$$

Finally, we substitute back for x:

$$2(v - \ln|v|) + C = 2(1 + u - \ln|1+u|) + C = 2 + 2\sqrt{x} - 2\ln|1+\sqrt{x}| + C\,_\square$$

Substitution in definite integrals

The next examples show two ways of evaluating *definite* integrals by substitution.

■ **Example 8.**   Find $\displaystyle\int_0^{\sqrt{\pi/2}} 2x\, \cos\left(x^2\right)\, dx$.

Solution: (**Antidifferentiate in x; plug in endpoints.**)   We'll write $f(x) = 2x\, \cos\left(x^2\right)$. If $F(x)$ is any antiderivative of $f(x)$, then➤ $F(\sqrt{\pi/2}) - F(0)$ is our   *By the FTC.*
answer.

We've already found a suitable F: using the substitution $u = x^2$, $du = 2x\, dx$, we saw above that $F(x) = \sin(x^2)$ is an antiderivative of f.➤ Thus   *We need only one antiderivative, so we set $C = 0$.*

$$\int_0^{\sqrt{\pi/2}} 2x\, \cos\left(x^2\right)\, dx = \sin\left(x^2\right)\Big]_{x=0}^{x=\sqrt{\pi/2}} = \sin(\pi/2) - \sin 0 = 1.$$ □

■ **Example 9.**   Find $\displaystyle\int_0^{\sqrt{\pi/2}} 2x\, \cos\left(x^2\right)\, dx$ again.

Solution: **(Substitute; create a new definite integral.)** As above, let $u = x^2$ and $du = 2x\,dx$. At the endpoints $x = 0$ and $x = \sqrt{\pi/2}$, $u = 0$ and $u = \pi/2$, respectively. Therefore, substituting for u, du, *and for the endpoints* gives

$$\int_0^{\sqrt{\pi/2}} 2x \cos\left(x^2\right) dx = \int_0^{\pi/2} \cos u \, du.$$

No need to substitute x back in.

The last integral can be calculated just as it stands:

$$\int_0^{\pi/2} \cos u \, du = \sin u \,\big]_0^{\pi/2} = \sin(\tfrac{\pi}{2}) - \sin 0 = 1,$$

as before.   □

Look again.   Notice carefully how we used substitution differently in the last two examples. In Example 8 we used substitution only as a temporary aid to find an antiderivative of the given integrand, with respect to the original variable x. In Example 9 we used substitution to transform the original definite integral into an entirely *new* definite integral, with new limits and a new integrand. The final result, of course, was the same in each case.

Substitution in general: why it works

In each example above, substitution "worked"—it led successfully to a correct antiderivative. To see *why* substitution works, we'll describe it in general language.

For any indefinite integral $\int f(x)\,dx$, substitution means finding a function $u = u(x)$ such that $f(x)$ can be rewritten in the form

$$f(x) = g(u(x))\,u'(x)$$

for some new function g. To say that substitution "works" is to say that

$$\int f(x)\,dx = \int g(u)\,du,$$

i.e., that if G is a function for which $G'(u) = g(u)$, then $G(u(x))$ is an antiderivative of f.

The chain rule guarantees that G "works":

$$\frac{d}{dx}\big(G(u(x))\big) = G'(u(x)) \cdot u'(x) = g(u(x)) \cdot u'(x) = f(x),$$

so $G(u(x))$ *is* the antiderivative we seek.

Substitution in *definite* integrals works for essentially the same reason. The fact that it does is sometimes known as the **change of variables theorem:**

> **Theorem 1.** Let f, u, and g be continuous functions such that for all x in $[a, b]$,
> $$f(x) = g\left(u(x)\right) \cdot u'(x).$$
> Then
> $$\int_a^b f(x)\,dx = \int_{u(a)}^{u(b)} g(u)\,du.$$

Proof: Let G be an antiderivative of g. By the FTC,

$$\int_{u(a)}^{u(b)} g(u)\,du = G(u(b)) - G(u(a)).$$

As we showed above,$^\bullet$ $G(u(x))$ is an antiderivative for $f(x)$. Therefore$^\bullet$

$$\int_a^b f(x)\,dx = G(u(b)) - G(u(a)),$$

Using the chain rule.

By the FTC again.

so the two integrals are equal.   □

Differentials: du, dx, and all that

The symbols du and dx are called the **differentials** of u and x, respectively. If $u = u(x)$, then du and dx are related in the expected way: $du = u'(x)\,dx$.

A rigorous (and rather subtle) theory of differentials exists in higher mathematics; we won't need it in this course. For us, the differential in an integral expression (dx in $\int f(x)\,dx$, for instance) is mainly a mnemonic device.$^\bullet$ In any integral, it reminds us of the variable of integration. In substitution problems, trading dx's for du's (or vice versa) is a useful bookkeeping device; it helps keep us honest with respect to Theorem 1.

I.e., an aid to memory.

Exercises ───────────────

Evaluate each of the following antiderivatives using the indicated substitution, then check your answers by differentiation.

1. $\displaystyle\int (4x+3)^{-3}\,dx \qquad u = 4x+3$

6. $\displaystyle\int \frac{\sin\left(\sqrt{x}\right)}{\sqrt{x}}\,dx \qquad u = \sqrt{x}$

2. $\displaystyle\int x\sqrt{1+x^2}\,dx \qquad u = 1+x^2$

7. $\displaystyle\int \frac{e^{1/x}}{x^2}\,dx \qquad u = 1/x$

3. $\displaystyle\int e^{\sin x}\cos x\,dx \qquad u = \sin x$

8. $\displaystyle\int x^3\sqrt{9-x^2}\,dx \qquad u = 9-x^2$

4. $\displaystyle\int \frac{(\ln x)^3}{x}\,dx \qquad u = \ln x$

9. $\displaystyle\int \frac{e^x}{1+e^{2x}}\,dx \qquad u = e^x$

5. $\displaystyle\int \frac{\arctan x}{1+x^2}\,dx \qquad u = \arctan x$

10. $\displaystyle\int \frac{x}{1+x}\,dx \qquad u = 1+x$

11. Example 6 says that $\displaystyle\int \frac{x}{\sqrt{2x+3}}\,dx = \frac{(2x+3)^{3/2}}{6} - \frac{3\sqrt{2x+3}}{2} + C$.
Verify this by differentiation.

For each of the following, find real numbers a and b so that equality holds, then evaluate the definite integral.

12. $\displaystyle\int_0^3 \frac{x}{\left(2x^2+1\right)^3}\,dx = \frac{1}{4}\int_a^b u^{-3}\,du$

13. $\displaystyle\int_{-\sqrt{\pi/2}}^{\sqrt{\pi}} x\cos\left(3x^2\right)dx = \frac{1}{6}\int_a^b \cos u\, du$

14. $\displaystyle\int_1^2 x^2 e^{x^3/4}\, dx = \frac{4}{3}\int_a^b e^u\, du$

15. $\displaystyle\int_{-2}^1 \frac{x}{1+x^4}\, dx = \frac{1}{2}\int_a^b \frac{1}{1+u^2}\, du$

16. Suppose that $\displaystyle\int_0^{12} g(x)\, dx = \pi$. Evaluate $\displaystyle\int_0^4 g(3x)\, dx$.

Evaluate each of the following antiderivatives. Check your answers by differentiation.

17. $\displaystyle\int \cos(2x+3)\, dx$

18. $\displaystyle\int \frac{2x^3}{1+x^4}\, dx$

19. $\displaystyle\int x\,(3x+2)^4\, dx$

20. $\displaystyle\int x\sqrt{3-x}\, dx$

21. $\displaystyle\int \tan x\, dx$

22. $\displaystyle\int xe^{x^2}\, dx$

23. $\displaystyle\int x^3 \left(x^4-1\right)^2 dx$

24. $\displaystyle\int \frac{x^2}{1+x^6}\, dx$

25. $\displaystyle\int \frac{x}{\sqrt{1-x^4}}\, dx$

26. $\displaystyle\int x^2\sqrt{4x^3+5}\, dx$

27. $\displaystyle\int \frac{x}{\sqrt{1+x^2}}\, dx$

28. $\displaystyle\int \frac{x+4}{x^2+1}\, dx$

29. $\displaystyle\int x\left(1-x^2\right)^{15} dx$

30. $\displaystyle\int (x+2)\left(x^2+4x+5\right)^6 dx$

31. $\displaystyle\int \frac{2x+3}{\left(x^2+3x+5\right)^4}\, dx$

32. $\displaystyle\int \frac{x+1}{\sqrt[3]{3x^2+6x+5}}\, dx$

33. $\displaystyle\int x\cos\left(1-x^2\right)dx$

34. $\displaystyle\int \frac{e^{2x}}{\left(1+e^{2x}\right)^3}\, dx$

35. $\displaystyle\int \frac{e^x}{(2e^x+3)^2}\, dx$

36. $\displaystyle\int \frac{e^{\sqrt{x}}}{\sqrt{x}}\, dx$

37. $\displaystyle\int \frac{2x+3}{(x+1)^2}\, dx$

38. $\displaystyle\int \frac{x^2}{x-3}\, dx$

39. $\displaystyle\int \sec x\tan x\, dx$

40. $\displaystyle\int \frac{\arcsin x}{\sqrt{1-x^2}}\, dx$

41. $\displaystyle\int \frac{\ln x}{x}\, dx$

42. $\displaystyle\int \frac{\sqrt{1+1/x}}{x^2}\, dx$

43. $\displaystyle\int \sec x\tan x\sqrt{1+\sec x}\, dx$

44. $\displaystyle\int \frac{5x}{3x^2+4}\, dx$

45. $\displaystyle\int \frac{\cos x}{1+\sin^2 x}\, dx$

46. $\displaystyle\int \frac{\sec^2 x}{\sqrt{1 - \tan^2 x}}\, dx$

47. $\displaystyle\int \sec^2 x \tan x\, dx$

48. $\displaystyle\int x \tan \left(x^2\right) dx$

49. $\displaystyle\int x^2 \sec^2 \left(x^3\right) dx$

50. $\displaystyle\int \frac{\cos x}{\sin^4 x}\, dx$

51. $\displaystyle\int x^4 \sqrt[3]{x^5 + 6}\, dx$

52. $\displaystyle\int \frac{e^{\tan x}}{1 - \sin^2 x}\, dx$

Evaluate each of the following definite integrals.

53. $\displaystyle\int_0^2 \frac{x}{\left(1 + x^2\right)^3}\, dx$

54. $\displaystyle\int_{-19}^8 \sqrt[3]{8 - x}\, dx$

55. $\displaystyle\int_1^e \frac{\sin(\ln x)}{x}\, dx$

56. $\displaystyle\int_e^{4e} \frac{dx}{x\sqrt{\ln x}}$

57. $\displaystyle\int_0^\pi \sin^3 x \cos x\, dx$

58. $\displaystyle\int_{-\pi/2}^\pi e^{\cos x} \sin x\, dx$

59. Let $I = \displaystyle\int \sec^2 x \tan x\, dx$.

   (a) Use the substitution $u = \sec x$ to show that $I = \frac{1}{2}\sec^2 x + C$.

   (b) Use the substitution $u = \tan x$ to show that $I = \frac{1}{2}\tan^2 x + C$.

   (c) The expressions for I in parts (a) and (b) look different, but both are correct. Explain this apparent paradox.   [HINT: For all x such that $\cos x \neq 0$, $\sec^2 x = 1 + \tan^2 x$.]

60. Use a u-substitution to show that $\displaystyle\int_0^1 x^n (1 - x)^m\, dx = \int_0^1 x^m (1 - x)^n\, dx$.

The substitution $u^n = ax + b$ can sometimes be used to help evaluate integrals involving expressions of the form $\sqrt[n]{ax + b}$. Use this substitution to evaluate each of the following antiderivatives.

61. $\displaystyle\int x\sqrt{2x + 1}\, dx$

62. $\displaystyle\int \frac{dx}{\sqrt{x} + \sqrt[3]{x}}$

6.3   Integral aids: tables and computers

And recreational . . .

Finding symbolic antiderivatives "from scratch" can be difficult. Still, there are excellent educational[*] reasons for doing so: finding the "right" substitution or other method may reveal satisfying order and pattern in apparent symbolic chaos.

Whatever its educational value, the process of symbolic antidifferentiation is difficult, tricky, and laborious enough that, in practice, calculus "users" resort freely to integral tables, computer software, or anything else that comes to hand.

Like all power tools, tables and mathematical software requires some care and caution in use. In this section we illustrate some of their possibilities and pitfalls.

Integral tables

"Antiderivative table" would be a better name.

We added the absolute value signs.

Standard scientific reference works contain extensive integral tables.[*] The *CRC Handbook of Chemistry and Physics, 48th Edition*, for instance, lists nearly 600 different antiderivative formulas on more than 40 pages. One of them reads like this:[*]

> **Fact: (Formula 403)**
>
> $$\int \frac{dx}{a + be^{px}} = \frac{x}{a} - \frac{1}{ap} \ln \left| a + be^{px} \right|$$

Notice the missing (but "understood") constant of integration; supplying one every time gets typographically tiresome.

Parameters and pattern-matching

Integral tables use **parameters**—letters *other* than the variable of integration—to stand for numerical constants that can vary from problem to problem. With this device a single integral formula can solve an entire family of similar problems. The next few examples concern the parameters in the formula above.

■ **Example 1.**   Verify Formula 403 by differentiation.

Watch the algebra carefully.

Solution: We differentiate the right side with respect to x—all other letters are constants:[*]

$$\left(\frac{x}{a} - \frac{1}{ap} \ln \left| a + be^{px} \right| \right)' = \frac{1}{a} - \frac{1}{ap} \frac{bpe^{px}}{a + be^{px}}$$

$$= \frac{1}{a} - \frac{1}{a} \frac{be^{px} + a - a}{a + be^{px}}$$

$$= \frac{1}{a} - \frac{1}{a} \left(1 - \frac{a}{a + be^{px}} \right)$$

$$= \frac{1}{a + be^{px}}.$$

That's what we wanted; the *CRC Handbook* got it right.   □

> **Getting it right.** Not every "equation" in published integral tables is correct. Integral tables change over time, as new antiderivative formulas come into favor and old ones fall into disuse. With each new version of an integral table, new typographical errors may creep in; some errors have lain undiscovered for years.
> Standard integral tables are sometimes used as benchmarks, to check mathematical software programs for accuracy. This process sometimes uncovers errors—not in the software but in the tables themselves. Some integral tables have been found to have error rates in the double-digit percentage range.

■ **Example 2.** Use Formula 403 to find $\displaystyle\int \frac{5}{3 - 2e^{-x}}\, dx$.

Solution: Writing

$$\int \frac{5}{3 - 2e^{-x}}\, dx = 5 \int \frac{dx}{3 - 2e^{-x}}$$

makes the parameter values directly apparent: $a = 3, b = -2, p = -1$.➤ The result:

Don't forget the 5 out front.

$$\int \frac{5}{3 - 2e^{-x}}\, dx = 5 \left(\frac{x}{3} + \frac{1}{3}\ln \left| 3 - 2e^{-x} \right| \right) + C. \qquad \square$$

Making the shoe fit

Antiderivatives that don't appear to fit an integral template can sometimes be made to do so. A u-substitution, some symbolic algebra, or a combination of both often effects this Cinderella-like transformation.➤

Sometimes the shoe just won't fit. Remember the Cinderella story?

■ **Example 3.** Find $\displaystyle I = \int \frac{\cos x}{5 + 2e^{3\sin x}}\, dx$.

Solution: As it stands the integral resembles Formula 403 vaguely, at best. However, substituting $u = \sin x$ and $du = \cos x\, dx$ reveals the familiar pattern:

$$\int \frac{\cos x}{5 + 2e^{3\sin x}}\, dx = \int \frac{du}{5 + 2e^{3u}}.$$

Setting $a = 5, b = 2$, and $p = 3$ completes the problem:

$$I = \frac{u}{5} - \frac{\ln \left| 5 + 2e^{3u} \right|}{15} + C = \frac{\sin x}{5} - \frac{\ln \left| 5 + 2e^{3\sin x} \right|}{15} + C. \qquad \square$$

■ **Example 4.** Find $\displaystyle I = \int \frac{e^x}{3e^{-x} + 2e^x}\, dx$.

Solution: A touch of exponential algebra➤ puts the integrand into the necessary form:

Multiplying top and bottom by e^{-x}.

$$\frac{e^x}{3e^{-x} + 2e^x} = \frac{e^x e^{-x}}{e^{-x}\left(3e^{-x} + 2e^x\right)} = \frac{1}{2 + 3e^{-2x}}.$$

Now we're ready for Formula 403, with $a = 2, b = 3, p = -2$:

$$I \doteq \int \frac{dx}{2 + 3e^{-2x}} = \frac{x}{2} + \frac{1}{4}\ln \left| 2 + 3e^{-2x} \right| + C. \qquad \square$$

Completing the square

Quadratic expressions (of the form $ax^2 + bx + c$) appear often in antiderivative problems. **Completing the square** often helps fit such integrands into an integral table's template. We illustrate with two examples.

■ **Example 5.** Find $I = \displaystyle\int \frac{dx}{x^2 + 4x + 5}$.

Such as the one at the end of this book.

Solution: Rummaging through an integral table[**] turns up two likely possibilities:

$$\int \frac{dx}{x^2 + a^2} = \frac{1}{a} \arctan\left(\frac{x}{a}\right); \quad \int \frac{dx}{x^2 - a^2} = \frac{1}{2a} \ln\left|\frac{x-a}{x+a}\right|.$$

Which one—if either—applies?

To decide, we'll complete the square in the denominator:

$$x^2 + 4x + 5 = (x^2 + 4x + 4) + (5 - 4) = (x+2)^2 + 1.$$

Easily checked by differentiation.

This, in turn, suggests the substitution $u = x + 2$, $du = dx$. Here's the result:[**]

$$\int \frac{dx}{x^2 + 4x + 5} = \int \frac{du}{u^2 + 1} = \arctan u + C = \arctan(x + 2) + C. \qquad \square$$

■ **Example 6.** Find $I = \displaystyle\int \frac{dx}{x^2 + 4x + 3}$.

Solution: The same two possibilities suggest themselves as in the previous example. Again, we complete the square:

$$x^2 + 4x + 3 = (x^2 + 4x + 4) + (3 - 4) = (x+2)^2 - 1.$$

The rest follows easily. Substituting $u = x + 2$ and $du = dx$ gives

$$\int \frac{dx}{x^2 + 4x + 3} = \int \frac{du}{u^2 - 1} = \frac{1}{2} \ln\left|\frac{u-1}{u+1}\right| + C = \frac{1}{2} \ln\left|\frac{x+1}{x+3}\right| + C. \qquad \square$$

A reality check: do the graphs look right?

Good sense dictates that, from time to time, results obtained by one method be tested by another. Symbolic antiderivatives especially deserve such attention. Functions, after all, are more than purely formal expressions: they're graphical and numerical objects as well. If symbolic operations (substitution, use of tables, and more to come) really make sense, they should stand up to graphical scrutiny. Symbolically-produced antiderivatives should, in other words, "look right." Now and then, therefore, we'll check that they do.

The last two examples invite such a look. Their symbolic results—

$$\int \frac{dx}{x^2 + 4x + 5} = \arctan(x + 2); \quad \int \frac{dx}{x^2 + 4x + 3} = \frac{1}{2} \ln\left|\frac{x+1}{x+3}\right|$$

are a bit mysterious. Although the integrands are almost identical, the antiderivatives look very different. Is there some mistake?

Let's look graphically. Here, first, are graphs of the two integrands:

Graphs of $f_1(x) = 1/(x^2 + 4x + 5)$ and $f_2(x) = 1/(x^2 + 4x + 3)$

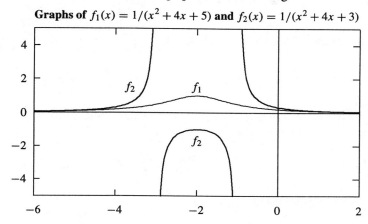

Now the mystery begins to dissolve. Despite their typographical similarity, the two integrands behave quite differently: f_1 behaves tamely, while f_2 has vertical asymptotes at $x = -3$ and $x = -1$. As $x \to \pm\infty$, on the other hand, both f_1 and f_2 tend to zero.

Plotting the antiderivatives F_1 and F_2 shows that they, too, behave as they should:

Graphs of $F_1(x) = \arctan(x + 2)$ and $F_2(x) = 0.5 \ln |(x + 1)/(x + 3)|$

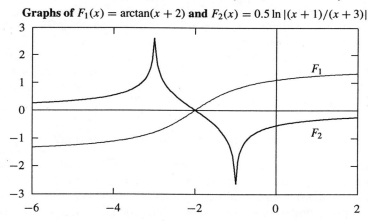

Notice:

Asymptotes. Like f_1 itself, the antiderivative F_1 is tamely behaved. Like f_2 itself, the antiderivative F_2 has vertical asymptotes at $x = -3$ and $x = -1$.➤➤   *But with different "directions."*

Long run behavior. Both F_1 and F_2 appear to tend toward the horizontal as $x \to \pm\infty$. This is as it should be: their respective derivatives both tend to zero in the long run.

Reduction formulas: one antiderivative in terms of another

Some formulas in integral tables have antiderivatives on *both* sides of an equation. Here's another entry from the *CRC Handbook*:

> **Fact: (Formula 311)**   If $n \neq 1$, then
> $$\int \tan^n x \, dx = \frac{\tan^{n-1} x}{n - 1} - \int \tan^{n-2} x \, dx.$$

Applying such a **reduction formula**—repeatedly, perhaps—transforms one integral problem into another, easier than the first.

■ **Example 7.** Find $\int \tan^4 x \, dx$.

Solution: Formula 311, with $n = 4$, says:

$$\int \tan^4 x \, dx = \frac{\tan^3 x}{3} - \int \tan^2 x \, dx.$$

With $n = 2$, this time.

The last integral may *itself* appear in an integral table. If so, we're done. If not, we apply Formula 311 again:◄

$$\int \tan^2 x \, dx = \frac{\tan x}{1} - \int \tan^0 x \, dx = \tan x - x.$$

Combining our results gives

$$\int \tan^4 x \, dx = \frac{\tan^3 x}{3} - \tan x + x + C.$$ $\square$

Antidifferentiation by machine

Old mathematical handbooks contain vast tables of trigonometric, exponential, and logarithm function values. Scientific calculators have rendered such tables, practically speaking, obsolete.

To a lesser (but significant) degree, sophisticated mathematical software effectively replaces integral tables. *Derive*, *Maple*, *Mathematica*, and other programs "know" not just integral formulas themselves, but—much more challenging—how and when to apply them.

Symbolic camouflage: different forms of the same result

Equivalent mathematical expressions can appear in radically different symbolic forms. Trigonometric expressions are especially tricky; the many "trigonometric identities" offer almost unlimited possibilities for disguise.

Computer symbolic operations often produce variant or unexpected forms of symbolic results. (What's convenient for a computer may be baffling to you and me—and vice versa.)

■ **Example 8.** According to one integral table,

$$\int \sin^2 x \, dx = \frac{x}{2} - \frac{\sin(2x)}{4}.$$

Maple says:

```
> int( sin(x)^2, x );

              - 1/2 cos(x) sin(x) + 1/2 x
```

Who's right?

Sometimes, it's much harder.

See for yourself.

Solution: Both are right (we'll forgive the omitted constant of integration). This time, the reason is simple.◄ Stirring the trigonometric identity $\sin(2x) = 2 \sin x \cos x$ into the first formula readily produces the second.◄ $\square$

When nothing works

Sometimes nothing seems to work. Symbolic methods, integral tables, algebraic tricks, and software may *all* fail to find an antiderivative in elementary form. Two reasons are possible: either we➤ haven't searched cleverly enough, or there *is* no elementary antiderivative. As we remarked at the beginning of this chapter, many elementary functions, even simple ones, don't have *elementary* antiderivatives.➤ None of these does, for instance:

$$e^{x^2}; \qquad \sin x^2; \qquad \frac{\sin x}{x}; \qquad \frac{x}{\ln x}.$$

Or a computer.

I.e., antiderivatives with formulas.

Like all continuous functions, these *have* antiderivatives, but not elementary ones. Notice how *Maple* handles one of them:

```
> int( sin(x)/x, x );
```

$$Si(x)$$

Unable to find an elementary answer, *Maple* returned the function Si, which is *defined* to be an antiderivative of $\sin(x)/x$. In effect, *Maple* punted.

The bright side. Even when symbolic methods fail, all is not lost. Numerical methods, such as those we'll study in the next chapter, often succeed. For example, numerical methods will nicely handle *definite* integrals of all the "problematic" functions above.

Exercises ──────────────

Evaluate each of the following integrals using a table of integrals or a software package. Making simple *u*-substitutions or completing the square *first* may help.

1. $\displaystyle\int \frac{dx}{3 + 2e^{5x}}$

2. $\displaystyle\int \frac{dx}{x(2x + 3)}$

3. $\displaystyle\int \frac{dx}{x^2(3 - x)}$

4. $\displaystyle\int_{e}^{4e} x^2 \ln x \, dx$

5. $\displaystyle\int \tan^3(5x) \, dx$

6. $\displaystyle\int \frac{dx}{x^2\sqrt{2x + 1}}$

7. $\displaystyle\int x \sin(2x) \, dx$

8. $\displaystyle\int x^2 e^{3x} \, dx$

9. $\displaystyle\int e^{2x} \cos(3x) \, dx$

10. $\displaystyle\int \frac{2x + 3}{4x + 5} \, dx$

11. $\displaystyle\int \frac{dx}{4 - x^2} \, dx$

12. $\displaystyle\int x^2\sqrt{1 - 3x} \, dx$

13. $\displaystyle\int \frac{4x + 5}{(2x + 3)^2} \, dx$

14. $\displaystyle\int \frac{dx}{x\sqrt{3x - 2}}$

15. $\displaystyle\int \frac{dx}{4x^2 - 1}$

16. $\displaystyle\int \frac{dx}{\left(4x^2 - 9\right)^2}$

17. $\displaystyle\int \frac{x+2}{2+x^2}\,dx$

18. $\displaystyle\int \frac{3}{\sqrt{6x+x^2}}\,dx$

19. $\displaystyle\int \frac{5}{4x^2+20x+16}\,dx$

20. $\displaystyle\int \frac{dx}{\sqrt{x^2+2x+26}}$

21. $\displaystyle\int x^3 \cos\left(x^2\right)\,dx$

22. $\displaystyle\int \frac{x}{\sqrt{x^2+4x+3}}\,dx$

23. $\displaystyle\int \frac{dx}{\left(x^2+3x+2\right)^2}$

24. $\displaystyle\int x^2\left(\cos x + 3\sin x\right)\,dx$

25. $\displaystyle\int \frac{e^x}{e^{2x}-2e^x+5}\,dx$

26. $\displaystyle\int \cos x \sin x \sin^2\left(2\cos^2 x + 1\right)\,dx$

27. $\displaystyle\int \sqrt{x^2+4x+1}\,dx$

28. $\displaystyle\int \frac{\cos x}{3\sin^2 x - 11\sin x - 4}\,dx$

29. $\displaystyle\int \frac{\cos x \sin x}{(\cos x - 4)(3\cos x + 1)}\,dx$

30. $\displaystyle\int \frac{e^{2x}}{\sqrt{e^{2x}-e^x+1}}\,dx$

31. $\displaystyle\int x \sin(3x+4)\,dx$

Chapter 7

Numerical Integration

7.1 The idea of approximation

The definite integral $\int_a^b f(x)\,dx$ is defined formally as a certain limit of approximating sums. This chapter is about different types of approximating sums—left sums, right sums, midpoint sums, and others. We'll study both how such sums approximate integrals and how much error they commit in the process.

Calculating approximating sums: help from technology

Calculating approximating sums is easy—in theory.➤ A right sum with 100 subdivisions for $\int_0^1 \sin x\,dx$, for instance, starts like this:

In practice, too, if n is small.

$$
\begin{aligned}
R_{100} &= \sum_{i=1}^{100} \sin\left(\frac{i}{100}\right) \cdot \frac{1}{100} \\
&= 0.01 \cdot [\sin(0.01) + \sin(0.02) + \sin(0.03) + \cdots + \sin(0.99) + \sin(1.00)].
\end{aligned}
$$

Evaluating such monsters numerically is no fit job for a human. For computers and calculators, it's simple.➤

Computers exist to do such things.

   How one uses a calculator or computer program to calculate approximating sums varies, of course, with the machine at hand. For example's sake, we'll illustrate how one program—*Maple*—does the job. We use *Maple* only as an illustration; the *ideas*, not the details of syntax, are our main concern. These ideas apply *regardless* of what form of technology➤ the reader may have at hand.

Other software programs, programmable calculators, etc.

   With *Maple* it's easy to calculate left, right, midpoint, and other approximating sums, even for large values of n.➤ Some of those one version➤ of *Maple* knows about are listed below, in *Maple* syntax:

Up to 100, say.

These commands don't work in all Maple versions—the idea is the point.

```
left(sin(x),x=0..1,10);        nleft(sin(x),x=0..1,10);
right(sin(x),x=0..1,10);       nright(sin(x),x=0..1,10);
midpoint(sin(x),x=0..1,10);    nmidpoint(sin(x),x=0..1,10);
trapezoid(sin(x),x=0..1,10);   ntrapezoid(sin(x),x=0..1,10);
simpson(sin(x),x=0..1,10);     nsimpson(sin(x),x=0..1,10);
```

(Commands beginning with n give *decimal* results.)

Maple tips: To simplify reading *Maple* talk, know these conventions (all are illustrated in this section):

- **Inputs** (what the user types) follow an "input prompt"—here, the "greater than" symbol > . Outputs (what *Maple* returns) are indented.

- To avoid roundoff errors, *Maple* doesn't write answers in decimal form unless told explicitly to do so.

■ **Example 1.**   Use *Maple* to calculate various approximating sums for $\int_0^1 \sin x \, dx$. Compare results with the "exact" answer.

Solution: The FTC gives the exact answer:

$$\int_0^1 \sin x \, dx = -\cos x \Big]_0^1 = -\cos 1 + \cos 0 \approx 0.4596976941.$$

Now, for comparison:

```
> right(sin(x),x=0..1,4);
        1/4 sin(1/4) + 1/4 sin(1/2) + 1/4 sin(3/4) + 1/4 sin(1)
```

The pattern looks correct. Let's see some numbers:

```
> nright(sin(x),x=0..1,10);
                    0.5013880981
> nleft(sin(x),x=0..1,10);
                    0.4172409996
> nmidpoint(sin(x),x=0..1,10);
                    0.4598892908
```

The results look reasonable; notice M_{10}'s impressive accuracy.   □

Tapping *Maple*. *Maple* is a package of powerful, useful, and convenient mathematical operations. We'll use it often in this book—but not for just anything. Typical uses of *Maple* include:

- Simple—but laborious—*numerical* operations.

  Example: Calculate L_{100} for the integral $\int_0^1 \sin x \, dx$.

- Simple—but laborious—*symbolic* operations.

  Example: Check by differentiation that

  $$\frac{x\sqrt{1+4x^2}}{2} + \frac{\ln(2x + \sqrt{1+4x^2})}{4}$$

  is an antiderivative for $\sqrt{1+4x^2}$.

- *Graphical* operations. Simple graphs are easy and quick to draw by hand. Graphs of complicated functions are no harder conceptually, but take more time and effort to draw accurately. We'll use *Maple* (or other technology).

  Example: Let $f(x) = \sin(x^2)$. Estimate the maximum value of $|f''|$ for $0 \le x \le 1$ using a graph of f.

Notice the common theme:

We use technology if the result—rather than the process—of calculation is of interest.

Error: a philosophical aside

How *well* do approximating sums—L_n, R_n, M_n, etc.—approximate an integral I? If n➤ is large we expect such approximations to be close to the true value of I. But *how* close? Within 0.01? Within 0.00001? Within 0.000000000001? How can we be *certain*?➤

The number of subdivisions

In Example 1 above we computed I exactly. In many cases I isn't known exactly.

To answer such questions, we'll look carefully at approximating sums and the sources of the errors they commit.

Why approximate?

Approximations—and the errors they commit—are important themes of this book. Here are three good reasons:

No closed form solutions. Many calculus problems, even simple-looking ones, can't be solved "in closed form," i.e., using "exact" methods, such as antidifferentiation and algebraic solution of equations.

Consider, for example, this problem:

Maximize $f(x) = x \sin x$ on the interval $[0, 2]$.

Harmless as the problem appears, symbolic methods are of little use. (Try it yourself➤ to see where symbolic methods fail.)

Differentiate, find roots, etc.

Accurate results. Approximate methods (graphing, numerical integration, etc.) usually "work" even if exact methods fail. Approximate methods, to be sure, yield "only" approximate results. If due care is taken, such estimates can be made highly accurate.

The maximum problem above, for instance, is easy to solve *approximately* by graphing.

Estimating error. An approximation to an unknown quantity is, by itself, of little value. A useful guess includes a "margin of error"—i.e., some assurance that the error committed is no worse than some computable amount.◂ Finding and using such "accuracy guarantees" for approximations to integrals is the main subject of this chapter.

Respectable opinion polls do this.

Exact answers, estimates, and accuracy guarantees

Approximating *known* quantities, such as $\int_0^1 \sin x\, dx$, is useful only as an exercise. Real approximation problems concern quantities◂ we *don't know exactly*. To "solve" real approximation problems, we need two things:

Such as $\int_0^1 \sin(x^2)\, dx$; see below.

1. an **estimate** for the unknown quantity;

2. an **error bound**—i.e., a bound on how much error the approximation can commit.

Error bounds: what to expect

We can't sensibly expect to compute *exactly* the error a given approximation commits—that would be tantamount to solving the original problem exactly. Error bounds, therefore, are usually *in*equalities.

For example, the error estimate

$$|I - L_n| \le \frac{K_1(b-a)^2}{2n}.$$

is typical. (L_n is the left sum with n subdivisions for $I = \int_a^b f(x)\, dx$; K_1 is a certain constant we'll discuss later.) Notice:

- The left side of the inequality—the absolute value of the error—is what we want to estimate. We hope it's small!

- Bounding the error's *absolute value* means that we care more about (or *know* more about) the *size* than about the *sign* of the error.◂

For the same reason, pollsters usually report error as, say, plus or minus 5%.

- The $\le$ inequality guarantees that the error is *no worse than* the (computable) quantity on the right. Error estimates represent conservative, worst-case scenarios; the *actual* error might well be less.

The bright side

Working out approximations and error estimates by hand is tedious. With computing, it's easy.

Still more errors: roundoff. Any approximation—L_{1000} for the integral $I = \int_0^1 x^2\, dx$, for instance—can commit several types of error. One source of error is the method itself. We know, for instance, that because the integrand is increasing, *every* left sum *underestimates* I.

Decimal rounding is another source of error. Even small roundoff errors may accumulate dangerously during delicate or repetitive calculations.

A careful treatment of roundoff errors is beyond our scope. As a rule, we'll control—though never completely avoid—roundoff errors by avoiding very large scale computations.

The simplest error bounds; monotone functions

If the integrand f happens to be *monotone* on $[a, b]$, left and right sums always *bracket* the integral $\int_a^b f$. This simple idea leads to a simple but useful error bound. We'll see how by example.

I.e., either increasing or decreasing, but not both.

■ **Example 2.** Let $I = \int_0^1 \sin(x^2)\, dx$. Also, let L_{100} and R_{100} denote left and right sum approximations of I, each with 100 subdivisions:

$$L_{100} \approx 0.30607; \qquad R_{100} \approx 0.31449.$$

Make a "best guess" at I. How far off could it be?

The integral I can't be evaluated using the FTC. Why not? Because (for deep reasons) the antiderivative of $\sin(x^2)$ is not an elementary function.

Check them if you like.

Solution: The integrand $\sin(x^2)$ *increases* everywhere on $[0, 1]$. Therefore:

For any n, L_n underestimates I; R_n overestimates I.

The picture shows the situation for *left* sums:

Left sums underestimate $\int_0^1 \sin(x^2)\, dx$

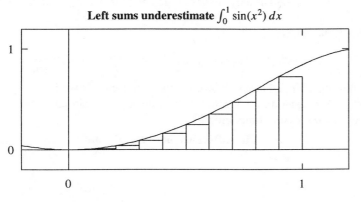

Thus I lies *somewhere* between L_{100} and R_{100}, i.e., in the interval $[0.30607, 0.31449]$. Lacking further information, we'll use the *midpoint*, 0.31028, to estimate I.

How much error could our guess possibly commit? Because the interval $[0.30607, 0.31449]$ has length 0.00842, the distance from I to the midpoint can't exceed the "radius," 0.00421. In symbols:

$$|I - 0.31028| < 0.00421.$$

We conclude:

The estimate $I \approx 0.31028$ holds with margin of error ± 0.00421.

□

A general error bound

See the next theorem.

The example illustrates a principle[◄] that applies to *any* monotone function f and integral $I = \int_a^b f$.

All the ingredients.   Let's assemble all our ingredients:

- The integrand, f, is *monotone* on $[a, b]$.

Equal-sized subintervals simplify computations; we'll always use them.

- The equally-spaced points $a = x_0 < x_1 < x_2 < \cdots < x_n = b$ *partition* $[a, b]$ into n equal subintervals,[◄] each of width $\Delta x = (b - a)/n$.

Check these carefully to review the notation.

- L_n and R_n are the left and right sums built from f, a, b, and n. In sigma notation:[◄]

$$L_n = \sum_{i=1}^{n} f(x_{i-1})\Delta x; \qquad R_n = \sum_{i=1}^{n} f(x_i)\Delta x.$$

See Section 5.4.

- L_n and R_n are both estimates for I. Because I lies *between* L_n and R_n, their *average*, $(L_n + R_n)/2 = T_n$,[◄] is another natural estimate for I.

Theorem 1 (Error bounds for left and right sums).
Suppose that f is monotone on $[a, b]$, and let $I = \int_a^b f(x)\,dx$. Then:

$$|I - R_n| \leq |f(b) - f(a)|\frac{(b - a)}{n}$$

$$|I - L_n| \leq |f(b) - f(a)|\frac{(b - a)}{n}$$

$$|I - T_n| \leq |f(b) - f(a)|\frac{(b - a)}{2n}$$

Proof: That I lies *between* L_n and R_n is the key idea. The crucial question is *how far apart* they can be—if they're close together, then L_n, R_n, and T_n are all close to I.

Here's the answer:

$$(7.1.1) \qquad |R_n - L_n| = |f(b) - f(a)|\,\Delta x = |f(b) - f(a)|\frac{(b - a)}{n}.$$

Convince yourself.

From this equation, all three parts of the theorem follow directly.[◄]

The proof for a decreasing f is almost identical.

Why does Equation 7.1.1 hold? A "picture proof" is simplest. For convenience, we'll use an *increasing* function f.[◄]

The difference between R_n and L_n

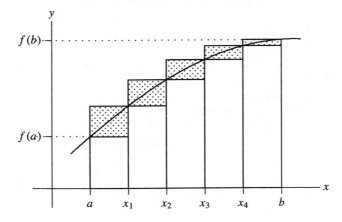

The shaded area represents the *difference* $|R_n - L_n|$—just what we want to measure. Notice:

- Each shaded box has *width* Δx.

- The *total height* of all n boxes is $|f(b) - f(a)|$.➤

Stack the shaded boxes vertically to see why.

The *total shaded area*, therefore, is what we claimed:

$$|R_n - L_n| = |f(b) - f(a)|\Delta x = |f(b) - f(a)|\frac{(b-a)}{n}. \qquad \square$$

Fine print: from pictures to equations

Pictures aren't proofs. For completeness, here's an analytic version (i.e., a translation into symbolic language) of what's above. As above, we assume that f is increasing on $[a, b]$.

$$
\begin{aligned}
|R_n - L_n| &= \sum_{i=1}^{n} f(x_i)\Delta x - \sum_{i=1}^{n} f(x_{i-1})\Delta x \\
&= \Delta x \sum_{i=1}^{n} (f(x_i) - f(x_{i-1})) \\
&= \Delta x \left[f(x_1) - f(x_0) + f(x_2) - f(x_1) + \cdots + f(x_n) - f(x_{n-1}) \right] \\
&= \Delta x \, (f(b) - f(a)) = \Delta x \, |f(b) - f(a)|.
\end{aligned}
$$

That's what we wanted to show.

Using the error bound formula: choosing n in advance

With an error bound formula, we can tell *in advance* how many subdivisions are needed for stipulated accuracy.

■ **Example 3.** How large must n be to assure that L_n estimates $I = \int_0^1 \sin x \, dx$ with error less than 0.01? Calculate such an L_n. Is it as good as claimed?

Solution: The error bound formula says that for this I,

$$|L_n - I| \le |\sin 1 - \sin 0| \cdot \frac{1}{n} \approx \frac{0.8415}{n}.$$

To assure that $|L_n - I| \le 0.01$, it's enough to insist that

$$\frac{0.8415}{n} < 0.01.$$

We're OK, therefore, if $n > 84.15$—any integer greater than 84 will do. Let's check with *Maple*:

```
> nleft( sin(x), x=0..1, 85 );
```

$$0.4547425626$$

The exact value of I, by comparison, is

$$\int_0^1 \sin x \, dx = -\cos x \Big]_0^1 = -\cos 1 + \cos 0 \approx 0.4596976941.$$

Thus L_{85} *under*estimates I by about 0.005—well within the error bound's prediction. $\qquad \square$

Exercises

1. Let $f(x) = x^3$ and $I = \int_0^1 f(x)\,dx$.

   (a) Evaluate I exactly using the FTC.

   (b) Compute L_4 and R_4 *by hand*. (You may use a scientific calculator.)

   (c) Compute the error bounds for L_4 and R_4 given by Theorem 1. How do the magnitudes of the *actual* approximation errors compare with the theoretical error bounds?

   (d) What is the actual error made by the approximation $I \approx T_4 = (L_4 + R_4)/2$? Is this consistent with Theorem 1?

   (e) If the error bounds in Theorem 1 are used, find the smallest value of n that guarantees that $|I - L_n| \le 0.005$ (i.e., that L_n approximates I to 2 decimal place accuracy). Justify your answer.

2. Let $f(x) = e^{-x^2}$ and $I = \int_0^3 f(x)\,dx$. [NOTE: The function f has no elementary antiderivative, so I can't be computed exactly using the FTC.]

   (a) Sketch a graph of f over the interval $[0, 3]$ and use it to estimate I.

   (b) Compute L_{20} and R_{20}. Does either of these values overestimate I? Which? Why?

   (c) If the error bounds in Theorem 1 are used, what is the smallest value of n that guarantees that $|I - R_n| \le 0.005$ (i.e., that R_n approximates I to 2 decimal place accuracy)? Explain.

For each of the integrals below, find the smallest value of n for which Theorem 1 guarantees that $|I - L_n| \le 0.000005$.

3. $\int_1^3 x\,dx$ 6. $\int_1^2 x^{-1}\,dx$

4. $\int_1^2 x^2\,dx$ 7. $\int_2^3 \sin x\,dx$

5. $\int_1^4 \sqrt{x}\,dx$

8–12. Repeat problems 3–7 using R_n rather than L_n.

13–17. Repeat problems 3–7 using T_n rather than L_n.

18. Let $I = \int_0^1 f(x)\,dx$. Suppose that f is a decreasing function on $[0, 1]$ such that $f(0) = 7$ and $f(1) = 4$, and that $L_{16} = 5.3172$,

    (a) Find a upper bound on the value of $|I - L_{16}|$.

    (b) Evaluate R_{16}.

    (c) Evaluate T_{16} and find an upper bound on the value of $|I - T_{16}|$.

19. Let $I = \int_0^\pi \sin x\,dx$.

    (a) Evaluate I exactly using the FTC.

    (b) Compute $|I - L_4|$.

    (c) Is the result in part (b) consistent with Theorem 1? Explain.

20. Let $I = \int_a^b f(x)\,dx$ and suppose that f is increasing on $[a, b]$. Explain why $L_n \leq M_n \leq R_n$ for any n.   [HINT: Draw a picture of such an f.]

21. Let $I = \int_a^b f(x)\,dx$ and suppose that f is decreasing on $[a, b]$. Rank the values L_n, T_n, and R_n in increasing order.   Why can you do this without knowing the value of n?

22. Let $I = \int_3^8 f(x)\,dx$. Indicate whether each of the following statements must be true, might be true, or cannot be true. Justify your answers.

   (a) If $L_{10} < I$ and $f'(x) > 0$ for all $x \in [3, 8]$, then $L_{20} < I$.

   (b) If f is monotone on $[3, 8]$, $f(3) = 5$, and $f(8) = 1$, then $|I - T_{1000}| \leq 0.05$.

   (c) If f is monotone on $[3, 8]$, $f(3) = 5$, and $f(8) = 1$, then $|I - L_n| > 0.1$ unless $n > 200$.

   (d) If $|R_{10} - I| < 0.05$, then $|R_{20} - I| < 0.05$.

23. Show that $R_n = L_n + \left[f(b) - f(a)\right] \cdot \dfrac{(b-a)}{n}$.

24. Show that $T_n = L_n + \frac{1}{2}\left[f(b) - f(a)\right] \cdot \dfrac{(b-a)}{n}$.

25. Show that if $f(a) = f(b)$, $L_n = R_n = T_n$.

26. Let f be an increasing function over the interval $[-2, 5]$ and define $I = \int_{-2}^5 f(x)\,dx$. Suppose that $L_{10} = 9.4132$ and $R_{10} = 9.5768$.

   (a) Evaluate T_{10}, the trapezoid approximation to I, and find an upper bound on $|I - T_{10}|$.

   (b) Suppose that $R_{50} = 9.5294$. Explain why $\frac{1}{2}(L_{10} + R_{50}) = 9.4713$ is an estimate of I that is guaranteed to be correct within $\pm\frac{1}{2}(R_{50} - L_{10}) = \pm\frac{1}{2}(9.5294 - 9.4132) = \pm 0.0581$.

27. Let $I = \int_a^b f(x)\,dx$ and suppose that f is monotone on the interval $[a, b]$. Show that $|I - T_n| \leq |f(b) - f(a)| \cdot \dfrac{(b-a)}{2n}$. (This is the third inequality in Theorem 1.)

28. Let $I = \int_a^b f(x)\,dx$ and suppose that f is monotone on the interval $[a, b]$. Show that $|I - M_n| \leq |f(b) - f(a)| \cdot \dfrac{(b-a)}{n}$.

29. Let $I = \int_0^2 \sin\left(x^2\right)\,dx$. Using a left, right, or trapezoid sum, estimate I with an error no greater than ± 0.01. Justify your answer.   [HINT: The integrand is not monotone over the interval $[0, 2]$.]

30. Let $I = \int_1^{12} f(x)\,dx$ where f is a decreasing function such that $f(1) = 10$, $f(2) = 2$, and $f(12) = 1$.

   (a) If the error bounds in Theorem 1 are used, what is the smallest value of n that guarantees that L_n approximates I to 2 decimal place accuracy (i.e., that $|I - L_n| \leq 0.005$)?

   (b) If the error bounds in Theorem 1 are used, what is the smallest value of n that guarantees that $|L_n - \int_1^2 f(x)\,dx| \leq 0.004$?

(c) If the error bounds in Theorem 1 are used, what is the smallest value of n that guarantees that $|L_n - \int_2^{12} f(x)\, dx| \le 0.001$?

(d) Show that the estimates computed in parts (b) and (c) can be combined to produce an estimate of I that is guaranteed to have 2 decimal place accuracy.

(e) The estimates in parts (a) and (d) both estimate I to 2 decimal place accuracy. How does the amount of computational effort necessary to compute the estimate in part (d) compare with that needed to compute the estimate in part (a)? (Compare the number of values of f that are used.)

31. For each of the following integrals,

    (i) Tabulate L_n and $I - L_n$ for $n = 2, 8, 32$, and 128. (Round answers to five decimal places.)

    (ii) How do the *actual* approximation errors computed in (i) compare with the bounds given by Theorem 1.

(a) $\int_1^2 x^2\, dx$ (c) $\int_1^2 x^{-1}\, dx$

(b) $\int_1^4 \sqrt{x}\, dx$ (d) $\int_2^3 \sin x\, dx$

32. Repeat parts (i) and (ii) of problem 31 using

(a) R_n rather than L_n.   (b) T_n rather than L_n.

33. (a) For each of integrals in problem 31, compute $\dfrac{L_{4n} - I}{L_n - I}$ for $n = 2, 8,$ and 32.

(b) Use the results you computed in part (a) to predict the approximation error made by L_{512}.

(c) Make a conjecture about how the magnitude of the approximation error depends on n.

34. Repeat the previous problem using

(a) R_n rather than L_n.   (b) T_n rather than L_n.

7.2   More on error: left and right sums and the first derivative

The simplicity of the error bound formulas in the previous section was purchased at a price: They apply to $\int_a^b f$ only if f is *monotone*.

Most integrands *aren't* monotone. It's important, therefore, to have⇥ more forgiving error bound formulas—ones that *don't* demand that f be monotone. In this section we'll derive such an error bound for left and right sums.⇥

As usual in mathematics, greater generality doesn't come for free: the new error bound formula is a little subtler. But we'll prevail . . . .

And understand.

For simplicity, we'll mainly discuss left sums; right sums behave almost identically.

Error bounds and derivatives: a continuing theme

In the rest of this chapter we'll derive several error bound formulas for approximations to $\int_a^b f$. Although these formulas differ from each other, each one depends, somehow or other, on a *derivative*⇥ of f. The connection between derivatives of f and the error committed by various rules is an important theme of this chapter.

As we'll see, error bounds for left and right sums are the simplest in this sense: they can be expressed in terms of the *first* derivative, f'.

First, second, or higher-order.

Left rule errors and the size of $|f'|$

The best case: $f' = 0$. The left rule "pretends," in effect, that f is *constant* on each subinterval. If f happens to *be* constant on $[a, b]$, then the left rule commits *zero* error in estimating $\int_a^b f$.⇥ The first derivative tells whether this happy situation applies: f is a *constant* function if and only if f' is the *zero* function.

Do you believe all this? Draw a picture.

The worst case: $|f'|$ is large. Alas, most functions aren't constant. The more f *differs* from a constant function, the worse the left rule behaves. Graphically speaking, the *steeper* the graph of f, the more error L_n commits. (The same principle holds whether f rises or falls; the only difference is whether L_n underestimates or overestimates.)

The quantity $|f'(x)|$ measures how *steeply* the f-graph rises or falls.⇥ If $|f'|$ is large on $[a, b]$, then f is "far from constant"; we'd expect L_n to behave poorly.⇥

The next example illustrates how $|f'|$ affects left rule errors for three linear functions.

The absolute value means we don't care, for now, whether f rises or falls.

If, say, $f'(x) = 100$, then $f(x) = 100x + C$—very far from a constant function.

■ **Example 1.** How well does L_4 approximate each of the three integrals $\int_0^1 x\, dx$, $\int_0^1 2x\, dx$, and $\int_0^1 (4 - 4x)\, dx$? Relate results to first derivatives.

Solution: Exact values of the integrals are easy to find:⇥

Use the FTC or just look.

$$\int_0^1 x\, dx = \frac{1}{2}; \qquad \int_0^1 2x\, dx = 1; \qquad \int_0^1 (4 - 4x)\, dx = 2.$$

Pictures show best how well or poorly L_4 works for each integral; shaded areas represent L_4 *errors*:

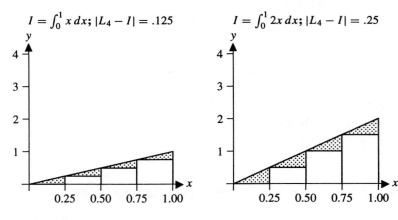

$$I = \int_0^1 x\,dx; \; |L_4 - I| = .125 \qquad I = \int_0^1 2x\,dx; \; |L_4 - I| = .25$$

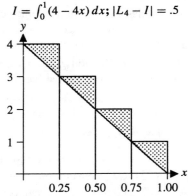

$$I = \int_0^1 (4 - 4x)\,dx; \; |L_4 - I| = .5$$

The pictures show that errors (whether overestimates or underestimates) increase as graphs get *steeper*, i.e., as f becomes "farther from constant." Derivatives tell the same story: for the three given integrands, $|f'| = 1$, $|f'| = 2$, and $|f'| = |-4| = 4$, respectively.

Maple's numbers agree with the pictures:

```
> nleft(x,x=0..1,4) , nleft(2*x,x=0..1,4) , nleft(4-4*x,x=0..1,4);
```

$$0.3750000000, \; 0.7500000000, \; 2.500000000$$

The table below shows everything:

L_4 **errors: how steepness matters**			
integral	$\int_0^1 x\,dx$	$\int_0^1 2x\,dx$	$\int_0^1 (4 - 4x)\,dx$
exact value	0.5	1.0	2.0
L_4	0.375	0.750	2.500
$I - L_4$	0.125	0.250	−0.500
$|I - L_4|$	0.125	0.250	0.500

The *relative size* of the errors is the main point.

    For linear functions f, the L_4 error is proportional *to $|f'|$.*   □

Left rule errors and the step size

With a smaller "step size" Δx,[>>] L_n presumably commits less error. How much less?   *I.e., with a larger n.*

■ **Example 2.**   Compare the errors committed by L_4 and L_8 in estimating $I = \int_0^1 x^2 \, dx$.

Solution: The exact value of I is $1/3$.[>>] Here are values for L_4 and L_8 *Convince yourself.*

$$L_4 = 0.2187500000; \qquad L_8 = 0.2734375000$$

and for the errors they commit:[>>]   *Use the fact that $I = 1/3$.*

$$|I - L_4| \approx 0.114583; \qquad |I - L_8| \approx 0.0598958.$$

The first error is about twice the second. Pictures agree; errors are shown shaded:

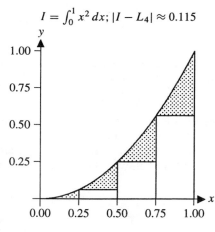
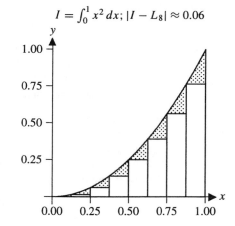

A careful look[>>] at the geometry shows, again, that halving the step size also (approximately) halves the error committed. Here's the main point: *Take one!*

> *The error L_n commits is (approximately) proportional to the step size Δx.*[>>]

I.e., inversely proportional to n.

□

Left rule error over *one* subinterval: a worst-case scenario

As the pictures above show, the error L_n commits in approximating an integral I is the sum of small errors committed over each subinterval.[>>] How large, at worst, can the error committed over *one* subinterval be? We answer this question precisely in the next example. Note, however, that the examples above suggest that the answer depends both on Δx[>>] and on $|f'|$.[>>]

In the best cases, such errors tend to cancel each other out; in the worst cases, they add up.

I.e., the stepsize.

I.e., the steepness of the f-graph.

■ **Example 3.**   Let $I = \int_a^b f$, and let K_1[>>] be any upper bound for $|f'|$ on $[a, b]$. (In other words, $|f'(x)| < K$ for all x in $[a, b]$.)

      How much error, at worst, can L_n commit over the single subinterval $[x_{i-1}, x_i]$? (One subinterval has width $\Delta x = (b - a)/n$.)

The subscript reminds us that we're bounding the first derivative.

Solution: Study the picture below:

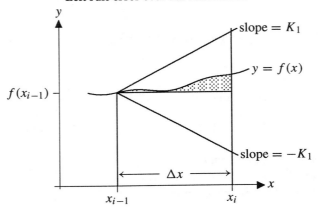

Left rule error over one subdivision

It shows:

As we said above.

What the left rule pretends. The left rule "pretends"◀ that f is constant on $[x_{i-1}, x_i]$. The shaded area represents the error attributable to this pretense.

What the derivative inequality means. The derivative inequality $|f'(x)| < K_1$ means◀ that the f-graph over $[x_{i-1}, x_i]$ must stay in the wedge *between* the two lines shown; they have slopes $\pm K_1$. (To "escape" this wedge, f would have to increase faster than K_1 allows.)◀

Think this through carefully.

A rigorous proof requires the mean value theorem. We studied it in Section 3.8 of Volume 1.

When the worst happens. The *worst possible error*—whether positive or negative— occurs if the f-graph is either the top line or the bottom line of the wedge. A careful look at the picture◀ shows that these errors correspond to the areas of the two right triangles, one above and one below the horizontal line. Each of these triangles has area◀

Take a careful look.

Check details.

$$
\frac{\text{base} \cdot \text{height}}{2} \;=\; \frac{\Delta x \cdot K_1 \cdot \Delta x}{2}
$$
$$
=\; K_1 \frac{\Delta x^2}{2}
$$
$$
=\; K_1 \frac{(b-a)^2}{2n^2}.
$$

The last expression answers our question.   □

A general error bound theorem

The example just above shows how much error L_n can commit over *one* subinterval of $[a, b]$. Over all n subintervals, therefore, L_n commits no more than n times as much error. We've just proved a useful theorem:

Theorem 2 (Left and right rule error bounds). Let
$I = \int_a^b f(x)\, dx$ and let L_n be the left approximating sum for I, with n equal subdivisions. Suppose that for all x in $[a, b]$, $|f'(x)| \leq K_1$. Then

$$
|I - L_n| \leq \frac{K_1(b-a)^2}{2n}.
$$

The same error bound applies for right sums.

■ **Example 4.**   Above we found the actual errors L_4 and L_8 commit in estimating $I = \int_0^1 x^2 \, dx$:

$$|I - L_4| \approx 0.114583; \qquad |I - L_8| \approx 0.0598958.$$

What does the theorem predict?

Solution: We need, first, a value for K_1. Because $f'(x) = 2x$, the inequality

$$|f'(x)| = |2x| \le 2$$

holds for every x in [0, 1]; we'll take $K_1 = 2$. The theorem says:

$$|I - L_4| \le 2\frac{1}{2 \cdot 4} = 0.25; \qquad |I - L_8| \le 2\frac{1}{2 \cdot 8} = 0.125.$$

We're OK—the actual errors are considerably less➤➤ than the theorem's guarantees.□   *About half.*

■ **Example 5.**   Use L_{20} and L_{200} to estimate $I = \int_0^3 \sin x^2 \, dx$.   What does the theorem guarantee about the error committed by each?

Solution: First we'll compute L_{20} and L_{200}:

```
> nleft(sin(x^2), x=0..3, 20);
```

$$0.7322735967$$

```
> nleft(sin(x^2), x=0..3, 200);
```

$$0.7703692456$$

The number K_1 can be, by definition, *any* upper bound for $|f'|$ on [0, 3]. How can we *find* such a K_1? Computing f' is easy, either by hand or by machine:

```
> diff(sin(x^2),x);
                     2
           2 x cos(x )
```

How large can $|2x \cos(x^2)|$ be if $0 \le x \le 3$? The simplest approach is graphical:

Graph of f': a value for K_1

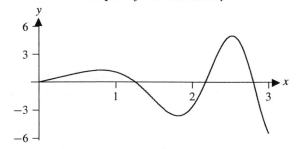

The graph shows that for all x in [0, 3], $|f'(x)| < 6$, so $K_1 = 6$ will do.

Now we have all our ingredients. By the theorem,

$$|I - L_{20}| \le \frac{6(3-0)^2}{40} = 1.35;$$

$$|I - L_{200}| \le \frac{6(3-0)^2}{400} = 0.135.$$

The second inequality means that I lies no farther than 0.135 from $L_{200} \approx 0.770$, i.e., somewhere in the interval [0.635, 0.905].

Both error bounds above are unimpressive, given the amount of computation involved. The main point is the *comparison*: a tenfold *increase* in n gives a tenfold *decrease*◄ in the error bound.   □

I.e., an improvement.

■ **Example 6.**   How much error do L_4 and L_8 commit in estimating $I = \int_0^1 10x \, dx$?◄ What does the error bound theorem predict?

By the FTC, I = 5.

Solution: First, the numbers:◄

Courtesy of Maple.

```
> int(10*x,x=0..1), nleft(10*x,x=0..1,4), nleft(10*x,x=0..1,
                        5, 3.75000000, 4.37500000
```

Thus $I = 5$, $L_4 = 3.75$, $L_8 = 4.375$. The *actual* errors, therefore, are

$$|I - L_4| = |5 - 3.75| = 1.25; \qquad |I - L_8| = |5 - 4.375| = 0.625.$$

What does the theorem say? Because f is *linear*, things are unusually simple: $f' = 10 = K_1$.◄ The theorem says, therefore, that

Usually we'd have to bound |f'|.

$$|I - L_4| \le 10\frac{1}{2 \cdot 4} = 1.25; \qquad |I - L_8| \le 10\frac{1}{2 \cdot 8} = 0.625.$$

In short, *our worst fears are realized*: L_4 and L_8 behave as badly as the theorem allows. The pictures (errors are shown shaded) illustrate this melancholy state of affairs:

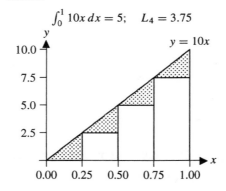
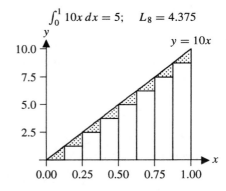

Morals

The theorem and examples suggest several general principles:

The size, not the sign, of the error.   The theorem bounds the *absolute value* of the error committed by L_n or R_n; it doesn't distinguish overestimates from underestimates. In particular examples it may be clear, perhaps from pictures, which is the case.

Finding K_1: are estimates OK?   The constant K_1 in the theorem can be *any* upper bound for $|f'|$ on $[a, b]$.

Finding such a bound can be hard, especially by hand. In practice, we'll usually resort (as we did in the example above) to graphical *estimates*. Using such estimates may seem illicit, but it isn't. *Over*estimating K_1 is harmless—the error bound "works" even if K_1 is much larger than necessary.◄ The only price paid for such sloppiness is a weaker error bound.

But don't underestimate K_1!

More and more subdivisions?   In theory, we can improve our L_n estimate by using thousands or millions of subdivisions. In practice, doing so is a bad idea. For one thing, roundoff errors would accumulate. More important, as we'll see soon, *other* approximating sums (e.g., midpoint sums) do much better, with far fewer computations.

Worst-case scenarios.   Error bound formulas, like those in this section, represent worst-case scenarios: they guarantee that the error a method commits is *no worse than* the error bound. In practice, the *actual error* is often much less than the *estimated error*—what the error bound formula predicts.

For $I = \int_0^3 \sin x^2 \, dx$ and $L_{200} \approx 0.770$, for example, the estimated error is about 0.135. It can be shown that $I = \int_0^3 \sin x^2 \, dx \approx 0.774$. The actual error, therefore, is only around 0.004.

Exercises

Compute left and right approximations for each of the following integrals using $n = 10$ equal subintervals. Compare these estimates with the exact values and check that the bound on the magnitude of the approximation error given in Theorem 2 is satisfied.

1. $\int_2^3 1 \, dx$

2. $\int_1^3 x \, dx$

3. $\int_1^2 x^2 \, dx$

4. $\int_1^4 \sqrt{x} \, dx$

5. $\int_1^2 x^{-1} \, dx$

6. $\int_2^3 \sin x \, dx$

For each of the following, find the smallest value of n for which Theorem 2 guarantees that L_n approximates the value within ± 0.005 (i.e., $|I - L_n| \le 0.005$). Justify your answers.

7. $\int_0^3 e^{-x^2} \, dx$

8. $\int_0^2 \sin \left(x^2\right) \, dx$

9. $\int_0^1 \left(1 + x^2\right)^{-1} \, dx$

10. Let $I = \int_1^4 e^x / x \, dx$. How does the bound on the approximation error made by L_n given by Theorem 2 compare with the bound given by Theorem 1?

11. Let $I = \int_0^2 \sqrt{4 - x^2} \, dx$.

   (a) Explain why $I = \pi$. [HINT: Draw a picture.]

   (b) Compute L_{10}. What is the *actual* approximation error (i.e., $\pi - L_{10}$)?

   (c) Why doesn't Theorem 2 provide a useful bound on the magnitude of the approximation error?

   (d) Compare the magnitude of the actual approximation error made by L_{10} with the bound given in Theorem 1.

12. Let $I = \int_a^b f(x)\,dx$ and suppose that the function graphed below is f'.

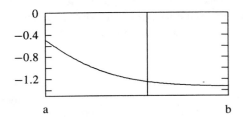

Rank the values I, L_n, and R_n in increasing order. Justify your answer.

13. Suppose that $f'(x) > 0$ and $f''(x) < 0$ when $x \in [a, b]$. Using a geometric argument (i.e., a picture), show that T_n is a more accurate estimate of $I = \int_a^b f(x)\,dx$ than L_n (i.e., $|I - T_n| < |I - L_n|$).

14. Let $I = \int_0^4 f(x)\,dx$ and suppose that f' is the function shown below.

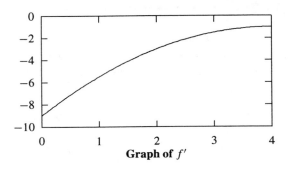

Graph of f'

(a) Find a value of n that guarantees that L_n, the left sum approximation to I with n equal subintervals, satisfies $|I - L_n| \le 0.0001$.

(b) Is $I < R_n$? Explain.

(c) Suppose that F is an antiderivative of f and that $f(2) = 0$. Find a value of n such that $|L_n - \int_2^4 F(x)\,dx| \le 0.01$.

15. Give an example of a function f such that R_n always underestimates the value of $\int_0^5 f(x)\,dx$ by the maximum amount allowed by Theorem 2.

16. Adapt the argument given in Example 3 to prove that $|I - R_n| \le \dfrac{K_1(b-a)^2}{2n}$.

17. Compute T_{10} and M_{10} for each of the integrals in problems 1–6.

(a) How do the approximation errors made by T_{10} and M_{10} compare with those made by L_{10}?

(b) How do the approximation errors made by T_{10} compare with those made by M_{10}?

(c) For which of the integrals does T_{10} make no approximation error? What is true about the first derivative of the integrand in each of these cases?

(d) Which of the integrals does T_{10} underestimate? What is true about about the second derivative of the integrand in each of these cases?
[HINT: Sketch each integrand over the interval of integration.]

(e) Which of the integrals does T_{10} overestimate? What is true about about the second derivative of the integrand in each of these cases?
[HINT: Sketch each integrand over the interval of integration.]

(f) For which of the integrals does M_{10} make no approximation error? What is true about the first derivative of the integrand in each of these cases?

(g) Which of the integrals does M_{10} underestimate? What is true about about the second derivative of the integrand in each of these cases?

(h) Which of the integrals does M_{10} overestimate? What is true about about the second derivative of the integrand in each of these cases?

18. Let $I = \int_a^b f(x)\,dx$ and suppose that $0 \le f'(x) \le K_1$ for all $x \in [a, b]$.

(a) Show that $0 \le I - L_n \le \dfrac{K_1\,(b-a)^2}{2n}$.

(b) Show that $-\dfrac{K_1\,(b-a)^2}{2n} \le I - R_n \le 0$.

(c) Show that if $f(x) = x$, the magnitude of $I - L_n$ is as large as allowed by parts (a) and (b).

(d) Parts (a) and (b) show that if f is nondecreasing on the interval $[a, b]$, L_n underestimates I and R_n overestimates I. Note that the error bounds have the same magnitude, but are opposite in sign. This suggests that the estimates L_n and R_n be combined to produce a better estimate of I. How can this be done? Where have you seen this approximation to I before?

19. Let f be a linear function. Show that T_n computes $\int_a^b f(x)\,dx$ *exactly* (i.e., with no approximation error).   [HINT: Draw a picture.]

20.   (a) Adapt the argument given in Example 3 to prove that $|I - M_n| \le \dfrac{K_1\,(b-a)^2}{4n}$.

(b) Prove that $|I - T_n| \le \dfrac{K_1\,(b-a)^2}{2n}$.   [HINT: Use the identity $T_n = (L_n + R_n)/2$.]

21. Consider the assertion:

If $0 < f'(x) < g'(x)$ for every $x \in [a, b]$, then $|L_{10} - \int_a^b f(x)\,dx| < |L_{10} - \int_a^b g(x)\,dx|$.

Is this assertion true or false? Justify your answer.

7.3 Trapezoid sums, midpoint sums, and the second derivative

Left and right sums, although simple conceptually, commit large errors. For most integrals the trapezoid and midpoint rules do better. In this section we see how *much* better and why.

The rules and their properties

First, formal definitions. Throughout this discussion $I = \int_a^b f(x)\, dx$ and the partition $a = x_0 < x_1 < x_2 < \cdots < x_n = b$ divides $[a, b]$ into n subintervals, each of length $\Delta x = (b - a)/n$.

Definition: (**The midpoint rule.**) A **midpoint sum** has the form

$$M_n = \sum_{i=1}^{n} f\left(\frac{x_{i-1} + x_i}{2}\right) \cdot \Delta x.$$

In the ith summand, f is evaluated at the *midpoint* of the ith subinterval $[x_{i-1}, x_i]$.
(**The trapezoid rule.**) A **trapezoid sum** has the form

$$T_n = \sum_{i=1}^{n} \frac{f(x_{i-1}) + f(x_i)}{2} \cdot \Delta x.$$

In the ith summand, the values $f(x_{i-1})$ and $f(x_i)$ are averaged.

(Pictures illustrating the ideas of trapezoid and midpoint sum appear in the following example, and also in Section 5.4.)

By the FTC, $I = 1/3$.

■ **Example 1.** Use M_4 and T_4 to approximate $I = \int_0^1 x^2\, dx$.◄ How much error does each commit?

We spare you the numerical computations, but check that the forms are right.

Solution: Using the FTC and the definitions above,◄

$$
\begin{aligned}
M_4 &= \left[f(1/8) + f(3/8) + f(5/8) + f(7/8)\right] \cdot \frac{1}{4} \\
    &= \left[(1/8)^2 + (3/8)^2 + (5/8)^2 + (7/8)^2\right] \cdot \frac{1}{4} \\
    &= \frac{63}{192};
\end{aligned}
$$

$$
\begin{aligned}
T_4 &= \left[\frac{f(0) + f(1/4)}{2} + \frac{f(1/4) + f(1/2)}{2} + \frac{f(1/2) + f(3/4)}{2} + \frac{f(3/4) + f(1)}{2}\right] \cdot \frac{1}{4} \\
    &= \left[f(0) + 2f(1/4) + 2f(1/2) + 2f(3/4) + f(1)\right] \cdot \frac{1}{8} \\
    &= \frac{66}{192};
\end{aligned}
$$

$$I = \frac{1}{3} = \frac{64}{192}.$$

The following pictures illustrate the situation:

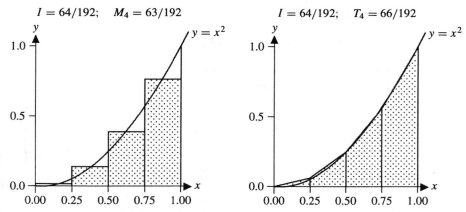

$I = 64/192; \quad M_4 = 63/192$ $I = 64/192; \quad T_4 = 66/192$

Notice, in particular, that M_4 *under*estimates I by $1/192$; T_4 *over*estimates I by exactly twice as much. □

M_n and tangent lines: another view

It's natural to think of M_n as a sum of *rectangular* areas.➤ In effect, we replace the graph of f over the ith subinterval (with midpoint m_i) with a *horizontal* line at height $y = f(m_i)$.

As in the picture above.

An alternative➤ view of M_n often helps in assessing the *sign* of the error:

But equivalent!

> *Replace the graph of f over the ith subinterval with the* tangent line at $x = m_i$.

The picture below illustrates the idea in a simple case:

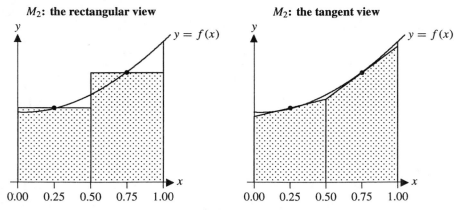

M_2: **the rectangular view** M_2: **the tangent view**

From this new viewpoint, M_n is a sum of *trapezoidal* areas—the top sides are determined by *tangent lines*.

That both approaches to M_n give the same answer has nothing especially to do with tangent lines. The fortunate fact is that the area of *any* trapezoid➤ is its base times its *height at the midpoint*.

With parallel vertical sides, as shown.

Concavity of f and the direction of error

It's graphically obvious➤ that if f is concave *up* on $[a, b]$, every trapezoid sum T_n necessarily *overestimates* $I = \int_a^b f$. The "tangent" interpretation of M_n shows that the opposite principle applies to midpoint sums. In short, the concavity of f determines the direction of T_n and M_n errors:➤

Isn't it?

It's easier to learn this principle than to memorize it.

If f is concave up *on* [a, b], *then* T_n *overestimates* I ; M_n *underestimates.*
If f is concave down, *the reverse holds.*

Error bounds for midpoint and trapezoid sums

The evidence so far suggests that trapezoid and midpoint sums usually do *better* than left and right sums at approximating integrals. How *much* better? Why?

What causes M_n and T_n errors? The role of f''

Pun intended.

Left and right sum approximations to $\int_a^b f$ commit *zero* error only if f is constant, i.e., if $f' = 0$. Thus left and right rule errors derive[←] from the *first* derivative: because $|f'|$ measures how far f *differs* from being constant, an upper bound K_1 for $|f'|$ plays a natural role in any "worst-case" error formula for L_n and R_n.

Pictures—draw one, if necessary—show why.

Midpoint and trapezoid sums do better than left and right sums for a simple but important reason:[←]

> M_n and T_n commit no *error if f is a* linear *function.*

Midpoint and trapezoid errors result, therefore, not from the steepness of the f-graph but from its *concavity*:

> *The more concave (upward or downward) the graph of f, the worse M_n and T_n behave.*

An error bound theorem for M_n and T_n

The magnitude of the second derivative.

"Nice" means that f has well-behaved derivatives f' and f'' on [a, b].

Since $|f''|$[←] measures the concavity of f, its appearance in the theorem below should be no surprise. The usual ground rules apply: $I = \int_a^b f$, M_n and T_n are midpoint and trapezoid sums with n equal subdivisions, and f is "nice."[←]

> **Theorem 3 (Midpoint and trapezoid rule error bounds).**
> Let $I = \int_a^b f(x)\,dx$, and let K_2 be an upper bound for $|f''|$ on $[a, b]$. Then
> $$|I - M_n| \leq \frac{K_2(b-a)^3}{24n^2}; \qquad |I - T_n| \leq \frac{K_2(b-a)^3}{12n^2}.$$

Notice:

The subscript reminds us of the second derivative.

- The number K_2[←] can be *any* upper bound for $|f''|$ on $[a, b]$, i.e., any number for which the inequality
$$|f''(x)| \leq K_2$$

From the theorem on left and right rule error bounds.

holds for every x in $[a, b]$. As with K_1,[←] K_2 can be estimated graphically. Ballpark estimates suffice; in fact, *over*estimating K_1 and K_2 is the conservative strategy.

- The precise ingredients of the error bounds—the constants 12 and 24, the power n^2, etc.—are, for the moment, mysterious. We'll justify them below.

But not more!

- Both formulas are *in*equalities. The actual errors (on the left sides) may be *less*[←] than the right sides predict.

- For a particular integral $I = \int_a^b f(x)\,dx$, the right sides of all our error formulas depend only on n—everything else is constant:

$$|I - M_n| \le \frac{C_1}{n^2}; \qquad |I - T_n| \le \frac{C_2}{n^2}; \qquad |I - L_n| \le \frac{C_3}{n}.$$

The first two estimates are "stronger" than the third because of the higher powers of n.

■ **Example 2.** Use M_{10}, T_{10}, and L_{10} to approximate $I = \int_0^1 \sin(x^2)\,dx$. What do the error bound theorems predict?

Solution: Here, first, are the numbers:➤ *With technology they're easy to find.*

$$M_{10} = 0.3098162947; \qquad T_{10} = 0.3111708112; \qquad L_{10} = 0.2690972619.$$

To find values for K_1 and K_2, we find derivatives and plot the results:➤

We could do all this by hand, but Maple simplifies the process. To compute a second derivative with Maple, say of $\sin(x)$, one can type `diff(sin(x),x,x);`.

```
> f1 := diff(sin(x^2),x);
                        2
              f1 := 2 cos(x ) x

> f2 := diff(sin(x^2),x,x);
                  2  2          2
        f2 := - 4 sin(x ) x  + 2 cos(x )

> plot( {f1,f2}, x=0..1);
```

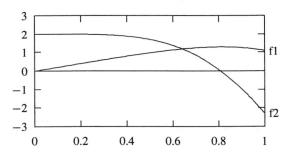

The graphs suggest reasonable values: $K_1 = 1.5$ and $K_2 = 2.5$ will do. Only arithmetic remains:

$$|I - M_{10}| \le \frac{K_2(b-a)^3}{24n^2} = \frac{2.5 \cdot 1^3}{2400} = 0.00104167;$$

$$|I - T_{10}| \le \frac{K_2(b-a)^3}{12n^2} = \frac{2.5 \cdot 1^3}{1200} = 0.00208333;$$

$$|I - L_{10}| \le \frac{K_1(b-a)^2}{2n} = \frac{1.5 \cdot 1^2}{20} = 0.075.$$

The maximum possible errors committed by M_n, T_n, and L_n are about 0.001, 0.002, and 0.08, respectively.   □

■ **Example 3.** How many subdivisions does each method require to approximate $I = \int_0^1 \sin(x^2)\,dx$➤ with assured error less than 0.0001?

The same integral as above.

Check details.

Solution: To answer, we set the right sides equal to 0.0001 and solve for n. For M_n, T_n, and L_n, respectively, we get:[*]

$$\frac{2.5}{24n^2} \le \frac{1}{10000} \iff n \ge 32.2749;$$

$$\frac{2.5}{12n^2} \le \frac{1}{10000} \iff n \ge 45.6435;$$

$$\frac{1.5}{2n} \le \frac{1}{10000} \iff n \ge 7500.$$

To be sure of approximating I to within 0.0001, the midpoint, trapezoid, and left rules need about 33, 46, and 7500 subdivisions, respectively. Notice the striking "efficiency" difference: the left rule takes over 200 times as much work as the midpoint rule!   □

The error bounds revisited: a closer look

That the error bounds above should *somehow* involve K_2 isn't, in context, surprising. Their particular forms, on the other hand, are hardly obvious from intuition. To end this section, we'll show informally[*] how the various constants [*] arise.

Formal proofs use the mean value theorem.

12, 24, etc.

Convince yourself!

A quadratic function: the worst offender.   The error bounds above apply to any function f on $[a, b]$ for which $|f''(x)| \le K_2$. Among *all* such functions, the polynomial $q(x) = K_2 x^2 / 2$ represents a sort of *extreme*, because $q''(x) = K_2$ for *all* x.[*] In other words, $|q''|$ is *always* as large as the theorem permits.

Given the "extreme" nature of q, we expect the maximum possible errors.

Check the details—we omitted some simple ones.

Errors on one subinterval.   How much error, at worst, can M_n and T_n commit over *one* subinterval? We'll answer by computing directly with the function q.[*] To simplify the algebra, we'll take $[0, h]$ as our subinterval.

The first step is to compare I_h (the *exact* integral of q over $[0, h]$) with M_h and T_h (the midpoint and trapezoid estimates over the same interval):[*]

$$I_h = \int_0^h q = \int_0^h \frac{K_2 x^2}{2}\, dx = \frac{K_2 x^3}{6}\bigg]_0^h = \frac{K_2 h^3}{6};$$

$$M_h = q(h/2) \cdot h = \frac{K_2 (h/2)^2}{2} \cdot h = \frac{K_2 h^3}{8};$$

$$T_h = \frac{q(0) + q(h)}{2} \cdot h = \frac{K_2 h^3}{4}.$$

The errors committed over $[0, h]$, therefore, are

$$I_h - M_h = \frac{K_2 h^3}{6} - \frac{K_2 h^3}{8} = \frac{K_2 h^3}{24};$$

$$I_h - T_h = \frac{K_2 h^3}{6} - \frac{K_2 h^3}{4} = -\frac{K_2 h^3}{12}.$$

As they should, the trapezoid and midpoint errors have opposite signs.

The mysterious constants 12 and 24 have appeared at last.[*]

Error over n subintervals.   Multiplying the quantities above by n gives the worst-case error over n subintervals. Replacing h with $(b - a)/n$ gives the error bound formulas of the theorem.

Exercises

1. Let $I = \int_0^1 e^{x^2} dx$.

   (a) Compute M_2 and T_2 *by hand*. (You may use a scientific calculator.)

   (b) Compute L_{10}, R_{10}, M_{10}, and T_{10}. Which of these approximations *underestimate* the exact value of I? Justify your answer.

   (c) Using the error bounds in Theorem 3, what is the smallest value of n that guarantees that $|I - M_n| \le 0.0005$ (i.e., that M_n approximates I to three decimal place accuracy)?

2. How large must n be to guarantee (using Theorem 3) that T_n estimates $\int_0^1 \sin x \, dx$ to within 10^{-10}? Will T_n underestimate or overestimate the exact answer? Why?

Compute T_{10} and M_{10} for each of the following integrals. Compare these estimates with the exact values and check that the bound on the magnitude of the approximation error given in Theorem 3 is satisfied.

3. $\int_2^3 1 \, dx$ 6. $\int_1^4 \sqrt{x} \, dx$

4. $\int_1^3 x \, dx$ 7. $\int_1^2 x^{-1} \, dx$

5. $\int_1^2 x^2 \, dx$ 8. $\int_2^3 \sin x \, dx$

9. Let f be the function graphed below.

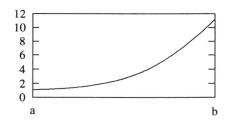

   Estimates of $\int_a^b f(x) \, dx$ were computed using the left, right, midpoint, and trapezoid rules, each with the same number of subintervals. The answers recorded were $8.52974, 9.71090, 9.74890$, and 11.04407. Which rule produced which estimate? Justify your answer and explain why it does not depend on knowledge of the value of n used to compute the estimates.

10. Let $I = \int_a^b f(x) \, dx$ where f is the function graphed below.

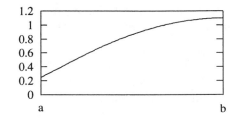

   Rank the values I, L_{30}, R_{30}, T_{30}, and M_{30} in increasing order. Justify your answer.

11. Let $I = \int_a^b F(x)\,dx$ where F is an antiderivative of the function f in the previous exercise. Does T_{100} underestimate or overestimate the value of I? Explain.

12. Let $I = \int_a^b f(x)\,dx$ and suppose that the graph below shows f''.

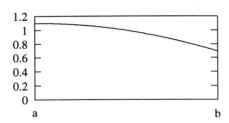

    Indicate whether each of the following inequalities must be true, might be true, or is false. Justify your answers.

    (a) $L_{100} \leq R_{100}$ (c) $M_{50} \leq L_{50}$

    (b) $T_{200} \leq M_{200}$

13. Let $I = \int_0^4 f(x)\,dx$ and suppose that f' is the function shown below.

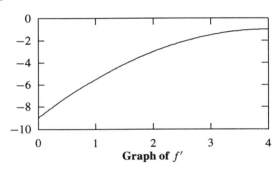

Graph of f'

    (a) Is $I < M_{10}$? Justify your answer.

    (b) Let F be an antiderivative of f. Find a value of n such that $|M_n - \int_1^2 F(x)\,dx| \leq 0.01$.

14. Let $I = \int_0^5 f(x)\,dx$ and suppose that each of the follow inequalities holds when $0 \leq x \leq 5$:

    (i) $-3 \leq f'(x) \leq -1$
    (ii) $2 \leq f''(x) \leq 6$

    Indicate whether each of the following statements **must** be true, **might** be true, or **cannot** be true. Justify your answers.

    (a) $I - T_{10} < 0.000005$

    (b) $M_{10} < R_{10}$

    (c) $|I - L_{10}| < |I - T_{10}|$.

15. Let $I = \int_{-1}^2 f(x)\,dx$ where f is a function with the following properties:

    (i) f is increasing on the interval $[-1, 2]$.

(ii) f'' is the function graphed below.

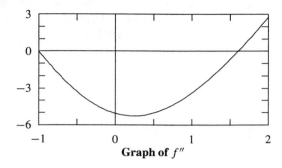

Graph of f''

(iii) f has the values tabulated below.

x	-1.00	-0.25	0.50	1.25	2.00
$f(x)$	2.0000	26.522	48.755	68.328	86.790

(a) Compute R_4 and an upper bound on the value of $|I - R_4|$.

(b) Compute T_4 and an upper bound on the value of $|I - T_4|$.

16. (a) Compute $(I - M_{10})/(I - T_{10})$ for each of the integrals in exercises 5–8. Are the results what you anticipated? Explain.

(b) Compute the M_{10} and T_{10} approximations to $\int_0^1 \sqrt{x}\,dx$, then compute $(I - M_{10})/(I - T_{10})$. Does the result surprise you? What's going on here?

Let $I = \int_0^{10} f(x)\,dx$. For each exercise below, give an example of a function f with the given property.

17. M_{200} underestimates I and the magnitude of the approximation error made is as large as Theorem 3 allows.

18. M_{200} overestimates I and the magnitude of the approximation error made is as large as Theorem 3 allows.

19. T_{200} underestimates I and the magnitude of the approximation error made is as large as Theorem 3 allows.

20. T_{200} overestimates I and the magnitude of the approximation error made is as large as Theorem 3 allows.

21. (a) Compare the error bound for M_n to the error bound for M_{10n}. Approximately how many additional decimal places of accuracy are gained by using ten times as many subintervals?

(b) Repeat part (a) for L_n and L_{10n}.

22. Let $I = \int_a^b f(x)\,dx$ and suppose that $0 \le f''(x) \le K_2$ for all $x \in [a, b]$.

(a) Show that $0 \le I - M_n \le \dfrac{K_2\,(b-a)^3}{24n^2}$.

(b) Show that $-\dfrac{K_2\,(b-a)^3}{12n^2} \le I - T_n \le 0$.

(c) Parts (a) and (b) show that if f is concave up on the interval $[a, b]$, M_n underestimates I and T_n overestimates I. This suggests that the estimates M_n and T_n can be combined to produce a better estimate of I. How can this be done?   [HINT: Combine the estimates in such a way that the worst case errors cancel.]

7.4 Simpson's rule

The story so far: integration rules in perspective

Simpson's rule, *aka* the **parabolic rule**, completes our menu of methods for approximating the value of an integral $I = \int_a^b f$. As the trapezoid and midpoint rules "improve" the simpler left and right rules, so Simpson's rule "improves" the midpoint and trapezoid rules.

The following hierarchy of rules shows how this improvement proceeds, step by step.

1. The **left and right rules** find I exactly➤ if f is any *constant function.* For arbitrary functions, therefore, L_n and R_n errors reflect the size of $|f'|$ (which measures how far f differs from being constant). So does the error bound formula:➤

$$|I - L_n| \leq \frac{K_1(b-a)^2}{2n}.$$

   I.e., commit zero error.

   The error bound for R_n is identical. K_1 is an upper bound for $|f'|$.

2. The **trapezoid and midpoint rules** find I exactly➤ if f is any *linear function.* For arbitrary functions, therefore, T_n and M_n errors reflect the size of $|f''|$ (which measures how far f differs from being linear). So do the error bound formulas:➤

$$|I - T_n| \leq \frac{K_2(b-a)^3}{12n^2}; \qquad |I - M_n| \leq \frac{K_2(b-a)^3}{24n^2}.$$

   I.e., commit zero error.

   K_2 is an upper bound for $|f''|$.

3. **Simpson's rule** finds I exactly➤ if f is any *quadratic function.* For arbitrary functions, therefore, we'd expect S_n errors to depend on the *third* derivative $|f'''|$ (which measures how far f differs from being quadratic).

   I.e., commits zero error.

By a lucky surprise, S_n does even better than expected. Although designed for *quadratic* integrands, Simpson's rule turns out to handle even *cubic* integrands exactly. Our resulting bonus➤ is that the S_n error bound formula involves an even *higher* derivative—the *fourth*—and an even higher power of n:➤

$$|I - S_n| \leq \frac{K_4(b-a)^5}{180n^4}.$$

Why is this a bonus? We'll see below.

K_4 is an upper bound for $|f^{(iv)}|$. Details later.

The last formula needs explanation, of course; we include it here for general comparison with the others. In the rest of this section we explain what its ingredients mean and show the rule in action.

Simpson's rule: definition and interpretations

Simpson's method of approximating $I = \int_a^b f$ starts with a partition $a = x_0 < x_1 < x_2 < \cdots < x_n = b$ of $[a, b]$ into an *even* number $n = 2m$ of subdivisions, each of length $\Delta x = (b-a)/n$.➤ We can now state the formal definition:

The other rules don't care whether n is even or odd.

> **Definition:** Simpson's approximation to I is the sum
>
> $$S_n = \big(f(x_0) + 4f(x_1) + 2f(x_2) + 4f(x_3) +$$
> $$2f(x_4) + \cdots + 4f(x_{n-1}) + f(x_n)\big)\frac{\Delta x}{3}.$$
>
> In sigma notation:
>
> $$S_n = S_{2m} = \frac{\Delta x}{3} \sum_{i=1}^{m} \big(f(x_{2i-2}) + 4f(x_{2i-1}) + f(x_{2i})\big).$$

But related!

The definition raises obvious questions: Why must n be even? What accounts for the coefficients 2, 3, and 4 in the approximating sum? What does any of this have to do with quadratic functions? We'll address these questions by interpreting Simpson's rule in two different[◄] ways.

Simpson sums as weighted averages of trapezoid and midpoint sums

We've already observed that for any n, the trapezoid sum T_n is the (ordinary) *average* $(L_n + R_n)/2$ of corresponding left and right sums. For a typical function over a small interval, the left and right rule estimates *straddle* the exact integral, so "splitting the difference" makes good sense.

In a similar way, Simpson's rule is built from the trapezoid and midpoint rules. Two ideas are key:

Which rule overestimates? Which underestimates? Concavity decides. See Section 7.3.

- For a typical function over a small interval, the trapezoid and midpoint methods *commit errors with opposite sign.*[◄]

- The trapezoid rule error bound is exactly *twice* that for the midpoint rule.

Put together, these facts suggest:

> *Over a small interval the trapezoid error is about* twice *as large as the midpoint error, and* opposite in sign.

Arrange for the errors made by T_n and M_n to cancel each other out.

This principle, in turn, suggests a natural strategy:[◄] Estimate I using the point 1/3 of the way from M_n to T_n. The result is precisely Simpson's estimate. In symbols:[◄]

S_{2n} deserves its subscript because it involves evaluating f at the endpoints of $2n$ subintervals.

$$S_{2n} = \frac{T_n + 2 \cdot M_n}{3}.$$

■ **Example 1.** Make sense of the formula above for $n = 3$ and $I = \int_0^1 \sin x \, dx$.

Check the algebra.

Solution: Is $S_6 = (T_3 + 2M_3)/3$? Let's check:[◄]

$$T_3 = \frac{1}{6}\left[\sin(0) + 2\sin(1/3) + 2\sin(2/3) + \sin(1)\right];$$

$$M_3 = \frac{1}{3}\left[\sin(1/6) + \sin(3/6) + \sin(5/6)\right];$$

$$\frac{T_3 + 2 \cdot M_3}{3} = \frac{1}{18}\left[\sin(0) + 2\sin(1/3) + 2\sin(2/3) + \sin(1)\right] + \frac{2}{9}\left[\sin(1/6) + \sin(3/6) + \sin(5/6)\right]$$

$$= \frac{1}{18}\left[\sin(0) + 4\sin(1/6) + 2\sin(2/6) + 4\sin(3/6) + 2\sin(4/6) + 4\sin(5/6) + \sin(1)\right]$$

$$= S_6.$$

Did you see the coefficients 2, 3, and 4 pop up in the computation?

We used Maple.

Things are as they should be.[◄]

The numbers work out nicely, too:[◄]

$$T_3 = 0.455433; \qquad M_3 = 0.461833; \qquad \frac{T_3 + 2M_3}{3} \approx 0.4597 = S_6. \qquad \square$$

Simpson's rule and approximation by parabolic arcs

A second, more geometric, approach to Simpson's rule concerns **parabolic approximation**. The other rules "pretend" that f is either constant (for left and right sums) or linear (for trapezoid and midpoint sums) on subintervals, but Simpson's rule pretends that f is *quadratic on successive pairs of subintervals.*

The process, more precisely, is in three steps:

1. Partition $[a, b]$ into an *even* number n of equal subdivisions; consider successive *pairs* of subdivisions.➤

We can do this because n is even.

2. Each such *pair* of subdivisions involves *three* equally-spaced partition points: x_{2i-2}, x_{2i-1}, and x_{2i}. Over each pair of subdivisions, "replace" the given f with a quadratic function q—the one whose graph (a parabola) passes through the three points $(x_{2i-2}, f(x_{2i-2}))$, $(x_{2i-1}, f(x_{2i-1}))$, and $(x_{2i}, f(x_{2i}))$.➤ For algebraic reasons, *one* such function q exists. (If the three points happen to be collinear, then the coefficient of x^2 is zero, so q is actually linear.)

In other words, q "agrees" with f at three consecutive partition points.

3. Find the (signed) area under➤ each small parabolic arc; the sum of all $n/2$ areas is Simpson's estimate S_n.

Or over, if $f < 0$.

The picture below illustrates the idea; the shaded area represents S_6:➤

Notice that S_6 involves three parabolas, not six.

Simpson's rule: the idea

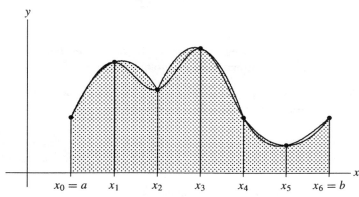

The last step above sounds daunting. Finding a new quadratic function for each of $n/2$ pairs of subintervals, for example, seems to portend lots of tedious algebra.

It's not that bad. As it happens, the area under the parabolic arc joining *any* three points $(x_{2i-2}, f(x_{2i-2}))$, $(x_{2i-1}, f(x_{2i-1}))$, and $(x_{2i}, f(x_{2i}))$ reduces, after the computational dust settles, simply to

$$\frac{\Delta x}{3}\left(f(x_{2i-2}) + 4f(x_{2i-1}) + f(x_{2i})\right),$$

precisely the ith summand in the definition above.

The "parabolic" approach to Simpson's rule has an immediate consequence: S_n *commits* zero error *for quadratic integrands.*

An error bound for Simpson's rule

Simpson's rule is exact—by design—for constant, linear, and quadratic polynomials.➤ By good luck, Simpson commits no error on *cubic* integrands, either.

I.e., all functions for which $f''' = 0$. Think about it.

■ **Example 2.** Use S_2 to estimate $I = \int_a^b x^3\, dx$. How much error is committed?

Solution: Let's calculate:➤

Here $x_0 = a$, $x_1 = (a+b)/2$, $x_2 = b$.

$$I = \int_a^b x^3\, dx = \frac{x^4}{4}\Bigg]_a^b = \frac{b^4 - a^4}{4};$$

$$S_2 = \frac{b-a}{6}\left(a^3 + 4\left(\frac{a+b}{2}\right)^3 + b^3\right).$$

Straightforward algebra shows that the two quantities are equal; hence S_2 commits *no* error. □

Added to everything above, Example 2 shows:

Both $f^{(iv)}$ and $f^{(4)}$ denote the fourth derivative of f.

> S_n commits zero error *for constant, linear, quadratic, and cubic integrands, i.e., for any function f with $f^{(4)} = 0$.* ◄◄

It's no surprise, then, that the error bound for Simpson's rule involves K_4, an upper bound for $f^{(4)}$ on $[a, b]$.

Theorem 4 (An error bound for Simpson's rule).
Let $I = \int_a^b f(x)\, dx$, and let K_4 be an upper bound for $|f^{(4)}|$ on $[a, b]$. Then

$$|I - S_n| \le \frac{K_4(b - a)^5}{180n^4}.$$

Notice:

- As with K_1 and K_2 in earlier theorems, K_4 may be *any* upper bound for $|f^{(4)}|$. As before, rough estimates are best found graphically.

- The parameter n appears to the *fourth* power in the denominator above; this accounts for Simpson's astonishing efficiency.

■ **Example 3.**   We calculated above that if $I = \int_0^1 \sin x\, dx$, then $S_6 = 0.45969967$. What does the error formula predict? How much actual error does S_6 commit?

Do it!

Solution: Because $f(x) = \sin x$, an easy computation ◄◄ shows that $|f^{(4)}(x)| = |\sin x| \le 1$; $K_4 = 1$ will do. Thus, by the theorem,

$$|I - S_n| \le \frac{K_4(b - a)^5}{180n^4} = \frac{1 \cdot (1 - 0)^5}{180 \cdot 6^4} \approx 0.0000043.$$

The predicted error, therefore, occurs in the *sixth* decimal place!

The *actual* error is even less:

$$I = \int_0^1 \sin x\, dx = -\cos x \Big]_0^1 = 1 - \cos 1 \approx 0.45969769,$$

so

$$|I - S_6| = |0.45969769 - 0.45969967| \approx 0.000002,$$

only about *half* what the theorem predicts. □

In our last example, we estimate an *unknown* quantity.

■ **Example 4.**   Use S_{10} to estimate $I = \int_0^1 \sin x^2\, dx$. How much error, at worst, might S_{10} commit? How many subdivisions would be needed to approximate I with error less than 10^{-8}?

Thankfully—imagine doing this by hand.

Solution: We'll use *Maple:* ◄◄

```
> f := sin(x^2);
                                    2
                         f := sin(x )


> nsimpson( f, x=0..1, 10);
```

$$0.3102602344$$

```
> f4 := diff(f,x,x,x,x);
```

$$f4 := 16 \sin(x^2) x^4 - 48 \cos(x^2) x^2 - 12 \sin(x^2)$$

(The command `diff(f,x,x,x,x);` calculates the fourth derivative; `diff(f,x$4);` does the same thing.)

Bounding the fourth derivative is easiest graphically:➤

> *It would be hard by symbolic methods.*

```
> graph(f4, x=0..1);
```

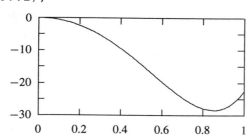

Given the graph, $K_4 = 30$ looks safe.

The rest is arithmetic. By the theorem,

$$|I - S_{10}| \leq \frac{30 \cdot (1 - 0)^5}{180 \cdot 10^4} \approx 0.000017.$$

The error, if any, enters around the fifth decimal place; 0.3103 is a meaningful estimate for I.

To assure error less than 10^{-8}, we need n so that

$$|I - S_n| \leq \frac{30 \cdot (1 - 0)^5}{180 \cdot n^4} < 10^{-8}.$$

Solving the last inequality for n gives

$$\frac{1}{6 \cdot n^4} < 10^{-8} \iff n^4 > \frac{10^8}{6} \iff n > \frac{100}{\sqrt[4]{6}} \approx 63.9.$$

Any even n over 63 will do.➤   □   *Why does n have to be even?*

Exercises

Show (by hand computation) that S_2 evaluates each of the following integrals exactly.

1. $\int_a^b 1 \, dx$ 3. $\int_a^b x^2 \, dx$

2. $\int_a^b x \, dx$ 4. $\int_a^b x^3 \, dx$

5. Show (by hand computation) that when S_2 is used to estimate $I = \int_a^b x^4 \, dx$, the magnitude of the approximation error is *equal* to the upper bound given in Theorem 4.

6. Let $I = \int_0^2 e^{x^2} \, dx$.

    (a) Compute S_4 using nothing more than a scientific calculator.

    (b) Bound the approximation error made by S_4.

    (c) Using the error bound formula for Simpson's rule given in Theorem 4, what is the smallest value of n necessary to guarantee that S_n approximates I to within 10^{-5}?

7. Let $I = \int_0^1 \cos(100x) \, dx$.

    (a) Compute the value of I exactly using the FTC.

    (b) Compute S_{10}. Show that the *actual* approximation error is less than 0.05.

    (c) Find an upper bound on the magnitude of the approximation error. Why is this upper bound so enormous?

8. Suppose S_{10} is the first Simpson sum guaranteed to approximate a certain definite integral I within 10^{-2}.

    (a) How much approximation error can S_{100} make?

    (b) What is the smallest value of n for which S_n must approximate I with an error no greater than $\pm 10^{-10}$?

9. Let $I = \int_{-1}^2 f(x) \, dx$ where f is a function such that $1 \le f^{(4)}(x) \le 8$ if $-1 \le x \le 2$, and f has the values tabulated below.

x	-1.00	-0.25	0.50	1.25	2.00
$f(x)$	2.0000	26.522	48.755	68.328	86.790

Compute S_4 and an upper bound on the value of $|I - S_4|$.

10. Let $I = \int_5^7 \cos x \, dx$.

    (a) Compute S_4, the Simpson's rule estimate of I with four equal subintervals, by hand. Use a calculator (*set in radian mode!*) to find the necessary values of the cosine function.

    (b) What does the error bound formula say about the error made when I is approximated by S_4?

    (c) Does S_4 underestimate or overestimate I? By how much?

11. Compute an estimate of $\int_0^2 xe^{-x^3} \, dx$ that is guaranteed to be correct within ± 0.001. Be sure to explain carefully why your estimate has the desired accuracy.

12. Let $I = \int_0^1 \sin(\sin(x)) \, dx$.

    (a) Compute bounds on the error made when I is approximated by M_n for $n = 4, 8$, and 16, respectively.

    (b) Compute bounds on the error made when I is approximated by S_n for $n = 4, 8$, and 16, respectively.

13. Let $I = \int_0^1 f(x)\,dx$ where some values of the function f are tabulated below:

x	0.00	0.25	0.50	0.75	1.00
$f(x)$	1.307	1.096	1.018	1.173	1.435

   (a) Compute estimates of I using the left, right, midpoint, trapezoid, and Simpson's rules.

   (b) A plot of the data makes it seem reasonable to assume that the graph of f is concave up on the interval $[0, 1]$. Use this assumption and your results from part (a) to find upper and lower bounds on I. Justify your choices.

14. Let $f(x) = e^{\sin x}$ and $I = \int_{-50\pi}^{150\pi} f(x)\,dx$.

   (a) It is straightforward to calculate that

   $$f^{(4)}(x) = \left(\cos^4 x - 6\cos^2 x \sin x - 4\cos^2 x + 3\sin^2 x + \sin x\right) e^{\sin x}.$$

   Use this to show that $\left|f^{(4)}(x)\right| \leq 15e \approx 41$ for any x.   [HINT: If x and y are real numbers, $|x + y| \leq |x| + |y|$.]

   (b) Use a graph of $f^{(4)}(x)$ to show that $\left|f^{(4)}(x)\right| < 11$ for any x.

   (c) What is the smallest value of n for which Theorem 4 guarantees that S_n approximates I within 0.001? Justify your answer.

   (d) Explain why $I = 100 \int_0^{2\pi} f(x)\,dx$.

   (e) What is the smallest value of n for which Theorem 4 guarantees that S_n approximates $\int_0^{2\pi} f(x)\,dx$ within 0.00001? Justify your answer.

   (f) Explain how the results in parts (d) and (e) can be combined to produce an estimate of I with error guaranteed to be less than 0.001.

7.5   Chapter summary

This chapter surveys and compares a variety of numerical methods for approximating definite integrals. These methods include the left and right rules, the trapezoid and midpoint rules, and Simpson's rule.

Numerical methods, although they give approximate answers, are generally easier to apply than symbolic methods of antidifferentiation. (The latter will be surveyed in the next chapter.)

Left and right rules.   The left and right rules for approximating $\int_a^b f(x)\,dx$ amount to pretending that the integrand function remains *constant* over short subintervals. The error committed by these simple rules, therefore, is relatively large. It depends, moreover, on the size of the *first* derivative f', which can be thought of as a measure of how much f differs from being constant on its domain.

For typical functions, the errors committed by the left and right rules have opposite sign—if one overestimates, the other underestimates. It's natural, therefore, to *average* the results of the left and right rules; doing so gives the trapezoid rule.

Trapezoid and midpoint rules.   The trapezoid and midpoint rules amount to pretending that the integrand f is *linear* over each subinterval—in this case, both rules give exact answers. The errors committed by these rules, therefore, are relatively smaller than those committed by the left and right rules. The errors they commit, moreover, depend on the size of the *second* derivative f'', which can be thought of as a measure of how much f differs from being linear on its domain.

For typical functions, the errors committed by the trapezoid and midpoint rules have opposite sign—if one overestimates, the other underestimates. The midpoint rule, moreover, generally commits error about one-half that of the trapezoid rule. With this in mind, it's natural to use a weighted average of the trapezoid and midpoint approximations, with the midpoint approximation given double weight. Calculating this average produces Simpson's rule.

Simpson's rule.   Simpson's rule amounts to pretending that the integrand f is *quadratic* over each pair of subintervals; thus Simpson's rule calculates integrals of quadratic functions exactly, with no error.

The errors committed by Simpson's rule are often much smaller than those committed by the other rules. These errors, moreover, depend on the size of the *fourth* derivative $f^{(4)}$. Estimating this size may be difficult by hand, but it's far easier with computing.

Chapter 8

Using the Definite Integral

8.1 Introduction

Measurement and the definite integral

Many important applications of calculus involve measuring something: the **area** of a plane region, the **volume** of a solid object, the **net distance** a moving object travels over an interval, the **length** of a curve from one point to another, the **work**➤ done against gravity in raising a satellite into orbit, the **present value**➤ of an income stream (allowing for interest and inflation), etc.

In the physicist's sense—see Section 8.4.

In the economist's sense—see Section 8.5.

In a few simple cases, admittedly, such quantities can be found by common sense alone. Measuring the area of a rectangular region, the length of a straight line segment, or the distance covered by an object moving at constant ~~speed~~ requires no big machinery from calculus. In practice, though, most regions aren't rectangular, curves aren't straight, and speeds aren't constant. In these more usual and more interesting situations, calculus tools—the definite integral, in particular—are indispensable.

Technical note: assuming good behavior. Throughout this section we'll assume, to avoid unhelpful distractions, that all the integrals $\int_a^b f(x)\,dx$ we meet make good mathematical sense. To guarantee this it's enough to assume that every integrand f is *continuous* on $[a, b]$; we'll do so from now on. In fact, discontinuous integrands *do* sometimes arise in practical applications. Even in such cases, however, the basic ideas of this section often apply, although perhaps in slightly different forms.

Definite integrals and area: reprise. The relation between the definite integral and area isn't new. From the very beginning➤ we've understood the definite integral geometrically in terms of area. For any continuous function f on $[a, b]$,

See the definition and pictures in Section 5.1.

$$\int_a^b f(x)\,dx = \text{signed area bounded by the } f\text{-graph for } a \leq x \leq b.$$

Remember: "Signed area" means that area under the x-axis counts as negative. Keeping track of positive and negative areas takes a little care, but the basic link between integrals and areas is by now familiar. Here's an easy example to bring the issue back to mind.

■ **Example 1.** Easy calculations➤ show that

Check for yourself.

$$I_1 = \int_0^\pi \sin x\,dx = 2; \qquad I_2 = \int_0^{2\pi} \sin x\,dx = 0.$$

Interpret these results in area language.

Draw one, or see Appendix J.

Solution: As its graph⁣ shows, $\sin x \geq 0$ on $[0, \pi]$, so I_1 measures the ordinary area—2 square units—under one arch of the sine curve.

The value of I_2 means that the "net" or "signed" area over the interval $[0, 2\pi]$ is zero. This makes good geometric sense: $\sin x \leq 0$ on the interval $[\pi, 2\pi]$, and the symmetry of the graph guarantees that the areas above and below the x-axis exactly cancel each other out. ☐

Definite integrals: not just for area anymore.   Any definite integral $\int_a^b f(x)\,dx$ *can* be interpreted as a signed area, as illustrated above. But if definite integrals measured *only* area, they wouldn't deserve the fuss we make over them. In fact, definite integrals can be used to measure or model many quantities *other than* area. Volume, arclength, distance, work, mass, fluid pressure, accumulated financial value, and other quantities can all be calculated as definite integrals. Choosing the *right* integral, and interpreting the result appropriately depends on the problem at hand.

■ **Example 2.**   We'll show in Section 8.3 that the **length of the curve** $y = f(x)$ **from** $x = a$ **to** $x = b$ is given by the integral

$$\int_a^b \sqrt{1 + f'(x)^2}\,dx.$$

Use this fact to interpret the integral

$$I = \int_0^1 \sqrt{1 + 4x^2}\,dx$$

in two different ways: (i) as the **length** of an appropriate curve; (ii) as an appropriate **area**. Evaluate these quantities numerically.

Important note: the curve we're measuring is NOT the integrand.

Solution: The integral I fits the arclength "template" if $f'(x) = 2x$. Clearly, $f(x) = x^2$ has this property. Therefore I gives the length of the curve $y = x^2$ from $x = 0$ to $x = 1$.⁣

Any definite integral measures the signed area bounded by its integrand. In this case, therefore, I measures the area under the curve $y = \sqrt{1 + 4x^2}$ from $x = 0$ to $x = 1$.

The pictures below illustrate both interpretations of I:

$\int_0^1 \sqrt{1 + 4x^2}\,dx \approx 1.479$: **length of a curve** $\int_0^1 \sqrt{1 + 4x^2}\,dx \approx 1.479$: **area under a curve**

How large, numerically, is I? Antidifferentiation is algebraically messy, so we'll use M_{20}, the midpoint rule with 20 subdivisions. Here's the result, to three decimal places:⁣

See the exercises for more on this calculation.

$$I \approx M_{20} = 1.479.$$

Thus *both* quantities (length and area) have the same approximate numerical value: 1.479 units of length and 1.479 units of area, respectively.➤➤   □

The respective units might be inches and square inches.

Two views of the definite integral

As the previous example suggests, definite integrals can be used to measure many disparate quantities. The key question, usually, is *what function* to integrate, and over *what interval*.

In applying the integral in various settings, it's useful to keep two different—but closely related—interpretations of a definite integral $\int_a^b f(x)\, dx$ in mind.

A limit of approximating sums: The integral is defined, formally, as a limit of approximating sums. Chapter 7 discusses and compares several kinds of approximating sums: left, right, trapezoid, midpoint, etc. Using **right sums**, for instance, we can write

$$\int_a^b f(x)\, dx = \lim_{n \to \infty} \sum_{i=1}^{n} f(x_i) \Delta x,$$

where the inputs x_i are the right endpoints of n equal-length subintervals of $[a, b]$. From this point of view, the integral "adds up" small contributions, each of the form $f(x_i) \Delta x$.

Accumulated change in an antiderivative: The fundamental theorem of calculus says that

$$\int_a^b f(x)\, dx = F(b) - F(a),$$

where, on the right, the function F can be any antiderivative of f on $[a, b]$.➤➤ The difference $F(b) - F(a)$ represents, in a natural way, the **accumulated change** (or "net change") in F over the interval $[a, b]$. In other words: To find the accumulated change in F over $[a, b]$, integrate f—the *rate function* associated to F—over $[a, b]$.

Remember: Antiderivative aren't unique. If F is one antiderivative of f, then so is $F + k$, where k is any constant.

Mathematically speaking, these two approaches to the integral are equivalent. The fundamental theorem of calculus says so; it guarantees that both methods give the same "answer." Having two *different* ways to think about the integral makes it more versatile in applications. Which viewpoint is "better" depends on the situation. The next example illustrates both viewpoints.

■ **Example 3.** The function $v(t) = 10 + 20t - 10t^2$ gives a car's *eastward velocity*, in miles per hour at time t hours, for $0 \le t \le 3$. Calculate $I = \int_0^3 v(t)\, dt$; interpret I both as a limit of sums and as accumulated change in an antiderivative.

Solution: Calculating I symbolically is easy:

$$I = \int_0^3 v(t)\, dt = 10t + 10t^2 - \frac{10t^3}{3} \Big]_0^3 = 30.$$

What does the answer mean? The pictures below give two different views of I:

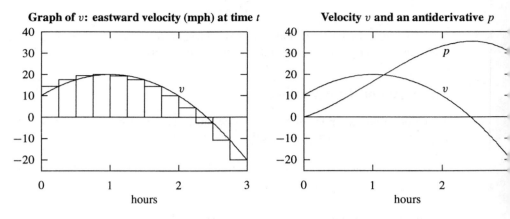

The left-hand picture suggests the limit-of-sums view of I. It shows R_{12}, the right approximating sum for I with 12 subdivisions. The signed area of each rectangle approximates the *eastward* distance covered by the car over that subinterval. (The last three rectangles contribute negative eastward distance—i.e., *westward* distance.) The right sum R_{12}, therefore, approximates the net accumulated eastward distance covered over the entire interval $0 \le t \le 3$. As the limit of such approximating sums, I represents the *exact* net eastward distance covered over [0, 3]—precisely 30 miles. Numerical evidence[◄◄] supports this idea:

We computed the right sums by machine, of course.

$$R_{12} \approx 25.938; \quad R_{50} \approx 29.082; \quad R_{200} \approx 29.774; \quad \ldots \quad I = 30.$$

The right-hand picture above illustrates the accumulated-change approach to I. Because the function v describes the car's eastward *velocity*, any antiderivative function p must describe the car's east-west *position*, given appropriate units and points of reference. Given our formula $v(t) = 10 + 20t - 10t^2$, we may as well choose the antiderivative function p defined by $p(t) = 10t + 10t^2 - 10t^3/3$. Since $p(0) = 0$, the car "starts" at position 0, and $p(t)$ represents the car's position at time t, measured in miles east of the starting point. Graphs of both v and p appear above right. As they show,

$$I = \int_0^3 v(t)\,dt = p(3) - p(0) = 30.$$

In words: Integrating the car's velocity over [0, 3] gives the accumulated change in position over the same interval.   □

Which view: sum or antiderivative?

Which view of the integral is "right" for a given application? Should we calculate an approximating sum, or look for an antiderivative?

The answer depends on the type and amount of information at hand. When an integrand presents itself in simple symbolic form, antidifferentiation is the obvious next step. But for data given graphically, or in tabular form, using approximating sums is a natural strategy. The next example illustrates both possibilities.

■ **Example 4.**   It's harvest time in the corn belt, and Farmer Brown is about to put her new Deere 12-row combine harvester through its paces. This machine is loaded. Farmer Brown has AC, AM, FM, CD, and 200 HP—but so does every other farmer in the county. What's really special on Brown's new Deere is something completely new: a continuous graphical readout, in real time, of the machine's instantaneous rate of harvesting, in bushels of corn per minute.

Farmer Jones, one farm south, can only dream of such conveniences. His two-year-old combine will take occasional instantaneous rate-of-harvest readings, but plotting them in real time is out of the question.

Brown and Jones, neighborly rivals, agree to compare their machines' harvesting performance over a 60-minute period, starting at $t = 0$. An hour later they pull over and compare their results—a plot and a table:

Farmer Brown's rate of harvest (bushels/minute)

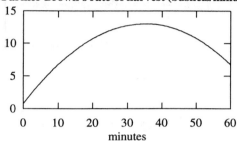

Farmer Jones's rate of harvest					
t (min)	5	15	25	35	45	55
bu/min	4	9	12	13	12	9

Who harvests more?

Solution: "Looks parabolic to me," says Farmer Brown. "That curve has a vertex at $(35, 13)$, and it goes through $(60, 27/4)$. Only one quadratic function has both of those properties, so my rate-of-harvest function r_B *must* have this formula:➤➤

Why is this the only possible quadratic formula? See the exercises.

$$r_B(t) = 13 - \frac{(t - 35)^2}{100}.$$

Integrating that will tell me how many bushels of corn I accumulated over the hour." Here's her result:

$$\int_0^{60} r_B(t)\, dt = \int_0^{60} \left(13 - \frac{(t - 35)^2}{100}\right) dt = 585.$$

Farmer Jones wants to integrate *his* rate-of-harvest function, r_J. But without an explicit formula for r_J, there's nothing to antidifferentiate. "No problem," thinks Jones. "Approximating sums are always available—with or without an explicit formula for the integrand. Since I know values of r_J at *midpoints* of six 10-minute intervals, I'll use M_6, the midpoint rule with six subdivisions." Here's his result:

$$\int_0^{60} r_J(t)\, dt \approx M_6 = (r_J(5) + r_J(15) + r_J(25) + r_J(35) + r_J(45) + r_J(55)) \cdot 10$$
$$= 590.$$

"Pretty close, but I'm up five bushels," says Jones. "Not so fast," says Brown. Why doesn't Brown concede? See the exercises.   □

Exercises ⎯⎯⎯⎯⎯⎯⎯⎯⎯⎯⎯⎯⎯⎯⎯⎯⎯⎯⎯⎯

1. In Example 4 Farmer Brown claimed that $r_B(t) = 13 - (t - 35)^2/100$ is the only quadratic function whose graph has a vertex at $(35, 13)$ and passes

through $(60, 27/4)$. Show this. (Hint: Any quadratic function can be written in the form $A + B(t - 35) + C(t - 35)^2$, for some constants A, B, and C. Assuming this, use the conditions given to find the appropriate values of A, B, and C.)

2. Why doesn't Farmer Brown concede in Example 4? To give one possible reason, find a quadratic function that "fits" Farmer Jones's data. (There is one, and only one.) Integrate this function to get *another* estimate for Farmer Jones's total harvest over the hour.

3. In the situation of Example 4, suppose we ignore the first and last 5 minutes, and consider only the interval $5 \le t \le 55$.

   (a) Use an integral to calculate how much Farmer Brown harvests from $t = 5$ to $t = 55$.

   (b) Use the trapezoid rule to estimate how much Farmer Jones harvests over the same period.

4. This problem is about Example 2, page 120, and the integral $I = \int_0^1 \sqrt{1 + 4x^2}\, dx$.

   (a) The left-hand picture in Example 2 illustrates the "arc length" interpretation of I. Use the picture to *estimate* the length of the curve $y = x^2$ from $x = 0$ to $x = 1$.

   (b) The right-hand picture in Example 2 illustrates the area interpretation of I. Use the picture to *estimate* the shaded area. Is your answer consistent with the previous part?

   (c) In Example 2 we used the midpoint approximation $I \approx M_{20} = 1.4788$. How much error, at worst, can M_{20} commit? (Use techniques of Chapter 7.)

   (d) The integral I can be calculated symbolically, using Formula #34 in the Table of Integrals. Do so. What's the "exact" answer?

5. Let f be any function for which f' is continuous on $[a, b]$. (This technical requirement guarantees that the length integral makes sense.) Let C denote the graph of $y = f(x)$ from $x = a$ to $x = b$. We said in Example 2 that

$$\text{length of } C = \int_a^b \sqrt{1 + f'(x)^2}\, dx.$$

We'll explain why this formula holds in Section 8.3. Here we'll *assume* that it holds and explore what it says.

   (a) Use the integral formula above to find the length of the straight line segment joining $(0, 0)$ and (a, b). (Hint: Start by finding a function f whose graph over $[a, b]$ is the straight line in question. Could you have found the same answer in an easier way?)

   (b) Write down an integral I that gives the length of the curve $y = x^2 + 1$ from $x = 0$ to $x = 1$. Estimate the value of I using M_{20}, the midpoint rule with 20 subdivisions.

   (c) Repeat the previous part for the two curves $y = \sin x$ and $y = 2 + \sin x$, in each case from $x = 0$ to $x = \pi$.

   (d) Let C be any real number, and f a well-behaved function as above. Use the integral formula above to explain why the length of the graph $y = f(x) + C$ from $x = a$ to $x = b$ does not depend on C. Then give a geometric explanation for the same fact.

6. A moral of Example 2 is that a given integral can be interpreted in various ways. In each part below, first calculate the given integral. Then interpret the numerical result in the stated context, using appropriate units of measurement.

(a) $\int_0^3 f(t)\,dt$, where the function $f(t) = 5t^2 - 20t + 50$ tells a certain car's speed, in miles per hour, t hours after midnight.

(b) $\int_0^3 f(t)\,dt$, where the function $f(t) = 5t^2 - 20t + 50$ tells a certain car's speed, in feet per second, t seconds after midnight.

(c) $\int_0^3 f(t)\,dt$, where the function $f(t) = 5t^2 - 20t + 50$ tells a certain car's acceleration, in feet per minute per minute, t minutes after midnight.

7. All parts of this problem are about the integral $I = \int_0^1 \sqrt{1 + x}\,dx$.

(a) Calculate I exactly, by antidifferentiation. Then give a decimal equivalent, rounded to 3 decimal places.

(b) I can be thought of as the area under some curve. Decide which curve, and draw the region in question. Does its area look about "right," numerically?

(c) I can also be thought of as the length of a graph $y = f(x)$, from $x = 0$ to $x = 1$. Find and plot such a function f. Does your graph appear to have about the "right" length?

(d) Is more than one function f possible in the previous part? If so, find and plot *another* possible function.

8.2   Finding volumes by integration

Calculus in the kitchen: slicing

Imagine a solid object—a long loaf of French bread, say—lying along the x-axis in the xy-plane, between $x = a$ and $x = b$. The usual way to *slice* such a loaf is with cuts *perpendicular* to the x-axis but, mathematically speaking, the "knives" that cut such slices are vertical planes, each of the form $x = k$ for some constant k. One such "cut," viewed from overhead, appears below:

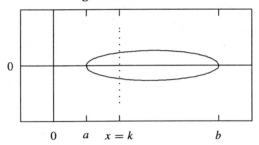

Slicing a loaf: an overhead view

The loaf's total *volume* is the sum of the volumes of all the slices. This obvious fact isn't much use in the kitchen. Mathematically, it lets us compute volumes by integration for a variety of solids.

Cross-sectional area and the volume of one slice

If the loaf is irregular, different slices have different volumes even if all have the same thickness. This difference arises from the fact that slices from the middle of the loaf have larger *cross-sectional area,*[◄] and hence more volume, than slices from the narrow ends.

Where the butter goes.

For any value of x between a and b, let $A(x)$ denote the **cross-sectional area** revealed by a knife cut at x (i.e., the area of the intersection of the loaf with the vertical plane at x, perpendicular to the x-axis). Clearly, $A(x)$ varies with x, rising and falling with the loaf's thickness.[◄]

For a loaf of white sandwich bread, A is essentially constant!

The *exact* volume of a slice of bread is hard to compute, because the cross section $A(x)$ varies with x. In the best possible case,[◄] $A(x)$ is constant. In that event, the following simple but important fact applies:

Mathematically, if not nutritionally.

> *If a slice, with thickness Δx, has* constant *cross-sectional area A, then the slice has volume $A \cdot \Delta x$.*

In a small x-interval (i.e., for a very *thin* slice) the cross-sectional area $A(x)$ is nearly constant.[◄] Therefore we can *approximate* the volume of a slice with thickness Δx by

Think about it. Do you agree?

$$\text{volume of one slice} \approx A(x)\Delta x,$$

where $A(x)$ is the cross-sectional area at any convenient value of x within the slice.

Reassembling Riemann's loaf: estimating volume

To estimate its volume, let's slice our loaf into n slabs of equal thickness $\Delta x = (b - a)/n$, with cuts[◄] at $x = x_0 = a, x = x_1, x = x_2, \ldots, x = x_n = b$. Here's an overhead view for $n = 4$:[◄]

Crumbless cuts, in this ideal world.

Cut on the dotted lines.

Four slices from the loaf

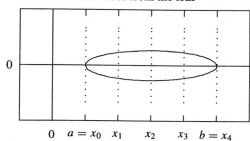

$$0 \quad a = x_0 \quad x_1 \quad x_2 \quad x_3 \quad b = x_4$$

Using the cross-sectional areas $A(x_1)$, $A(x_2)$, ... , $A(x_n)$ for the slabs leads to this estimate for total volume:

$$\text{total volume} \approx A(x_1)\Delta x + A(x_2)\Delta x + \cdots + A(x_n)\Delta x.$$

Here comes the crucial observation:

> *This volume estimate is an approximating sum*➡ *for the integral*

$$\int_a^b A(x)\,dx$$

> *of the area function.*

A right sum, to be precise.

Now we play the usual limit game: as n tends to infinity, the slab thickness Δx tends to zero so the volume *estimate* tends to the true volume of the loaf. Meanwhile, the approximating sums tend to the true value of the integral.

We'll state our conclusion as a general theorem, applicable to *all* well-behaved solids:➡

Edible or not.

> **Theorem 1.** Suppose that a solid lies with its base on the xy-plane, between the vertical planes $x = a$ and $x = b$. For all x in $[a, b]$, let $A(x)$ denote the area of the cross section at x, perpendicular to the x-axis. If $A(x)$ is a continuous function, then
>
> $$\text{volume} = \int_a^b A(x)\,dx.$$

Note:

Why the integral exists. Requiring that A vary *continuously* with x assures that the integral exists. For most ordinary, smooth, solid objects, A *is* continuous. More exotic objects can often be handled by breaking them up (mentally, of course) into simpler objects to which the formula above applies.

Looking for a formula. Using The theorem requires an explicit formula for A. Finding one may be difficult. For the simple solids studied here, a formula for A is easily found.

The simplest case. For solids with *constant* cross section (e.g., sticks of butter, glasses of milk, etc.) the theorem merely restates elementary volume formulas.➡

Convince yourself of this, say for an ordinary circular cylinder.

Using the theorem: solids of revolution

And have for centuries.

Among all solids, calculus users consistently prefer[◂◂] **solids of revolution**, i.e., solids formed by revolving some region in the xy-plane about an axis, often the x-axis. The figure below shows the solid formed by revolving the region under the curve $y = x^2$, from $x = 0$ to $x = 1$, about the x-axis:

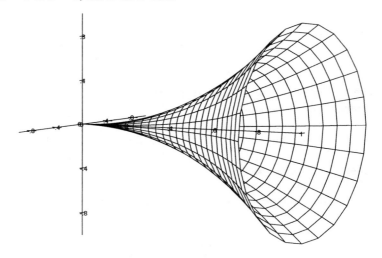

The revolutionary advantage

The picture above illustrates this.

Solids of revolution share one great advantage: their cross sections perpendicular to the axis of rotation are *circular*.[◂◂] Slices of such solids, in other words, are **disks**. This attribute leads to simple, convenient volume computations.

It's the same solid shown above. Does either picture suggest a numerical answer?

■ **Example 1.**   Find the volume of the solid formed by revolving the region shown below about the x-axis.[◂◂]

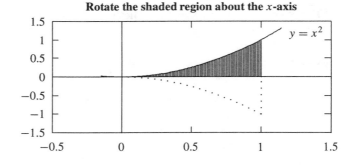

Rotate the shaded region about the x-axis

Solution: For any x in [0, 1], the cross section at x is a circle with radius $y = x^2$. Thus

$$A(x) = \pi \left(x^2\right)^2 = \pi x^4.$$

By Theorem 1,

$$\text{volume} = \int_0^1 A(x)\, dx = \int_0^1 \pi x^4\, dx = \frac{\pi}{5} \approx 0.628. \qquad \square$$

A computation similar to the previous one applies whenever the region under the graph of a positive function $y = f(x)$, from $x = a$ to $x = b$, is revolved around the x-axis.

■ **Example 2.** Find the volume of the sphere of radius r.

Solution: The sphere is the result of revolving the area under the curve $f(x) = \sqrt{r^2 - x^2}$, from $x = -r$ to $x = r$, about the x-axis.➤➤ The cross section at x, therefore, is a circle with radius $f(x)$. Hence

Draw your own picture!

$$A(x) = \pi(f(x))^2 = \pi\left(r^2 - x^2\right),$$

and so➤➤

Check algebra.

$$\begin{aligned}
\text{volume} &= \pi \int_{-r}^{r} \left(r^2 - x^2\right) dx \\
&= \pi \left[r^2 x - \frac{x^3}{3}\right]_{-r}^{r} \\
&= \frac{4\pi r^3}{3}.
\end{aligned}$$

□

Other axes. Rotation around the x-axis is not the only possibility. Whatever the axis, the principle of Theorem 1 applies: volume is found by integrating the cross sectional area.

■ **Example 3.** What's the volume of the solid formed if the region of Example 1 is rotated about the axis $y = -1$.

Solution: The solid formed in this way is a hollow "horn." For any x in [0, 1] the cross section at x is a **washer**, with inner radius 1 and outer radius $1 + x^2$.➤➤ The area $A(x)$ of such a washer is

Draw your own picture of a typical "washer."

$$\begin{aligned}
A(x) &= \pi\,(\text{outer radius})^2 - \pi\,(\text{inner radius})^2 \\
&= \pi\,(1 + x^2)^2 - \pi\,1^2 = \pi\,(2x^2 + x^4).
\end{aligned}$$

By the theorem,

$$\text{volume} = \int_0^1 A(x)\,dx = \pi \int_0^1 (2x^2 + x^4)\,dx = \frac{13\pi}{15} \approx 2.723.$$

Non-circular cross sections

For solids with non-circular cross sections, $A(x)$ may be harder to find; otherwise, the idea is exactly the same.

■ **Example 4.** A pyramid of height one has square horizontal cross sections; at the bottom, the edge length is one. Find the volume.➤➤

Draw your own picture.

Solution: Let the x-axis run vertically, from $x = 0$ to $x = 1$. Since the edge length decreases *linearly* from 1 (at the base) to 0 (at the apex), the edge length $l(x)$ at height x is $1 - x$. Thus the cross-sectional area at height x is

$$A(x) = l(x)^2 = (1 - x)^2 = 1 - 2x + x^2,$$

and the volume is

$$\text{volume} = \int_0^1 A(x)\,dx = \int_0^1 \left(1 - 2x + x^2\right) dx = \tfrac{1}{3}.$$

□

Exercises ─────────────────────────────

Sketch the region in the first quadrant bounded by the given curves and find the volume of the solid formed when this region is revolved around the x-axis.

1. $y = x^3$, $y = 0$, $x = 8$ 3. $y = x + 6$, $y = x^3$, $x = 0$

2. $y = x^4$, $y = 1$, $x = 0$ 4. $y = x^2$, $y = x^3$

Sketch the region bounded by the given curves and find the volume of the solid of revolution formed when the region is revolved around the y-axis.

5. $y = \sqrt{x}$, $y = 0$, $x = 4$ 7. $y = x^2$, $y = 2^x$, $x = 0$

6. $y = \sqrt{x}$, $y = x^2$ 8. $y = e^x$, $y = 0$, $x = 0$, $x = 1$

Sketch the region bounded by the given curves and find the volume of the solid of revolution formed when the region is revolved around the line $y = a$.

9. $y = \sqrt{x}$, $y = 0$, $x = 1$, $a = 1$ 11. $x = y^2 - 4$, $y = 2 - x$, $a = 2$

10. $y = x^2 - 1$, $y = x + 1$, $a = -1$ 12. $y = \sqrt{x}$, $y = x^2$, $a = -2$

13. Find a formula for the volume of a cone with radius r and height h.
    [HINT: Cross-sections parallel to the base are circles.]

14. Find a formula for the volume of a pyramid with height h and a square base each edge of which has length ℓ. [HINT: Cross-sections parallel to the base are squares.]

15. The base of a certain solid is the region enclosed by $y = 1/x$, $y = 0$, $x = 1$, and $x = 4$. Every cross section of the solid taken perpendicular to the x-axis is an isosceles right triangle with its hypotenuse across the base. Find the volume of the solid.

16. The circumference (in inches) of a pole at several heights (in feet) is given in the table below. Use numerical integration (e.g., the trapezoid rule or Simpson's rule) to estimate the volume of the pole. (Assume that cross sections of the pole taken parallel to the ground are circles.)

height	0	10	20	30	40	50	60
circumference	16	14	10	5	3	2	1

17. Find the volume of the solid formed when the region bounded by the curves $y = \arctan x$, $y = 0$, and $x = 1$ is revolved around the y-axis.

18. A drinking glass has circular cross-sections. The glass has height 5 inches, bottom diameter 2 inches, and top diameter 3 inches. How much liquid can the glass hold?

19. Assume that the earth is a sphere with radius $r = 24{,}900$ miles.

(a) Find the volume of the earth north of latitude 45°.

(b) Find the volume of the earth between the equator and latitude 45°.

20. A cylindrical gasoline tank with radius 4 feet and length 25 feet is buried on its side underneath a service station (i.e., its flat ends are perpendicular to the surface above). If the gasoline in the tank is 6 feet deep, what is the volume of gasoline in the tank?

21. Let V be the volume of the solid formed when the region bounded by the curves $y = \arctan x$, $y = 0$, $x = -3$, and $x = -1$ is revolved around the x-axis.

    (a) Express V as a definite integral.

    (b) If the integral in part (a) is estimated using a left sum with $n = 10$ subintervals, does the result underestimate or overestimate V? Justify your answer.

22. A fuel oil tank is 10 feet long and has flat ends that are perpendicular to the ground. Cross-sections parallel to the flat ends have the shape of the ellipse $\frac{x^2}{9} + \frac{y^2}{36} = 1$. If the fuel oil in the tank is 9 feet deep, what is the volume of the fuel oil in the tank?

The following exercises provide an introduction to the method of **cylindrical shells**.

23. Let $R > 0$ be a constant, $\Delta r = R/n$, and $r_k = k\Delta r$.

    (a) Show that the area of an annulus (i.e., a ring) with inner radius r and outer radius $r + \Delta r$ is $\pi(2r + \Delta r)\Delta r$.

    (b) Explain geometrically why $\sum_{k=0}^{n-1} \pi(2r_k + \Delta r)\Delta r = \pi R^2$.
    [HINT: The interior of a circle of radius R can be divided into a circle of radius Δr and $n - 1$ concentric annuli, each with thickness Δr.]

    (c) Explain why $\lim_{n\to\infty} \sum_{k=0}^{n-1} \pi(2r_k + \Delta r)\Delta r = \int_0^R 2\pi r\, dr$.   [HINT: First show that $\lim_{n\to\infty} \sum_{k=0}^{n-1}(\Delta r)^2 = 0$.]

    (d) Explain geometrically why $\sum_{k=1}^{n} \pi(2r_k - \Delta r)\Delta r = \pi R^2$.

24. Suppose that f is continuous and that $f(x) \geq 0$ when $a \leq x \leq b$. Let V be the volume of the solid obtained when the region bounded by the curve $y = f(x)$ and the lines $x = a$, $x = b$, and $y = 0$ is rotated about the y-axis.

    (a) Let $n \geq 1$ be an integer, $\Delta x = (b - a)/n$, and $x_k = a + k\Delta x$. Explain why $V \approx \sum_{k=0}^{n-1} \pi(2x_k + \Delta x)f(x_k)\Delta x$.   [HINT: Think of the expression as the sum of the volume of a solid cylinder and the volumes of $n - 1$ concentric hollow cylinders.]

(b) Show that $\displaystyle\sum_{k=0}^{n-1} 2\pi x_k f(x_k)\Delta x$ is a left Riemann sum approximation to the

integral $\displaystyle\int_a^b 2\pi x f(x)\,dx$.

(c) Show that $\displaystyle\lim_{n\to\infty}\sum_{k=0}^{n-1}\pi(2x_k + \Delta x)f(x_k)\Delta x = \int_a^b 2\pi x f(x)\,dx$.

[HINT: Show that $\displaystyle\lim_{n\to\infty}\sum_{k=0}^{n-1} f(x_k)(\Delta x)^2 = 0$.]

(d) Explain why it is reasonable to believe that $V = \displaystyle\int_a^b 2\pi x f(x)\,dx$.

25. Suppose that f is continuous and that $f(x) \geq 0$ when $a \leq x \leq b$. Let V be the volume of the solid obtained when the region bounded by the curve $y = f(x)$ and the lines $x = a$, $x = b$, and $y = 0$ is rotated about the y-axis. Also, let $n \geq 1$ be an integer, $\Delta x = (b - a)/n$, and $x_k = a + k\Delta x$.

(a) Suppose that f is decreasing on the interval $[a, b]$. Explain why

$$\sum_{k=1}^{n}\pi(2x_k - \Delta x)f(x_k)\Delta x \leq V \leq \sum_{k=0}^{n-1}\pi(2x_k + \Delta x)f(x_k)\Delta x.$$

(b) Suppose that f is decreasing on the interval $[a, b]$. Show that
$$V = \int_a^b 2\pi x f(x)\,dx.$$

(c) Show that if f is increasing over the interval $[a, b]$, then
$$V = \int_a^b 2\pi x f(x)\,dx.$$

26. Suppose that f is continuous and that $f(x) \geq 0$ when $a \leq x \leq b$. Let V be the volume of the solid obtained when the region bounded by the curve $y = f(x)$ and the lines $x = a$, $x = b$, and $y = 0$ is rotated about the y-axis.

(a) Let r and Δx be numbers. Show that $\pi(r + \Delta x)^2 - \pi r^2 = 2\pi\Delta x(r + \frac{1}{2}\Delta x)$.
[HINT: $x^2 - y^2 = (x - y)(x + y)$.]

(b) Let $\Delta x = (b - a)/n$ and $x_k = a + k\Delta x$. Explain why

$$V \approx \sum_{k=0}^{n-1} 2\pi\left(x_k + \tfrac{\Delta x}{2}\right) f\left(x_k + \tfrac{\Delta x}{2}\right)\Delta x.$$

(c) Explain why $\displaystyle\lim_{n\to\infty}\sum_{k=0}^{n-1} 2\pi\left(x_k + \tfrac{\Delta x}{2}\right) f\left(x_k + \tfrac{\Delta x}{2}\right)\Delta x = \int_a^b 2\pi x f(x)\,dx$.

27. Suppose that f is continuous and that $f(x) \geq 0$ when $a \leq x \leq b$. Let V be the volume of the solid obtained when the region bounded by the curve $y = f(x)$ and the lines $x = a$, $x = b$, and $y = 0$ is rotated about the y-axis.

   (a) Suppose that $|f'(x)| \leq K$ when $a \leq x \leq b$ and that $a \leq c < d \leq b$. Explain why $f(c) - K(d-c) \leq f(z) \leq f(c) + K(d-c)$ when $c \leq z \leq d$.

   (b) Let $\Delta x = (b - a)/n$ and $x_k = a + k\Delta x$. Use the result in part (a) to explain why

   $$\sum_{k=0}^{n-1} \pi(2x_k - \Delta x)\big(f(x_k) - K\Delta x\big)\Delta x \leq V \leq \sum_{k=0}^{n-1} \pi(2x_k + \Delta x)\big(f(x_k) + K\Delta x\big)\Delta x.$$

   (c) Show that $V = \displaystyle\int_a^b 2\pi x f(x)\, dx$.

28. Let V be the volume of the solid of revolution formed when the region bounded by the curve $y = x^{1/3}$, and the lines $y = 0$, $x = 0$, and $x = 8$ is revolved around the y-axis.

   (a) Compute V using the method of disks and washers.

   (b) Compute V using the method of cylindrical shells.

29. Let V be the volume of the solid of revolution formed when the region bounded by the curve $y = x\sqrt{1 - x^2}$, and the lines $y = 0$, $x = 0$, and $x = 1$ is revolved around the y-axis.

   (a) Use the method of disks and washers to show that

   $$V = \int_0^{1/2} \frac{\pi}{2}\left(1 + \sqrt{1 - 4y^2}\right) dy - \int_0^{1/2} \frac{\pi}{2}\left(1 - \sqrt{1 - 4y^2}\right) dy.$$

   (b) Use the method of cylindrical shells to show that $V = \displaystyle\int_0^1 2\pi x^2 \sqrt{1 - x^2}\, dx$.

   (c) Show that $V = \pi^2/8$.

8.3   Arclength

Integrals can be used to measure geometric quantities other than area and volume. **Arclength** offers another a good example, one that shows clearly the role of approximating sums.

■ **Example 1.** How long is the curve $y = \sin x$ from $x = 0$ to $x = \pi$, shown below?

Part of the sine curve: how long is it?

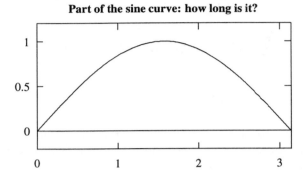

(Before continuing, *somehow* guess the length. Use eyeball, string, a ruler, whatever.)

Use the distance formula.

Solution: Straight line segments are easy to measure.[*] *Curves* pose more of a problem. In this chapter's spirit, a natural strategy suggests itself:

Approximate the curve with straight line segments and add up their lengths.

Below is one such piecewise-linear approximation to the sine curve, built from four segments.

Approximation by a piecewise-linear arc

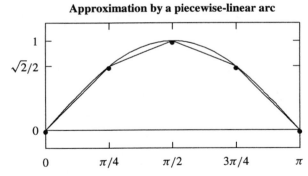

They're shown.

It's OK to skip the numerical computation, but is the answer reasonable?

From the coordinates of the endpoints[*] it's a routine (though tedious) matter to calculate the total length; here, it's about 3.79.[*]

With more and shorter line segments, we'd expect the piecewise-linear "curves" to approach the sine curve itself. In particular, the "polygonal" length should approach L, the length of the sine curve. In other words, L should be the *limiting* length of polygonal approximations to the curve.

To show this principle in action numerically, we've tabulated below the lengths of several piecewise-linear approximations to our elusive L. Each corresponds to subdividing the x-interval $[0, \pi]$ into n equal subintervals:

Approximating the length of a curve					
number of segments	4	8	16	31	100
polygonal length	3.7901	3.8125	3.8183	3.8197	3.8201

For now, 3.82 is our best guess.   □

Generic problem, generic solution.   The generic arclength problem reads like this:

> *Find the length of the graph of f from x = a to x = b.*

The previous example suggests a generic strategy:

> *Approximate with piecewise-linear arcs and take a limit.*

Great, but how does it all work in practice? How is the integral involved? *Which* integral is involved?

To describe the situation more precisely we need a picture:

Approximating a curve C with a polygonal arc C_n

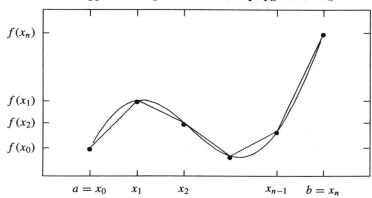

The picture shows these ingredients:

1. C, the graph of f from $x = a$ to $x = b$.

2. A partition of $[a, b]$ into n equal subintervals, with endpoints $x_0, x_1, \ldots, x_n$. ➤➤   *In all, $n + 1$ of them.*

3. A polygonal arc C_n, made from n line segments, joining the points on C above $x_0, x_1, \ldots x_n$.

Now we can put the idea succinctly: length of $C = \lim_{n \to \infty}$ (length of C_n).

How long is C_n? A useful estimate.   To measure C_n, we add the lengths of its n segments. The i-th segment joins $(x_{i-1}, f(x_{i-1}))$ to $(x_i, f(x_i))$, so its length is

$$\sqrt{(f(x_i) - f(x_{i-1}))^2 + (x_i - x_{i-1})^2}.$$

The total length of C_n, therefore, is

$$(8.3.1) \qquad \text{length of } C_n = \sum_{i=1}^{n} \sqrt{(f(x_i) - f(x_{i-1}))^2 + (x_i - x_{i-1})^2}.$$

We'll rewrite Equation 8.3.1 as an approximating sum for an integral. The first step is pure algebra:

$$
\begin{aligned}
\text{length of } C_n &= \sum_{i=1}^{n} \sqrt{(f(x_i) - f(x_{i-1}))^2 + (x_i - x_{i-1})^2} \\
&= \sum_{i=1}^{n} \sqrt{\left(\frac{f(x_i) - f(x_{i-1})}{x_i - x_{i-1}}\right)^2 + 1} \cdot (x_i - x_{i-1}).
\end{aligned}
$$

This clumsy expression can be simplified. By the definition of derivative, the approximation

$$\frac{f(x_i) - f(x_{i-1})}{x_i - x_{i-1}} \approx f'(x_i)$$

holds if $x_i - x_{i-1}$ is small; it *is* small here if n is large.

Putting the pieces together, and writing Δx_i for $(x_i - x_{i-1})$, we get

$$\text{length of } C_n = \sum_{i=1}^{n} \sqrt{\left(\frac{f(x_i) - f(x_{i-1})}{x_i - x_{i-1}}\right)^2 + 1} \cdot (x_i - x_{i-1})$$

$$\approx \sum_{i=1}^{n} \sqrt{(f'(x_i))^2 + 1} \cdot \Delta x_i.$$

The last line is our payoff: the quantity

$$\sum_{i=1}^{n} \sqrt{(f'(x_i))^2 + 1} \, \Delta x_i$$

is an *approximating sum* for the integral

$$\int_a^b \sqrt{(f'(x))^2 + 1} \, dx.$$

Our conclusion follows:

Fact: **(Arclength by integration.)** The integral

$$\int_a^b \sqrt{(f'(x))^2 + 1} \, dx$$

gives the length of the f-graph from $x = a$ to $x = b$.

■ **Example 2.** Write the length of the sine curve from $x = 0$ to $x = \pi$ as an integral. Estimate its value.

Solution: By the fact above, the length is the value of the integral

$$I = \int_0^\pi \sqrt{(\cos x)^2 + 1} \, dx.$$

The FTC won't help us evaluate I. Like many integrands that involve square roots, this one has no convenient antiderivative. Lacking an antiderivative, we'll estimate I using the midpoint rule with 100 subdivisions. According to our computer $I \approx 3.8202$—a reassuring result, given earlier work.   □

Do curves have length? In other words, do piecewise-linear approximations successfully approximate a curve's arclength as we claimed above?

With modern computer graphics, the answer seems obvious. Almost every "curve" in this book, for instance, is really a collection of short line segments—shorter than the eye can perceive individually–strung end to end. In this case, appearances don't deceive: all but the very worst-behaved curves *can* be approximated in the sense discussed above.

Curves *do* have length.

Exercises

1. Compute the length of the line $y = x$ from $x = 0$ to $x = 1$. Does the answer agree with the usual distance formula?

2. Compute the length of the line $y = mx + b$ from $x = 0$ to $x = 1$. Does the answer agree with the usual distance formula?

3. Let C be the curve described by the equation $y = \dfrac{1}{3}x^3 + \dfrac{1}{4x}$ over the interval $1 \le x \le 3$.

   (a) Sketch the curve C and estimate its length.

   (b) Compute the length of C exactly.   [HINT: Do your algebra *very* carefully. If you do it correctly, the square root disappears.]

4. Let C be the curve $y = \left(e^x + e^{-x}\right)/2$ from $x = 1$ to $x = 4$. Find the length of C *exactly*.   [HINT: $\left(\frac{1}{2}\left(e^x + e^{-x}\right)\right)^2 = 1 + \left(\frac{1}{2}\left(e^x - e^{-x}\right)\right)^2$.]

5. Explain in words why the length approximation **increases** as the number of linear segments increases.

6. Find the length of the curve $y = x^2$ from $x = 1$ to $x = 2$ within ± 0.005.

7. Find the length of the curve $y = e^x$ from $x = 0$ to $x = 1$ within ± 0.005.

8. Let J be the length of the curve $y = f(x)$ from $x = 0$ to $x = 4$ where f is a function such that f' is the function shown below.   [NOTE: The graph of f', not f, is shown below.]

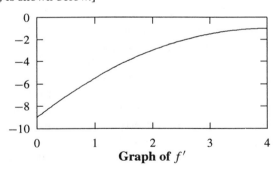

Graph of f'

   (a) Use the information in the table below to compute L_4, the left sum approximation to J using four equal subintervals.

x	0	1	2	3	4
$f'(x)$	-9	-5.5	-3	-1.5	-1

   (b) Is $L_4 < J$? Justify your answer.

8.4 Work

How much work does a rocket do in lifting a satellite into orbit? How much work does a car do on the 934-mile drive from Perham, Minnesota to Durham, North Carolina? How much work does a horse do in plowing the back 40? Is it more work to raise a 60-pound bucket from a 60-foot well or a 50-pound bucket from a 70-foot well?

All good questions, with thoroughly practical consequences. Doing any kind of work takes "fuel" (hydrogen, gasoline, hay, cheeseburgers, ...)—it's important to know how much is needed "on board." **Work** is the common ingredient (and the common word) in every situation above. We'll need a mathematically useful definition.

The meaning of work

To a physicist, digging ditches is a lot of work; filing tax returns is almost none.

The everyday meaning of "work" is all too familiar; "work" covers everything from completing tax returns to digging ditches. Physicists have a narrower but more precise definition.◄

Except, perhaps, in lifting the globe.

Work is done when a *force acts through a distance*. Without movement *no* work is done; a house does no work by holding up its roof. Even poor Atlas, the Titan forced to bear the world on his shoulders, accomplished no *work*,◄ fatiguing though his labors must have been.

Simplifying assumptions: are they reasonable?

Throughout this section we assume:

> *A force (either constant or variable) acts along a straight line; movement occurs along the same line.*

These assumptions are convenient, but are they reasonable? Few real-life physical phenomena, after all, are this simple. Forces may act *at angles* to the direction of motion, and motion may occur along *curves*. Moving cars, for instance, encounter *many* forces—gravity, air drag, rolling resistance, crosswinds, etc.—all of them varying over the course of a trip. Given the complex nature of physical reality one might expect *nothing* useful to survive our simplifying assumptions.

As we'll see.

On moving cars, for example.

In fact much *does* survive. Some interesting physical phenomena *do* conform to our "straight-line" assumptions.◄ A complicated combination of forces,◄ moreover, can often be understood as a *sum* of simpler forces, each acting as assumed above.

Work done by a constant force

In the simplest instance the force is *constant* along its line of action. In this case the "obvious" definition applies:

$$\text{work} = \text{force} \times \text{distance}.$$

The **units of work** reflect this definition. Here in the U.S., lifting a 30-pound toddler 3 feet to hip height, for instance, takes

$$30 \text{ pounds} \times 3 \text{ feet} = 90 \text{ foot-pounds}$$

of work. In Canada,◄ lifting the same kid to the same hip takes

In the metric system the basic unit of force is the newton—approximately 4.5 pounds.

$$133.44 \text{ newtons} \times 0.914 \text{ m} \approx 121.964 \text{ newton-meters}$$

of work. Other work units, such as *inch-ounces* and *mile-pounds*, are possible, but all have the same "dimension": units of *distance* times units of *force*.

Work done by a variable force

In typical physical situations work is performed by a force that *varies* along its line of action. Even hoisting a toddler can be construed this way: on the way up, the child moves away from the center of the earth and so becomes "lighter." (For *practical* purposes this effect is small enough to ignore.)➤ In more interesting situations the force varies more dramatically:

But see Example 4 below.

Springs. The farther a spring is stretched from its "natural length," the greater the force with which it pulls back. According to **Hooke's law** an ideal spring's "restoring force" $F(x)$ is *proportional* to x, the distance the spring is stretched. Thus

$$F(x) = kx$$

for some constant k, called the **spring constant.**➤

The numerical value of k depends on the units of measurement and on the "stiffness" of the spring.

Gravity. An object's weight decreases as it moves away from the earth. **Newton's law of universal gravitation** quantifies this commonplace observation:➤

Commonplace now—not in Newton's time.

> *The force of earth's gravity on an object is inversely proportional to the square of the distance from the object to the center of the earth.*

Thus the force F required to lift an object (i.e., to counteract gravity) is

$$F(x) = \frac{k}{x^2}$$

if the object is x miles distant from the center of the earth. The numerical value of k➤ depends on the object and on the units of measurement.

k is the proportionality constant.

Buckets and ropes. The force required to draw water from a well *decreases* as the bucket rises—the higher the bucket, the less rope to be pulled up. We'll pursue this observation in a later example.

Pumping iron. One way to "press" a heavy barbell is to exert a *constant* upward force➤ all the way from shoulder level to the top of one's reach. Although *physically* simple, this strategy is not *physiologically* ideal. Mechanics of the human body make it "easier" to exert upward force at some levels of arm extension than at others. To the extent that rules permit, therefore, a human weightlifter would exert a *varying* upward force through a barbell's travel. (But note: *whatever* the weightlifter's strategy, the work of lifting is the same.)➤

Equal to the barbell's weight.

Be sure you understand why.

Such examples argue for a definition that permits *variable* forces.

Definition: (Work) Let F be a continuous function. If the force $F(x)$ acts along an axis from $x = a$ to $x = b$, then the work done is

$$\int_a^b F(x)\, dx.$$

After a simple example we'll discuss why the definition is reasonable.

■ **Example 1.** Stretching a spring 1 foot beyond its natural length requires 10 pounds of force. How much work is done in stretching the spring from rest to 1 foot? From rest to a feet? How does the answer depend on a?

Solution: By Hooke's law, $F(x) = kx$. The given conditions say that $F(1) = 10$, so $k = 10$. By the definition above:

$$\text{work to stretch 1 ft} \;=\; \int_0^1 10x\,dx = 10\frac{x^2}{2}\Big]_0^1 = 5 \text{ foot-pounds;}$$

$$\text{work to stretch } a \text{ ft} \;=\; \int_0^a 10x\,dx = 10\frac{x^2}{2}\Big]_0^a = 5a^2 \text{ foot-pounds.}$$

The last answer reveals an interesting relationship between force and work:

> *The* force *required to stretch a spring a feet from rest is proportional to a; the* work *done in the process is proportional to* a^2.

$\square$

Work as an integral: why the definition makes sense

Why is it "right" to define work as an integral? Does the integral really represent a physically meaningful quantity, one that deserves the name "work?"

Any sensible definition of work done by a *variable* force should at least "agree with" the simpler definition of work done by a *constant* force. Ours does: if $F(x) = k$, then

$$\text{work} = \int_a^b F(x)\,dx = \int_a^b k\,dx = k(b-a) = \text{force} \times \text{distance.}$$

So far, so good; but why is the definition "right" for variable forces? We'll give two arguments.

Adding small contributions: the integral as a sum

Left endpoints or midpoints would have done as well.

The work integral $\int_a^b F(x)\,dx$ can be thought of as a limit of **approximating sums**. Specifically, let's slice the interval $[a, b]$ into n equal subintervals, each of length Δx, and let x_i be the *right* endpoint of the i-th subinterval. Then

$$\int_a^b F(x)\,dx = \lim_{n\to\infty} \sum_{i=1}^n F(x_i)\,\Delta x.$$

Similar pictures appear in Section 5.4.

The following picture should look familiar:

Approximating work: a right sum

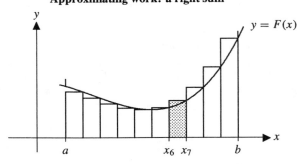

The sum $\sum_{i=1}^n F(x_i)\,\Delta x$ is the key quantity. Each summand $F(x_i)\,\Delta x$ represents the work done by the *constant* force $F(x_i)$ over the subinterval $[x_{i-1}, x_i]$. (The shaded area represents the work done over the subinterval $[x_6, x_7]$.) The sum *adds* these small contributions. (The entire polygonal area represents this total.) Here's the punchline:

Each approximating sum is an estimate to the total work done over the full interval $[a, b]$.

As $n \to \infty$, therefore, the sums tend (by definition of the integral) to $\int_a^b F(x)\,dx$. The integral, therefore, plausibly measures the "true" work done.

Increments and elements: a mnemonic argument

As further evidence for the integral definition of work we offer the following mnemonic argument, using **increments** and **elements**.

Let x be any point in the interval $[a, b]$. Consider the tiny interval $[x, x + \Delta x]$, with **increment of distance** Δx. Because F remains essentially constant over any tiny interval,➡ the work ΔW done over the interval $[x, x + \Delta x]$ satisfies

Every continuous function has this property.

$$\Delta W \approx F(x)\,\Delta x.$$

Letting $\Delta x \to 0$ now gives $dW/dx = F(x)$, or

$$dW = F(x)\,dx;$$

dW is called the **element of work.** "Adding up" these elements—by integration—gives the total work done from $x = a$ to $x = b$:

$$\text{Total work} = W = \int dW = \int_a^b F(x)\,dx.$$

(The notation $W = \int dW$ should be understood informally: it means simply that the integral "adds" up small contributions.)

Calculating work: miscellaneous examples

The integral definition of work is simple enough. Deciding what to integrate, and over what interval, is usually the hardest part of applying the formula.

Working against gravity

Earth's gravity affects even far distant objects, never dying out completely. How, then, is putting objects into orbit possible? Calculus gives the answer.

■ **Example 2.**   Communications satellites weigh around 1000 pounds and orbit the earth at "altitudes" of about 15,000 miles. How much work is done against earth's gravity in lifting such a satellite into orbit?

Solution: By Newton's law of gravitation➡ the (variable) force F needed to counteract earth's gravity is

Discussed earlier in this section.

$$F(x) = \frac{k}{x^2},$$

where k is a constant and x denotes the distance from earth's center. That the satellite weighs 1000 pounds on earth's surface—about 4000 miles from earth's center—means that (allowing for approximation)

$$1000 = \frac{k}{4000^2} \implies k = 1.6 \times 10^{10} \implies F(x) = \frac{1.6 \times 10^{10}}{x^2}.$$

(Force is measured in pounds.) Therefore the work done (measured in mile-pounds) in moving from $x = 4000$ to $x = 19000$➡ is found by an easy integral calculation:

$x = 19000$ corresponds to an "altitude" of 15000 miles.

$$\int_{4000}^{19000} \frac{1.6 \times 10^{10}}{x^2} = -\frac{1.6 \times 10^{10}}{x}\bigg]_{4000}^{19000} = 3,158,000 \text{ mile-pounds.}$$

That's heavy lifting, but not impossibly so—a loaded Boeing 747 accomplishes about the same work against gravity to reach cruising altitude. □

■ **Example 3.** How much *additional* work is required to lift the satellite another 85,000 miles, to an altitude of 100,000 miles?

Solution: The next 85,000 miles of altitude come almost free. The extra work done in raising the satellite from $x = 19000$ to $x = 104000$ miles is a relatively trifling

$$\int_{19000}^{104000} \frac{1.6 \times 10^{10}}{x^2} \, dx = -\frac{1.6 \times 10^{10}}{x}\bigg]_{19000}^{104000} \approx 688,259 \text{ mile-pounds,}$$

less than 22% of the work required to gain the first 15,000 miles of altitude.

A similar computation gives the work $W(a)$ done in lifting the satellite from the earth to *any* altitude a:

$$W(a) = \int_{4000}^{a} \frac{1.6 \times 10^{10}}{x^2} = -\frac{1.6 \times 10^{10}}{x}\bigg]_{4000}^{a} = 10^6 \left(4 - \frac{1600}{a}\right).$$

A close look at the last expression reveals a surprising conclusion, with important consequences for space flight:

Lifting the satellite to any *altitude a requires no more than* 4×10^6 *mile-pounds of work.*

Plotting W against a shows that W is bounded above—and suggests why space flight is possible:

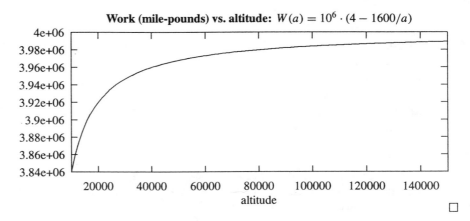

■ **Example 4.** Gravity's force diminishes—however slightly—even when lifting a 30-pound toddler 3 feet to hip height. We argued earlier in this section that this effect is negligible. Is it?

Solution: Ignoring Newton's law gave the rough answer of 90 foot-pounds of work. Now let's take Newton's law into account. All the computations are similar to those above; only the numbers are different. We'll need these facts:➤➤

Check details if you wish.

- At $x = 4000$, $F(x) = k/x^2 = 90$, so $k = 4.8 \times 10^8$.

- Converting feet to miles gives 3 feet ≈ 0.0005681 miles. Thus the toddler's vertical travel is from $x = 4000$ to $x = 4000.0005681$ miles.

- One mile-pound equals 5280 foot-pounds.

All the ingredients are now in place. The total work done, in foot-pounds, is

$$5280 \int_{4000}^{4000.0005681} \frac{4.8 \times 10^8}{x^2}\, dx = -5280 \left. \frac{4.8 \times 10^8}{x} \right]_{4000}^{4000.0005681} \approx 89.99998719.$$

The "savings" from diminishing gravity, therefore, is only 0.00001281 foot-pounds—about the work done by a medium-sized raindrop falling one foot. □

■ **Example 5.** Is it more work to raise a 70-pound bucket of water from a 50-foot well or a 50-pound bucket from a 70-foot well? Rope weighs 0.25 pounds per foot.➤➤

Guess before computing anything!

Solution: Let x denote "height," i.e., the distance (in feet) from either bucket to the *bottom* of its well. As the first bucket travels from $x = 0$ to $x = 50$, the force required to raise it varies with x. At height x, this force must counteract both the weight of the bucket and the weight of $50 - x$ feet of rope.➤➤ Thus, for the first bucket, the force function is

Draw a diagram illustrating the various quantities involved.

$$F(x) = 70 + 0.25 \cdot (50 - x) = 82.5 - 0.25x;$$

the work done is

$$\int_0^{50} F(x)\, dx = \int_0^{50} (82.5 - 0.25x)\, dx = 3812.5 \text{ foot-pounds.}$$

For the second bucket the computation is almost identical. This time

$$F(x) = 50 + 0.25 \cdot (70 - x) = 67.5 - 0.25x;$$

the work done is

$$\int_0^{70} F(x)\, dx = \int_0^{70} (67.5 - 0.25x)\, dx = 4112.5 \text{ foot-pounds.}$$

The answer stands to reason—the same amount work is done on each bucket,➤➤ but drawing up the longer rope takes more work. □

Because work = force × distance.

Exercises ————————————————

1. Suppose that the spring in a bathroom scale is stretched by 0.06 inches when a 115-lb person stands on the scale.

   (a) How much work is done on the spring when a 115-lb person steps on the scale?

(b) How much work is done on the spring when a 175-lb person steps on the scale?

2. The cable attached to an elevator car weighs 5 lb/ft. If 200 feet of cable must be wound onto a pulley to lift the car from the bottom to the top floor of a building, how much work is done just lifting the cable?

3. A cylindrical tank with radius 5 feet and height 10 feet is half-filled with water. How much work must be done to pump all the water over the top rim of of the tank? (The density of water is 62.4 pounds per cubic foot.)   [HINT: Slice the tank into layers each of thickness Δx. The layer of water with its bottom at height x needs to be pumped a distance of $10 - x$ feet.]

4. A cylindrical gasoline tank with radius 4 feet and length 15 feet is buried on its side underneath a service station. The top of the tank is 10 feet underground; its flat ends are perpendicular to the surface above. Find the amount of work needed to pump all the gasoline in the tank to the nozzle which is 3 feet above the ground. (Gasoline weighs 42 pounds per cubic foot.)   [HINT: Each horizontal cross-section of the tank is a rectangle, one side of which has length 15 feet.]

5. Is it more work to raise a 60-pound bucket from a 60-foot well or a 50-pound bucket from a 70-foot well? Rope weighs 0.25 pounds per foot.

6. A bucket that weighs 80 lbs when filled with water is lifted from the bottom of a well that is 75 feet deep. However, the bucket has a hole in it, so it weighs only 40 lbs when it reaches the top of the well. Assuming that water leaks from the bucket at a constant rate and that the rope weighs 0.65 lb/ft, find the work required to lift the bucket from the bottom of the well to the top.

7. According to Hooke's Law, the force $F(x)$ necessary to stretch or compress an ideal spring x units away from its natural length is *proportional* to x, i.e., $F(x) = kx$. (The value of k—called the spring constant—depends on the spring.)

   A certain spring has natural length 18 inches; a force of 10 pounds is enough to compress it to a length of 16 inches.

   (a) What is the value of the spring constant k?

   (b) How much work is done in compressing the spring from 16 inches to 12 inches?

8. It takes $40x$ pounds of force to keep a certain spring stretched x feet from rest.

   (a) Find the work done in stretching the spring 2 feet, starting from rest.

   (b) Find the work done in stretching the same spring s feet from rest. (The answer depends on s, of course.)

   (c) Farmer Ole's aging plow horse, Sven, can do only 10000 foot-pounds of work on his daily oats ration. Ole hitches Sven to the spring described above. How far can Sven pull?

9. A certain spring exerts $4x$ pounds of force when stretched x feet from rest. One end is fixed to the ceiling. A chain, 10 feet long, weighing 2 pounds per foot, hangs from the other end. The end of the chain just brushes the floor. Find the work done in pulling down on the chain a distance of 2 feet. Don't ignore the weight of the chain.

10. Redo Example 2 in this section, but assume that the orbit is 20,000 miles from earth's center.

11. An object weighing k pounds slides without friction along the graph of $y = f(x)$. (The x and y units are feet.) A horizontal force $F(x)$ is applied to move the object from $x = a$ to $x = b$. Physical intuition says that at x, the horizontal force $F(x)$ required is proportional to the *slope* of the graph. (Right? What "should" $F(x)$ be if the graph is horizontal at x? Vertical?) In other words, $F(x) = kf'(x)$.

   (a) Find the work done if $k = 10$, $a = 0$, $b = 1$, $f(x) = x$.

   (b) Find the work done if $k = 10$, $a = 0$, $b = 1$, $f(x) = x^3$.

   (c) Find the work done if $k = 10$, $a = 0$, $b = 1$, $f(x) = x^n$, for n any positive integer.

   (d) Explain the relation between your answers to the three previous parts.

12. It takes kx pounds of force to keep a spring stretched x feet from rest.

   (a) How much work is done in stretching the spring 10 feet from rest? (The answer depends on k, of course.)

   (b) How much work is done in stretching the spring from $x = a$ feet to $x = a + 10$ feet. (The answer depends on both k and a.)

8.5   Present value

The idea: money grows

Anyone, given the choice, would prefer a dollar today to a dollar tomorrow. This makes excellent sense: a dollar tomorrow is worth *less* than a dollar today, because today's dollar, prudently invested, can earn some interest by tomorrow.

But *how much* less is a dollar tomorrow worth than a dollar today? What is an advertised $1 million lottery prize really worth if it's paid in 20 yearly $50,000 installments? Given the choice, should I collect my salary in annual, monthly, or weekly installments, or doesn't it matter? How much should I deposit now to cover 4 years of college tuition, starting in 10 years?

We'll answer some below.

All these questions[←] concern what economists call **present value**, i.e., the value *now* of a future payment or "stream" of payments. Present value calculations—using integral calculus—are vital in economic decisions, e.g., in choosing among competing investment options.

The present value of one future payment

And most important—don't skip it.

The first[←] example illustrates the idea in the simplest case: the present value of *one* future payment. Notice especially that exponential functions are involved, and how present value depends on an interest rate.

■ **Example 1.**   A savings bond will pay $1000 on its maturity date, 10 years from today. What is it worth now, assuming that 5% compound interest is available? What if 10% interest were available?

Solution: Let $V(t)$ denote the bond's value at time t years from now. We're given that $V(10) = 1000$; we want to find $V(0)$.

To start, recall this important fact: Under continuously compounded interest, the value V grows *exponentially*. In other words, $V(t)$ has the form

$$V(t) = V(0)e^{rt},$$

where r is the annual interest rate. (Here, $r = 0.05$ or $r = 0.10$.)

> Remember why? Here's a quick refresher. Growth at continuously compounded interest rate r, means, in differential equation language, that for all t,
>
> $$V'(t) = rV(t).$$
>
> It's easy to check that for any constant C, the function $V(t) = Ce^{rt}$ solves this DE.[←] Since $V(0) = Ce^0 = C$, it follows that $V(0)$ is the "right" value for C.

See for yourself by differentiation that $V' = rV$.

Knowing that $V(10) = 1000$, we can solve for $V(0)$. For any interest rate r,

$$V(10) = V(0)e^{10r} = 1000 \implies V(0) = 1000e^{-10r}.$$

Thus if $r = 0.05$, $V(0) = 1000e^{-0.5} \approx 606.53$; if $r = 0.1$, $V(0) = 1000e^{-1.0} \approx 367.88$.

In economic terms: the **present value** of a single $1000 payment, 10 years in the future, is about $607 assuming 5% annual interest; it's about $368 assuming 10% interest.   □

The example prompts three general remarks:

A negative exponential. The example shows that the present value, PV, of one future payment depends on three things: P, the size of the payment; t, the time until the payment is made; and r, the interest rate. We found, using this notation, that

$$PV = Pe^{-rt}.$$

The general form of this expression will reappear soon. The *negative* exponent reflects the fact that the present value of a future payment is normally *less* than the future payment itself.

Backward in time. The present value of a future payment P can be thought of as the amount to be deposited *now*, at interest rate r, to accrue to value P after time t. Equivalently, one can think of following an investment's value *backward* in time, from value P at time t to value PV at time zero.

Simplifying assumptions. Present value calculations are simplest if we assume (as we did implicitly above) that (i) the interest rate r remains constant over time; and (ii) that interest is compounded *continuously*. (The first assumption is just for convenience; the second lets us use calculus.) We'll assume both (i) and (ii) in most of what follows.

The discrete case: present value of several future payments

Many cases of interest involve *several* payments, occurring at several future times. The *total* present value of a sequence of payments is found, naturally, by addition.

■ **Example 2.** A lottery jackpot, although advertised as \$1,000,000, is paid in 20 annual installments of \$50,000. What's the present value of the prize, assuming 6% annual interest? What if 8% interest were available?

Solution: By the formula above, the present value of *one* \$50,000 payment k years in the future is $50000e^{-0.06k}$. Thus the total present value of all 20 payments is

$$50000 \left(e^{-0.06 \times 0} + e^{-0.06 \times 1} + e^{-0.06 \times 2} + \cdots + e^{-0.06 \times 19} \right) \approx \$599,983.$$

That's much less than the advertised million. At 8% interest the computation is similar: the present value is

$$50000 \left(e^{-0.08 \times 0} + e^{-0.08 \times 1} + e^{-0.08 \times 2} + \cdots + e^{-0.08 \times 19} \right) \approx \$519,033. \quad \square$$

Several payments: the general form. The example illustrates the general picture. If r is the interest rate, and payments $P_1, P_2, \ldots, P_n$ are made at times $t_1, t_2, \ldots, t_n$, then the present value of all n payments is found by summation:

(8.5.1) $$PV = \sum_{i=1}^{n} P_i e^{-rt_i}.$$

The continuous case: present value of an income stream

In real life, financial transactions occur at specific moments—the beginning of a month, the end of a pay period, 11:59 pm on April 15, etc. Nevertheless, economists often picture income as an unbroken "stream," flowing continuously in time.** The next example—in which all three graphs are drawn as continuous, unbroken curves—illustrates this point of view. Notice, too, that the graphs show the *rates* of income flow as functions of time.

Hence, e.g., the term "cash flow."

Each one represents an income stream.

■ **Example 3.**   The graphs below[*] show varying rates of daily income, over a 360-day "year," for three idealized workers: a tax consultant, an ice cream vendor, and a factory worker. (Which is which? Day 0 is January 1; U.S. tax returns are due April 15.)

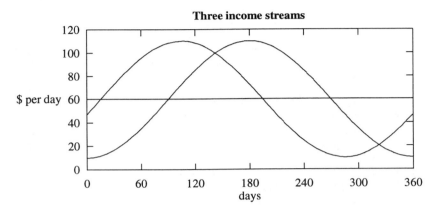

Three income streams

How much does each worker earn over the whole year? Who's best off? Why?

Solution: First, which graph is which? Judging from their "peaks" (around April 15 (day 105) and July 1 (day 180)), the two sinusoidal graphs represent the tax consultant and the ice cream vendor.[*] The straight-line graph, therefore, represents the factory worker's constant \$60 daily wage, day in and day out. Label the graphs T, I, and F.

Our fable occurs north of the equator. How do the graphs show this?

The factory worker's total annual income is easiest to calculate. At \$60/day for a 360-day year, the factory worker earns \$21,600—the area under the F-graph from $t = 0$ to $t = 360$. The other two workers' total income can also be found as areas, under the T- and I-graphs, respectively. A close look reveals a mild surprise. All three graphs enclose the *same* total area,[*] so all three workers earn the same total annual income: \$21,600.

Take a close look now; convince yourself.

Nevertheless, the workers are *not* (quite) equally well off. The tax consultant[*] is luckiest. Compared to I and F, T receives more income *earlier* in the year. For now that's all we can say. Below we'll *quantify* T's advantage, by calculating the present value of each income stream at $t = 0$.   □

Who else?

Fair assumptions? We made several simplifying assumptions in Examples 1– 3: that interest rates remain constant over time, that interest is compounded continuously, that daily earnings follow simple patterns, etc.

None of these assumptions need be literally true. For instance, few factory workers are paid by the day—including weekends. Similarly, ice cream sales rise and fall with temperature, not just with time of the year.

Do such simplifications render our results useless? Certainly not. *All* economic decisions require some combination of calculation and guesswork. To some extent the future is always unknowable. In practice, making sensible simplifying assumptions is inevitable—and it's standard operating procedure.

Why treat income as a stream?   Treating income as a continuous stream lends itself nicely to calculus: functions, derivatives, and integrals all arise naturally. From this point of view, an income stream is a function p of time: at time t, $p(t)$ is the (instantaneous) *rate* at which income "flows" to the recipient. It follows, then, that the integral $\int_a^b p(t)\,dt$ is the *total* income received over the time interval $a \le t \le b$. (That's why we looked at area under the curves in Example 3.)

Present value as an integral: the definition.   Building on the previous examples, let's define➤➤ the present value of a continuous income stream. Three ingredients are involved:

- a time scale t; $t = 0$ corresponds to the *present time;*➤➤

- an interest rate r; usually taken as a constant;➤➤

- a continuous income-stream function $p(t)$, defined for $a \le t \le b$.➤➤ At time t, $p(t)$ tells the *rate of income flow with respect to time.*

Finally, an important➤➤ caution on compatible *units of measurement*:

t, r, and p must use the same time unit.

For example, if t measures *days*, then r is the *daily* interest rate, and $p(t)$ is measured in money units per *day*.

> **Definition:** Let t, r, and p be as above. The **present value of the income stream** p is defined by the integral
>
> $$PV = \int_a^b p(t)e^{-rt}\,dt.$$

First we'll use the definition, then defend it as sensible. In the meantime, notice the similarity to Equation 8.5.1.

■ **Example 4.**   Assuming a 5% annual interest rate, find the present value (at the beginning of the 360-day year) for each of the three workers T, I, and F of Example 3. How different are the results if 10% interest is available?

Solution: Let t denote time in days; then $t = 0$ is the "present" time, and the income flow lasts from $t = 0$ to $t = 360$.

   Because t involves *days*, r should measure the *daily*➤➤ interest rate. Here, therefore, $r = 0.05/360$.

   The three stream functions—we'll call them p_T, p_I, and p_F—are shown graphically➤➤ on page 148. To find our present values we'll need *formulas* for p_T, p_I, and p_F. One formula can be "read" easily from the graphs: $p_F(t) = 60$. The other two graphs look periodic; with some effort we could produce their formulas from trigonometric ingredients.➤➤ Here we'll just state them:

$$p_T(t) = 50\cos\left(\pi \cdot \frac{t - 105}{180}\right) + 60; \qquad p_I(t) = 50\cos\left(\pi \cdot \frac{t - 180}{180}\right) + 60.$$

   The rest is straightforward (but messy) calculation. Here are the results:➤➤ With $r = 0.05/360$,

$$PV_F = \int_0^{360} 60e^{-0.05t/360}\,dt \approx 21068.89;$$

$$PV_T = \int_0^{360} \left(50\cos\left(\pi \cdot \frac{t - 105}{180}\right) + 60\right) e^{-0.05t/360}\,dt \approx 21203.55;$$

$$PV_I = \int_0^{360} \left(50\cos\left(\pi \cdot \frac{t - 180}{180}\right) + 60\right) e^{-0.05t/360}\,dt \approx 21067.78.$$

After stating and using the definition, we'll say why it's sensible.

The time for which present value is calculated.

In practice r is not always constant, of course.

I.e., the income stream starts at time a and ends at time b.

Really important!

Not yearly.

Look again.

See the exercises.

We spare you the calculation—but see the exercises.

Tax consultants usually seem to.

As predicted, T comes out on top,[**] about \$135 ahead of the others.

At a 10% annual interest rate—i.e., $r = 0.1/360$—the tax consultant's advantage is a little greater:

$$PV_F \;=\; \int_0^{360} 60e^{-0.1t/360}\,dt \approx 20555.12;$$

$$PV_T \;=\; \int_0^{360} \left(50\cos\left(\pi \cdot \frac{t-105}{180}\right) + 60\right) e^{-0.1t/360}\,dt \approx 20817.26;$$

$$PV_I \;=\; \int_0^{360} \left(50\cos\left(\pi \cdot \frac{t-180}{180}\right) + 60\right) e^{-0.1t/360}\,dt \approx 20550.78. \qquad \Box$$

Why the integral definition makes sense.   The typographical similarity between

$$\sum_{i=1}^{n} P_i e^{-rt_i} \quad \text{and} \quad \int_a^b p(t)e^{-rt}\,dt,$$

(the formulas for present value in the discrete and continuous cases, respectively) is no accident. Roughly speaking, the integral is the *continuous* version of the *discrete* sum.

To relate the integral and the sum more precisely, we'll approximate a continuous income stream by a discrete sequence of payments. Suppose, then, that for $a \le t \le b$, $p(t)$ describes a continuous income stream, and let $a = t_0 < t_1 < t_2 < \cdots < t_n = b$ partition the interval $[a, b]$ into n short subintervals.[**] We'll approximate the income stream p with a finite sequence of payments $P_1, P_2, \ldots P_n$, occurring at times $t_1, t_2, \ldots t_n$, respectively.[**]

Remember partitions? See, e.g., Section 5.4.

Payments P_i occur at the end of each subinterval.

How large should each payment P_i be? Well, over a (short) subinterval $[t_{i-1}, t_i]$, the stream function p doesn't change much,[**] so the estimate $p(t) \approx p(t_i)$ is reasonable.[**] Thus the *total income* received over the time interval $[t_{i-1}, t_i]$ is approximately $p(t_i)\Delta t_i$. In our (approximate) discrete version, therefore, it's natural to use $P_i = p(t_i)\Delta t_i$. To summarize: we approximate the continuous stream p, defined for $a \le t \le b$, by n *separate* payments: at time t_i, the payment is $P_i = p(t_i)\Delta t_i$. As n tends to infinity, moreover, we expect our approximation to improve.

Because p is continuous.

We pretend, in effect, that $p(t)$ is constant on each subinterval.

We saw above that the total present value of these n payments, given interest rate r, is

$$\sum_{i=1}^{n} P_i e^{-rt_i} = \sum_{i=1}^{n} p_i e^{-rt_i}\,\Delta t_i.$$

aka, Riemann sum.

Here's the key point: the last sum is an approximating sum[**] for $\int_a^b p(t)e^{-rt}\,dt$—just the integral we've been waiting for. As n tends to infinity, therefore, the approximation tends both to the desired integral *and*, plausibly, to the the desired present value.

Exercises ———————————————————————————————

1. Al and Bob, each 42 years old, hope to retire as millionaires. To this end, they'll deposit money *now* to accrue to \$1 million by age 65.

   (a) How much should Al deposit now if 6% interest is available?

   (b) How much should Al deposit now if 8% interest is available?

(c) Bob has $100,000 on hand. What interest rate will he need to meet his retirement goal?

2. Chris, now 20, wants to retire at 65 with a $1 million nest egg. How much should Chris deposit now at 6%? At 8%? In speculative junk bonds, paying 20%?

3. **(Real interest rates.)** Inflation—the rise of prices over time—reduces the buying power of future money. In effect, inflation imposes *negative* interest on a deposit. Economists "control for inflation" by defining the **real interest rate** as the difference

$$\text{real interest rate} = \text{nominal interest rate} - \text{inflation rate}.$$

Predicting future inflation necessarily involves guesswork; so, therefore, must the real interest rate. Historically, real interest rates have tended to hover in the range 2%–4%.

Anne and Betty, now 42 years old, hope to retire as *real* millionaires, i.e., as millionaires *after* inflation. They can accomplish this by depositing enough money now to accrue—at *real interest* rate r—to $1 million by age 65.

(a) How much should Anne deposit now if 2% real interest is available?

(b) How much should Anne deposit now if 4% real interest is available?

(c) Betty has $200,000 on hand. What real interest rate will she need to meet her retirement goal?

4. Christine, now 20, wants to retire at 65 with a $1 million *real* nest egg. How much should Christine deposit now at 2% real interest? At 4%? In speculative junk bonds, paying 15% real interest?

5. A child born in 1993 will start college in 2011 and graduate in 2015. Nobody really knows what college tuitions will be by then. However, St. Lena College offers an "early decision" guarantee for the class of 2015: enroll as an infant and lock in yearly tuition payments of $40,000, $42,000, $44,000, and $46,000, payable in 2011, 2012, 2013, and 2014.

How much should new parents deposit in 1993 to cover these future payments, if 6% annual interest is available? What if 8% interest were available?

6. Consider again the college tuition scenario in the previous problem. This time, we'll take a "continuous" point of view, as follows. Over the four years 2011–2015, total tuition paid is $172,000. Suppose, now, that this money is paid "continuously," at the *constant* rate of $43,000 per year.

How much difference does the continuous point of view make? To decide, answer the same questions as in the previous problem.

7. In each part below, find the present value of the given income stream. In each case, treat income as a continuous stream. (Thus, e.g., think of yearly income of $12,000 as flowing in at the *constant rate* of $12,000/year.)

(a) A yearly income of $12,000, beginning 10 years in the future and lasting for 10 years. Assume the yearly interest rate $r = 0.06$.

(b) A yearly income of $12,000, beginning 10 years in the future and lasting for 10 years. Assume annual interest rate $r = 0.08$.

    (c) A yearly income of \$12,000, beginning 10 years in the future and lasting for 10 years. Assume annual interest rate $r = 0$. (This investor doesn't trust banks; he "deposits" the money under the floorboards.)

    (d) At time t years, $10 \le t \le 20$, income flows at the rate $p(t) = 12{,}000e^{0.04t}$. Assume annual interest rate $r = 0.08$. (Note: In practice, many income streams *do* increase over time, e.g., to correct for inflation. This one rises at a 4% instantaneous rate.)

    (e) As in the previous part, but assume an annual interest rate of $r = 0.1$.

8. Consider carefully the two wavy graphs in Example 3, page 148. We claimed above that their formulas are

$$p_T(t) = 50\cos\left(\pi\,\frac{t-105}{180}\right) + 60; \qquad p_I(t) = 50\cos\left(\pi\,\frac{t-180}{180}\right) + 60.$$

    (a) Which graph goes with which formula? Explain briefly in words how you know.

    (b) The functions $\cos t$, p_T, and p_I are all **periodic functions**, i.e., they repeat themselves on time intervals of a fixed length. How long is the period of each of these functions?

    (c) The formulas for p_T and p_I are "built" from $\cos t$ using various constants: 50, 60, 105, 180, etc. Discuss briefly what effect each of these constants has on the corresponding graph.

9. Symbolically calculating the integrals in Example 4, page 149, is a rather messy affair. Instead, use the midpoint rule with 50 subdivisions to *estimate* the first three integrals there.

10. Symbolically calculating the integrals in Example 4, page 149, is messy, but it *can* be done. One way is to use this antiderivative formula (which appears in the endpapers):

$$\int e^{ax}\cos(bx)\,dx = \frac{e^{ax}}{a^2+b^2}\bigl(a\cos(bx) + b\sin(bx)\bigr)$$

    (a) Verify by differentiation that the antiderivative formula really holds.

    (b) Use the integral formula (with appropriate choices for constants) to show symbolically that (as claimed in Example 4)

$$\int_0^{360}\left(50\cos\left(\pi\cdot\frac{t-180}{180}\right) + 60\right)e^{-0.1t/360}\,dt \approx 20{,}550.78.$$

8.6   Chapter summary

This chapter surveys—very briefly—several standard applications of the integral to geometric and physical problems.

Measurement and the definite integral.   In practical applications a definite integral usually *measures* something: area, volume, the length of a curve, a physical or economic quantity, etc. This chapter tells how to use integrals to measure various quantities.

The hardest question, usually, is deciding *what integral* measures a given quantity. In choosing the "right" integral it's useful to keep two different (but equivalent) interpretations of the integral in mind: (1) $\int_a^b f(x)\,dx$ as a limit of approximating sums; (2) $\int_a^b f(x)\,dx$ as the accumulated change in an antiderivative. Section 8.1 describes and compares these points of view.

Volumes.   Calculating **volumes** offers another geometric application of the integral. For complicated three-dimensional figures, such calculations require multivariable calculus, but for many simple solids, ordinary calculus integrals suffice. The principle behind typical volume calculations is **slicing**. A convenient x-axis is chosen; the solid lies along the axis from $x = a$ to $x = b$. The solid is then sliced (in imagination, of course) with cuts perpendicular to the axis. If $A(x)$ denotes the cross-sectional area revealed by such a cut at position x along the axis, then the solid's volume is the integral

$$\text{volume} = \int_a^b A(x)\,dx.$$

As with other applications, the main challenge is usually setting up, rather than evaluating, the integral. For this purpose **solids of revolution** offer a key advantage: all cross sections perpendicular to the axis of rotation are *circles*. As a result, finding the integrand $A(x)$ is usually easy for such solids.

Arc length.   The length of a curve C is defined as the limit, as n tends to infinity, of the lengths of "polygonal approximations" to C. If the curve is the graph of $y = f(x)$ from $x = a$ to $x = b$, then each polygonal approximation can be thought of as an approximating sum for the integral $\int_a^b \sqrt{1 + f'(x)^2}\,dx$. The value of this integral, therefore, gives the length of C.

Work.   Calculating **work** offers a physical application of the integral. Work (in the physicist's sense) is done when a force acts through a distance. In the simplest case, where the force is constant, work is the *product* of the force and the distance through which it acts. Calculus methods are needed if the force is *variable*—as it is in most applications of genuine interest.

The force of **gravity**, for instance, is essentially constant near the surface of the earth, but varies significantly, as described by Newton, over astronomical distances. By combining Newton's law of gravitation with elementary integral techniques, one can compute, for example, the work done against gravity in lifting satellites to various levels of orbit.

Present value.   Suppose that $t = 0$ represents the present time, that a function $p(t)$ represents the rate at which a stream of income "flows" into an account during a time interval $a \le t \le b$, and that interest is continuously compounded, at the constant rate

r. Then, as shown in Section 8.5, the integral $\int_a^b p(t)e^{-rt}\,dt$ gives the **present value** of the income stream. The integral takes account of the fact that future payments are worth less, because of interest effects, than present payments.

Chapter 9

More Antidifferentiation Techniques

9.1 Integration by parts

Introduction

In this chapter we return to a project begun in Chapter 6: to find, for a given elementary function f, an elementary *antiderivative function*, i.e., a function F for which $F' = f$. (Recall: an "elementary" function is one built from the standard basic "elements"—power functions, exponential and logarithm functions, trigonometric functions, etc.)

As we said in Chapter 6, finding elementary *antiderivatives* is, as a rule, harder than finding *derivatives*. Sometimes the problem is actually impossible: some elementary functions—even simple-looking ones—have *no* elementary antiderivatives.➤

See the end of this section for several examples.

Still, many important elementary functions *do* have elementary antiderivatives. Searching for them is a worthy goal in its own right, and an excellent vehicle for understanding various classes of functions.

This chapter presents methods for finding antiderivative formulas for various types of integrands. We make no pretense of exhausting the surprisingly deep and difficult subject of antidifferentiation. However, if we combine the symbolic methods discussed here with a moderate-sized integral table, we can handle most integrals that arise in practical applications. With extra help from modern mathematical software, we can do better still.

Integration by parts and the product rule

Substitution is the reverse version of the *chain rule*. In a similar sense, integration by parts reverses the *product rule* for derivatives.

The product rule for *derivatives* says that for differentiable functions u and v,

$$(u \cdot v)' = u \cdot v' + v \cdot u'.$$

From the *antiderivative* point of view, the product rule says that uv is an *antiderivative* of $uv' + u'v$. In symbols:➤

We added the input variable x, this time.

$$\int \big(u(x) \cdot v'(x) + v(x) \cdot u'(x) \big) \, dx = u(x) \cdot v(x).$$

Equivalently,

$$\int u(x) \cdot v'(x)\, dx = u(x) \cdot v(x) - \int v(x) \cdot u'(x)\, dx.$$

We trade uv' for $u'v$.

This last identity is worth a closer look. Its simple but crucial idea—swapping derivatives for antiderivatives[◄] inside an integral sign—is surprisingly useful throughout mathematics. In elementary calculus, the formula is known as the **integration by parts formula**. As we'll see, the formula sometimes lets us trade a hard integration problem for an easier one.

If we write $u = u(x)$, $v = v(x)$, $du = u'(x)\, dx$, and $dv = v'(x)\, dx$, the formula looks simpler still. We'll state the result as a theorem; it applies both to definite and indefinite integrals.[◄]

The FTC explains why it applies to indefinite integrals.

> **Theorem 1 (Integration by parts).** If u and v are differentiable functions, then
>
> - $\displaystyle \int u\, dv = uv - \int v\, du.$
>
> - $\displaystyle \int_a^b u\, dv = uv \Big]_a^b - \int_a^b v\, du.$

Here's how the method works under ideal conditions.

■ **Example 1.** Find $\displaystyle \int x\, e^x\, dx$ and $\displaystyle \int_0^1 x\, e^x\, dx$.

Solution: Setting $u = x$ and $dv = e^x\, dx$ "fits" the template above:

$$\int x\, e^x\, dx = \int u\, dv.$$

Finding v from dv is another antiderivative problem; sometimes that's hard.

To use the formula, we'll need values for du and v. The first is simple: since $u = x$, $du = dx$. Finding a suitable v *can* be hard,[◄] but here it's easy: if $v = e^x$, then $dv = e^x\, dx$, as desired.

Now the formula kicks in:

$$\int x\, e^x\, dx = \int u\, dv = uv - \int v\, du = x\, e^x - \int e^x\, dx = x\, e^x - e^x + C.$$

Checking the answer involves—of all things—the product rule:

$$(x\, e^x - e^x + C)' = e^x + x\, e^x - e^x = x\, e^x.$$

Our candidate does what it should.

Evaluating the *definite* integral is now routine:

$$\begin{aligned}
\int_0^1 x\, e^x\, dx &= x\, e^x \Big]_0^1 - \int_0^1 e^x\, dx \\
&= x\, e^x - e^x \Big]_0^1 \\
&= 1.
\end{aligned}$$

□

■ **Example 2.** Find $\displaystyle \int x^2\, \ln x\, dx$.

Solution: If we set $u = \ln x$, then $dv = x^2\,dx$, $du = \dfrac{1}{x}\,dx$, $v = \dfrac{x^3}{3}$, and everything works nicely:

$$
\begin{aligned}
\int x^2 \ln x\,dx &= \int u\,dv = uv - \int v\,du \\[2mm]
&= \frac{x^3}{3}\ln x - \int \frac{x^2}{3}\,dx \\[2mm]
&= \frac{x^3}{3}\ln x - \frac{x^3}{9} + C.
\end{aligned}
$$

It's now a routine matter to check, by differentiation, that the answer is correct. ➼ □

Do so. Which differentiation rule is involved?

How to choose u and dv "successfully" isn't always obvious. Setting $u = \ln x$ and $dv = x^2\,dx$ in Example 2, for instance, was natural enough, but not inevitable. We might instead have tried, say, $u = x^2$ and $dv = \ln x\,dx$. (The latter choices, though not incorrect, turn out not to be helpful.)

Sometimes, surprising choices of u and dv turn out to "work."

■ **Example 3.** Find $\displaystyle\int \ln x\,dx$.

Solution: At first glance the problem doesn't seem to fit the mold at all. Yet it does. If we set $u = \ln x$, then $dv = dx$, $du = \frac{1}{x}\,dx$, $v = x$ and the formula says: ➼

Verify that everything "fits."

$$
\int \ln x\,dx = x\,\ln x - \int 1\,dx = x\,\ln x - x + C.
$$

As usual, the answer is easy to check by differentiation—using the product rule! ➼ □

Convince yourself.

Tricks of the trade: wise and foolish choices

Integration by parts trades one indefinite integral expression, $\int u\,dv$, for another, $uv - \int v\,du$. The bargain is worth making, of course, only if the second expression is simpler or more tractable than the first.

When things go wrong

In practice, such satisfaction can't be guaranteed. On the contrary, choosing u and dv unwisely may make things *worse*, not better.

■ **Example 4.** (A step in the wrong direction.) What's wrong ➼ with choosing $u = e^x$ and $dv = x\,dx$ for the indefinite integral $\int x\,e^x\,dx$?

In hindsight, anyway.

Solution: Our choices aren't *wrong*; they just don't help. Setting $u = e^x$, $dv = x\,dx$, $du = e^x\,dx$, and $v = \dfrac{x^2}{2}$ gives

$$
\int x\,e^x\,dx = uv - \int v\,du = e^x \cdot \frac{x^2}{2} - \int \frac{x^2}{2}\cdot e^x\,dx.
$$

The new integral looks harder than the old. This time, we struck out.   □

Once u and dv are chosen, we need values for du and v. Finding du is *always* straightforward; only differentiation is involved. Finding v, by contrast, is an *antidifferentiation* problem; it may be hard or, worse, impossible.

■ **Example 5.** Find $\displaystyle\int 2x\, e^{x^2}\, dx$.

Solution: The integrand is certainly a product, and so perhaps a candidate for integration by parts. Plunging ahead, let's try $u = 2x$ and $dv = \exp(x^2)\, dx$. Then $du = 2\, dx$, but what's v? The answer: nothing useful; $\exp(x^2)$ has no elementary antiderivative. We can't even write v down, let alone use it to solve our problem.

Two possible conclusions follow: either we chose u and dv poorly, or integration by parts was the wrong tool for this job. In this case, the latter is correct. The *ordinary* substitution➤ $u = x^2, du = 2x\, dx$ works nicely:➤

cf. Section 6.2

Yet another moral: Try easy methods—substitution, especially—first.

$$\int 2x\, e^{x^2}\, dx = \int e^u\, du = e^{x^2} + C. \qquad \square$$

When things go right

Mishaps are possible, but for surprisingly many antiderivatives—even ones that don't appear to be products—integration by parts succeeds. The trick is to choose u and dv successfully. In practice, "successfully" means

- dv can be antidifferentiated to give v;

- $\int v\, du$ is simpler or more familiar than $\int u\, dv$.

As we've seen, not every choice of u and dv succeeds. But if one choice fails, another may work. Intuition for good and bad choices comes with practice.

LIATE—a mnemonic for choosing u and dv. Whether an attempt at integration by parts succeeds or fails depends, usually, on the "right" choices of u and dv.

In his article *A technique for integration by parts* (American Mathematical Monthly 90, 1983, pp. 210–211) Herbert E. Kasube proposes the "LIATE" rule for choosing u. (Choosing u is enough; given u, dv must be "everything else.")

As Kasube observes, most integrands are built from functions of 5 types: (L) logarithmic; (I) inverse trigonometric; (A) algebraic; (T) trigonometric; (E) exponential. The LIATE rule says:

Choose u in the order LIATE.

In other words, u should be a function of the *first* available type in the list L,I,A,T,E; dv is "everything else." In Example 1 above, for example, the choice $u = x$ is of type A (algebraic), because neither logarithmic (L) nor inverse trigonometric (I) functions were present.

Not every indefinite integral succumbs to integration by parts. For those that do, the LIATE method seems almost always to work.

Mixing methods

Some problems require more than one method.

■ **Example 6.**   Find $\int \arctan x \, dx.$ ➤   *What does LIATE say here?*

Solution: Let $u = \arctan x$ and $dv = dx$. Then $du = \dfrac{1}{1+x^2} \, dx$, $v = x$, and, by
"parts": ➤   *Short for "integration by parts."*

$$\int \arctan x \, dx = uv - \int v \, du = x \arctan x - \int \frac{x}{1+x^2} \, dx.$$

We're not done, but nearly so: the last integral succumbs to the ordinary substitution
$u = 1 + x^2$, $du = 2x \, dx$: ➤   *Watch the 2's.*

$$\begin{aligned}
x \arctan x - \int \frac{x}{1+x^2} \, dx &= x \arctan x - \frac{1}{2} \int \frac{du}{u} \\[2mm]
&= x \arctan x - \frac{1}{2} \ln|u| + C \\[2mm]
&= x \arctan x - \frac{1}{2} \ln\left(1 + x^2\right) + C. \qquad \square
\end{aligned}$$

Integration by parts for experts

Several useful variations on the basic theme of integration by parts exist. We'll
illustrate two by example.

■ **Example 7.**   **(Repeated integration by parts.)** Find $\int x^2 e^x \, dx$.

Solution: Let $u = x^2$, then $dv = e^x \, dx$, $du = 2x \, dx$, $v = e^x$, and

$$\int x^2 e^x \, dx = x^2 e^x - 2 \int x e^x \, dx.$$

Did we get anywhere? Yes—we've already seen that the last integral can be handled
by *another* integration by parts. ➤ We won't redo the problem; here's the result: *With $u = x$, $dv = e^x \, dx$.*

$$\int x^2 e^x \, dx = x^2 e^x - 2 \int x e^x \, dx = x^2 e^x - 2 \left(x e^x - e^x\right) + C. \qquad \square$$

■ **Example 8.**   **(Integrate twice, then solve.)** Find $\int e^x \sin x \, dx$.

Solution: If we set $u = \sin x$, then $dv = e^x \, dx$, $du = \cos x \, dx$, $v = e^x$, and ➤   *Watch carefully, especially for signs.*

$$I = \int e^x \sin x \, dx = e^x \sin x - \int e^x \cos x \, dx.$$

Not much progress yet; let's try parts *again* on the last integral. If $u = \cos x$, then
$dv = e^x \, dx$, $du = -\sin x \, dx$, and $v = e^x$, so

$$I = e^x \sin x - \int e^x \cos x \, dx = e^x \sin x - \left(e^x \cos x + \int e^x \sin x \, dx\right).$$

The original integral, I, has reappeared! Are we chasing our tail? No. We can *solve*
the last equation for I:

$$I = e^x \sin x - \left(e^x \cos x + I\right) \quad \Longrightarrow \quad 2I = e^x \sin x - e^x \cos x$$

$$\Longrightarrow \quad I = \frac{1}{2}\left(e^x \sin x - e^x \cos x\right) + C. \qquad \square$$

Reality check

That "wraparound" trick in the previous example seems almost too good to be true. Let's check results graphically. Here, for comparison, are graphs of both $f(x) = e^x \sin x$ and $F(x) = (e^x \sin x - e^x \cos x)/2$, supposedly an antiderivative:

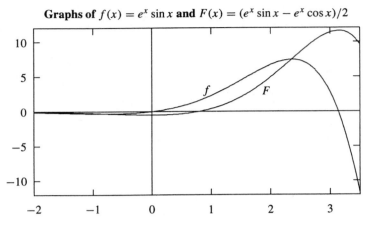

Graphs of $f(x) = e^x \sin x$ and $F(x) = (e^x \sin x - e^x \cos x)/2$

The graphs look all right:

Asymptotes. The integrand $f(x)$ tends quickly to zero as $x \to -\infty$. Properly, therefore, the antiderivative F appears essentially horizontal toward the left of the picture.

Wild swings in the long run. As $x \to \infty$, $f(x)$ changes sign at each multiple of π, oscillating between larger and larger positive and negative values. The antiderivative F, therefore, also oscillates dramatically, with successive peaks and valleys at multiples of π. The graphs bear this out.

Reduction formulas: stepping down to a solution

It holds for any integer $n > 0$.

The equation^{**}

$$\int x^n e^x \, dx = x^n e^x - n \int x^{n-1} e^x \, dx$$

I.e., "reduced."

is called a **reduction formula**. It expresses the left-hand indefinite integral in terms of *another*, slightly simpler^{**} integral.

■ **Example 9.** Use the reduction formula above to find $\int x^3 e^x \, dx$.

Solution: The reduction formula says, for $n = 3$:

$$\int x^3 e^x \, dx = x^3 e^x - 3 \int x^2 e^x \, dx.$$

In Example 7 above we found a value for the last integral; plugging that result in above gives our answer:

$$
\begin{aligned}
\int x^3 e^x \, dx &= x^3 e^x - 3 \int x^2 e^x \, dx \\
&= x^3 e^x - 3\left(x^2 e^x - 2x e^x + 2 e^x + C\right) \\
&= x^3 e^x - 3x^2 e^x - 6x e^x + 6 e^x - 3C.
\end{aligned}
$$

Since C is an arbitrary constant, so is $-3C$; therefore we'll write, simply,

$$\int x^3 e^x \, dx = x^3 e^x - 3x^2 e^x - 6x e^x + 6 e^x + C.$$

A routine check➻ shows that the answer is right.   □   *Details left to you.*

The previous example shows that the reduction formula above works for $n = 3$, but why does it hold for *any* integer n?

Integration by parts explains why. If we set $u = x^n, dv = e^x \, dx, du = nx^{n-1} \, dx$, and $v = e^x$, then the integration by parts formula says precisely what we want:

$$\int x^n e^x \, dx = x^n e^x - n \int x^{n-1} e^x \, dx.$$

Integral tables➻ contain many reduction formulas; most of them are proved by   *cf. the next section* integration by parts.

Is a function antidifferentiable in closed form?

As we know well by now, the answer is "not necessarily." Our favorite examples of functions without elementary antiderivatives are $\sin(x^2)$, $\cos(x^2)$, and $\exp(x^2)$, but there are many➻ more. Integration by parts sometimes helps us recognize such   *Many, many more.* functions.

■ **Example 10.** For a given non-negative integer n, does $x^n \exp(x^2)$ have an elementary antiderivative?

Solution: For $n = 0$, no—in that case, the integrand is $\exp(x^2)$, which we discussed above. For $n = 1$, yes—in that case, the simple substitution $u = x^2$ produces the antiderivative $\exp(x^2)/2$.➻ We might guess, then, that the answer is "yes" for   *Check that it's right!* *odd n* and "no" for *even n*.

Our guess is correct. To see why, consider the integral

$$\int x^n \exp(x^2) \, dx$$

for $n \geq 2$. Integration by parts, with $u = x^{n-1}$ and $dv = x e^{x^2} \, dx$, leads➻ to the   *Convince yourself.* reduction formula

$$\int x^n \exp(x^2) \, dx = x^{n-1} \frac{\exp(x^2)}{2} - \frac{n-1}{2} \int x^{n-2} e^{x^2} \, dx.$$

Observe two things:

- The left-hand integrand has an elementary antiderivative *if and only if* the one on the right does.

- Applying the reduction formula repeatedly knocks down the power of x, by *two* each time. We'll reach, eventually, *either* zero or one, depending on whether n is odd or even.

These two facts, combined with what we know for $n = 0$ and $n = 1$, show what we claimed:

$x^n \exp(x^2)$ *is antidifferentiable in closed form if and only if n is odd.*

□

Exercises ——————————————

Find each of the following antiderivatives using integration by parts and the suggested u and dv. Check your answers by differentiation.

1. $\displaystyle\int xe^{2x}\,dx \qquad u = x, \quad dv = e^{2x}\,dx$

2. $\displaystyle\int x\sin(3x)\,dx \qquad u = x, \quad dv = \sin(3x)\,dx$

3. $\displaystyle\int x\sec^2 x\,dx \qquad u = x, \quad dv = \sec^2 x\,dx$

4. $\displaystyle\int \sqrt{x}\ln x\,dx \qquad u = \ln x, \quad dv = \sqrt{x}\,dx$

5. $\displaystyle\int x\sqrt{1+x}\,dx \qquad u = x, \quad dv = \sqrt{1+x}$

6. $\displaystyle\int \arcsin x\,dx \qquad u = \arcsin x, \quad dv = dx$

7. Evaluate $\displaystyle\int x\cos^2 x\,dx$ using integration by parts.
   [HINT: $\cos^2 x = \frac{1}{2}\big(1 + \cos(2x)\big)$.]

8. Evaluate $\displaystyle\int x\sin x\cos x\,dx$ using integration by parts.
   [HINT: $\sin^2 x = \frac{1}{2}\big(1 - \cos(2x)\big)$.]

9. (a) Use integration by parts to show that $\displaystyle\int \sin^2 x\,dx = -\sin x\cos x + \int \cos^2 x\,dx$

   (b) Use part (a) and the trigonometric identity $\sin^2 x + \cos^2 x = 1$ to show that $\displaystyle\int \sin^2 x\,dx = \frac{1}{2}(x - \sin x\cos x)$.

Evaluate each of the following definite integrals using integration by parts. Check your answers by comparing them to a midpoint rule estimate computed with $n = 2$.

10. $\displaystyle\int_0^\pi x\cos(2x)\,dx$

11. $\displaystyle\int_0^1 xe^{-x}\,dx$

12. $\displaystyle\int_1^e x\ln x\,dx$

13. $\displaystyle\int_{\pi/4}^{\pi/2} x\csc^2 x\,dx$

14. $\displaystyle\int_{-1}^{\sqrt{2}/2} x^2\arctan x\,dx$

15. $\displaystyle\int_1^4 e^{3x}\cos(2x)\,dx$

Find each of the following antiderivatives. Check your results graphically.

16. $\int x^2 \cos x \, dx$

17. $\int \arccos x \, dx$

18. $\int \sqrt{x} \ln \left(\sqrt[3]{x} \right) dx$

19. $\int (\ln x)^2 \, dx$

20. $\int \dfrac{\ln x}{x^2} \, dx$

21. $\int x e^x \sin x \, dx$

22. $\int \arctan(1/x) \, dx$

23. $\int x \arctan x \, dx$

24. $\int e^{\sqrt{x}} \, dx$

25. $\int x^5 \sin \left(x^3 \right) dx$

26. $\int \sin (\ln x) \, dx$

27. $\int \sqrt{x} e^{-\sqrt{x}} \, dx$

28. $\int \sin \left(\sqrt{x} \right) dx$

29. $\int \dfrac{\arctan \left(\sqrt{x} \right)}{\sqrt{x}}$

30. $\int \sqrt{x} \arctan \left(\sqrt{x} \right)$

31. Let n be a positive integer such that $n \geq 1$ and define $I_n = \int x \, (\ln x)^n \, dx$.

   (a) Evaluate I_1.

   (b) Use integration by parts to derive the following **reduction formula**:

$$I_n = \frac{x^2}{2} (\ln x)^n - \frac{n}{2} I_{n-1}.$$

   (c) Use the reduction formula in part (b) to evaluate I_2 and I_3.

   (d) Explain why $\left(\int x \, (\ln x)^n \, dx \right)' = x \, (\ln x)^n$.

   (e) Verify the reduction formula in part (b) by showing that

$$x \, (\ln x)^n = \frac{d}{dx} \left(\frac{x^2}{2} (\ln x)^n - \frac{n}{2} \int x \, (\ln x)^{n-1} \, dx \right).$$

32. Let n be a positive integer such that $n \geq 1$.

   (a) Verify the reduction formula

$$\int x^n e^x \, dx = x^n e^x - n \int x^{n-1} e^x \, dx$$

   by differentiation.

   (b) Use integration by parts to derive the reduction formula in part (a).

33. Let n be a positive integer such that $n \geq 1$. Use integration by parts to show that

$$\int (\ln x)^n \, dx = x \, (\ln x)^n - n \int (\ln x)^{n-1} \, dx.$$

34. Suppose that $r \neq -1/2$. Use integration by parts to show that

$$\int \left(x^2 + a^2\right)^r = \frac{x \left(x^2 + a^2\right)^r}{2r + 1} + \frac{2ra^2}{2r + 1} \int \left(x^2 + a^2\right)^{r-1} \, dx.$$

35. We said in this section that none of the functions $\sin\left(x^2\right)$, $\cos\left(x^2\right)$, and $\exp\left(x^2\right)$ has an elementary antiderivative. Use these facts, if necessary, in the following problems.

   (a) The function $x \sin\left(x^2\right)$ has an elementary antiderivative. Find it.

   (b) The function $x^3 \sin\left(x^2\right)$ has an elementary antiderivative. Find it. [HINT: Try $u = x^2, dv = x \sin(x^2)$.]

   (c) The function $x^3 \cos\left(x^2\right)$ has an elementary antiderivative. Find it.

   (d) Show that the function $x^2 \cos\left(x^2\right)$ has no elementary antiderivative. [HINT: Set $u = x$, $dv = x \sin\left(x^2\right)$; then imitate the argument in Example 10, page 161.]

36. In Example 10, page 161, we showed that for positive integers n, $x^n \exp(x^2)$ has an elementary antiderivative if n is odd, but has no elementary antiderivative if n is even. Imitate the arguments there to show that the same result holds for the function $x^n \sin(x^2)$.

37. Suppose that f is a continuous function, that $f(0) = 2$, and that $\int_0^\pi f(x) \sin x \, dx + \int_0^\pi f''(x) \sin x \, dx = 6$. Find $f(\pi)$.

38. Suppose that f is a continuous function, that $\int_{-\pi/2}^{3\pi/2} f'(x) \, dx = 1$, and that $\int_{-\pi/2}^{3\pi/2} f''(x) \cos x \, dx = 4$. Evaluate $\int_{-\pi/2}^{3\pi/2} f(x) \cos x \, dx$.

9.2 Partial fractions

Rational functions and their antiderivatives

Recall that a **rational function** is any function that can be written as the *quotient*➤
$p(x)/q(x)$ of two polynomials. All three of the following expressions define rational
functions:

➤ *I.e., ratio.*

$$\frac{2}{1-x^2}; \qquad \frac{2+5x+3x^2+3x^3}{x(1+x^2)(x+2)}; \qquad \frac{x^3}{1+x^2}.$$

This section is about the problem of antidifferentiating rational functions. Notice
first that there *is* a problem. Polynomials themselves are easy to antidifferentiate, but
quotients of polynomials are another matter entirely. Even *derivatives* of quotients
are sometimes sticky to compute; *anti*derivatives can be worse still.

The **method of partial fractions** is a systematic technique for antidifferentiating
rational functions. The basic idea is to "divide-and-conquer"—using some flashy
algebra, we rewrite a given rational function as a sum of *simpler* rational functions—
ones we can easily antidifferentiate.

First examples: the method in action

Before describing the method in full theoretical regalia, we'll illustrate it concretely
with several examples.➤

➤ *Read the examples carefully; they motivate the general discussion that follows.*

■ **Example 1.** Find $\displaystyle\int \frac{2}{1-x^2}\, dx$.

Solution: As it stands, the problem looks hard. The integrand doesn't directly
resemble any of our standard basic forms, and neither u-substitution nor integration
by parts looks promising.

The trick is to *rewrite* the integrand as a sum of two simpler terms:

$$\frac{2}{1-x^2} = \frac{2}{(1+x)(1-x)} = \frac{1}{1+x} + \frac{1}{1-x}.$$

The last two summands are the **partial fractions** from which the method takes
its name.➤ We'll explain later how we found them; for the moment, the point
is that rewriting the problem as above brightens our outlook considerably. Each
partial fraction is easy to antidifferentiate separately; adding the results completes
the problem. Here are the details:➤

➤ *Summing the **partial** fractions gives the **total** fraction.*

➤ *Watch the minus signs.*

$$\int \frac{2}{1-x^2}\, dx = \int \frac{dx}{1+x} + \int \frac{dx}{1-x} = \ln|1+x| - \ln|1-x| + C.$$

The answer, as always,➤ is easy to check by differentiation. Let's do so:

➤ *Well, almost always.*

$$(\ln|1+x| - \ln|1-x| + C)' = \frac{1}{1+x} - \frac{-1}{1-x} = \frac{2}{1-x^2},$$

as claimed. □

The gap. The gap in the solution above concerns how we *found* the useful equation

$$\frac{2}{(1+x)(1-x)} = \frac{1}{1+x} + \frac{1}{1-x}.$$

We'll fill this gap soon. Notice, however, that once written down, the equation is
easy to *check* algebraically, e.g., by finding a common denominator for both terms
on the right.

■ **Example 2.** Find $\displaystyle\int \frac{2 + 5x + 3x^2 + 3x^3}{x(1 + x^2)(x + 2)}\, dx$.

By manipulating the right hand side.

Solution: It's easy to check[◄] that

$$\frac{2 + 5x + 3x^2 + 3x^3}{x(1 + x^2)(x + 2)} = \frac{1}{x} + \frac{1}{1 + x^2} + \frac{2}{x + 2}.$$

As in the previous example.

With the integrand written[◄] as a sum of partial fractions, the rest is easy:

$$\int \frac{dx}{x} + \int \frac{dx}{1 + x^2} + \int \frac{2}{x + 2}\, dx = \ln|x| + \arctan x + 2\ln|x + 2| + C. \quad \square$$

■ **Example 3.** Find $\displaystyle\int \frac{x^3}{1 + x^2}\, dx$.

I.e., the highest power of x.

Check this for yourself.

Solution: This time, the denominator can't be factored. However, since the **degree**[◄] of the numerator is higher than that of the denominator, **long division** is possible. Here's the result:[◄]

$$\frac{x^3}{1 + x^2} = x - \frac{x}{1 + x^2}.$$

As the sum of a polynomial and a partial fraction.

For the second integral, substitute $u = 1 + x^2$.

Once again, rewriting the integrand[◄] makes antidifferentiation comparatively easy:[◄]

$$\int x\, dx - \int \frac{x}{1 + x^2}\, dx = \frac{x^2}{2} - \frac{1}{2}\ln\left|1 + x^2\right| + C. \quad \square$$

The forest for the trees: antiderivatives and their ingredients

$\sin(x^2)$ doesn't, for example.

As we know, not every elementary function has an elementary antiderivative.[◄] Must every *rational* function have one? We found a nice antiderivative for each function above, but that could be just chance.[◄]

Textbook authors have been known to rig problems.

Happily, every rational function *has* an elementary antiderivative. A rigorous proof of this theoretical fact involves deep mathematics; we won't attempt one. The key idea, however, is that of partial fractions. In particular, it can be shown that the antiderivative of *any* rational function involves *only* the ingredients seen in the examples above: logarithm, arctangent, and rational functions.

Knowing that an antiderivative formula exists is one thing; finding one explicitly is quite another. The hardest problem, usually, is algebraic—rewriting the given rational function as a sum of partial fractions. Since the method itself[◄] is our main interest, we'll gerrymander problems to keep the algebra simple.

Not its possible complications.

Rational numbers, rational functions, and partial fractions

The analogy continues in higher mathematics, as those who study abstract algebra will see.

Rational *functions* are closely akin to rational *numbers.*[◄] Both are quotients of simpler objects; many of the standard ideas and operations that apply to ordinary fractions apply as well to rational functions.

Proper *vs.* improper

For instance, a positive rational number p/q is called **improper** if $p \geq q$. Any improper fraction can be written as the sum of an *integer* and a *proper* fraction. Thus, for example,

$$\frac{29}{6} = 4 + \frac{5}{6};$$

the numbers 4 and 5 are found by long dividing 29 by 6.➤➤

4 is the quotient; 5 is the remainder.

Rational functions behave similarly. A rational function $r(x) = p(x)/q(x)$ is **proper** if p has lower degree than q; otherwise, r is **improper**.➤➤ The analogy with rational numbers continues:

The integrand in Example 3 is improper.

> *Any improper rational function can be written as the sum of a polynomial and a proper rational function.*

For instance:

$$\frac{x}{x+1} = 1 - \frac{1}{x+1}; \qquad \frac{x^2}{x+1} = x - 1 + \frac{1}{x+1}.$$

As with ordinary fractions, the right sides of these equations can be found by long division.➤➤

Convince yourself.

Even Miss Manners would approve. Rational functions, no matter how "improper," commit no breach of morals or manners. Using such terminology, quaint as it may be, is part of the fun of mathematics.

Partial fractions—of numbers and of polynomials

Any rational *number m/n* can be written as the sum of fractions of a special type: fractions whose denominators are the **prime factors** of n, or powers thereof. For example, 6 has prime factors 2 and 3, and

$$\frac{5}{6} = \frac{5}{2 \cdot 3} = \frac{1}{2} + \frac{1}{3}; \qquad \frac{5}{12} = \frac{5}{2^2 \cdot 3} = \frac{3}{4} + \frac{-1}{3}.$$

The idea for rational *functions* is similar. Any rational function $p(x)/q(x)$ can be written as the sum of **partial fractions**—i.e., other rational functions whose denominators are the **irreducible factors** of $q(x)$, or powers thereof. We used this idea above when we wrote

$$\frac{p(x)}{q(x)} = \frac{2 + 5x + 3x^2 + 3x^3}{x(1+x^2)(x+2)} = \frac{1}{x} + \frac{1}{1+x^2} + \frac{2}{x+2}.$$

Notice especially the denominators on the right: each one is an **irreducible factor** of $q(x)$. (An **irreducible** polynomial is one that can't be factored any further.)➤➤

Irreducible factors are analogous to prime numbers: neither can be factored.

Partial fractions: only two possible forms.   All this jargon may sound formidably abstract. Indeed, the algebra of polynomials and rational functions is a huge and growing theoretical subject. We'll need to see only the tiniest tip of that immense iceberg. For us, only these facts matter:

Factoring. Any polynomial can be factored as a product of *linear* and *irreducible quadratic* factors.

What's irreducible? Linear polynomials are automatically irreducible. A quadratic polynomial $ax^2 + bx + c$ may or may not have linear factors. The quadratic formula tells which is the case: If $b^2 - 4ac < 0$, then no further factoring is possible.

Proper behavior. Any *proper* rational function can be written as a sum of *proper* partial fractions.

The situation boils down, for us, to the fact that any proper rational function can be written as a sum of partial fractions of just *two basic types*:

$$\frac{A}{(ax+b)^n} \quad \text{and} \quad \frac{Ax+B}{(ax^2+bx+c)^n}.$$

(In each case n is a positive integer; a, b, c, A, and B are constants.) Antidifferentiating *any* rational function, therefore, reduces to antidifferentiating these two basic types.

Antiderivatives of the basic forms

Type I antiderivatives.   Partial fractions of the first type are always easy to antidifferentiate. The substitution $u = ax+b$ shows◀ that

Do you agree?

$$\int \frac{A}{(ax+b)^n}\,dx = \begin{cases} A\dfrac{\ln|ax+b|}{a} & \text{if } n = 1; \\[2ex] \dfrac{A}{a(1-n)(ax+b)^{n-1}} & \text{if } n > 1. \end{cases}$$

The general formula may look formidable. In most specific cases, though, antidifferentiation is easy.

■ **Example 4.**   Find $\displaystyle\int \frac{6}{2x+1}\,dx$ and $\displaystyle\int \frac{6}{(2x+1)^5}\,dx.$

Check the easy details.

Solution: For both integrals, substituting $u = 2x+1$ and $du = 2dx$ does the trick:◀

$$\int \frac{6}{2x+1}\,dx = \int \frac{3}{u}\,du = 3\ln|2x+1| + C;$$

$$\int \frac{6}{(2x+1)^5}\,dx = \int \frac{3}{u^5}\,du = -\frac{3}{4(2x+1)^4} + C. \qquad \square$$

Type II antiderivatives.   Antidifferentiating partial fractions of the second form,

$$\frac{Ax+B}{(ax^2+bx+c)^n},$$

can be quite messy, especially if $n > 1$. In that case reduction formulas can be used to knock the exponent n down.◀

We'll usually avoid such cases, but see the exercises for an example.

For $n = 1$, **completing the square** in the denominator simplifies things considerably. Let's see some examples.

■ **Example 5.**   Find $\displaystyle\int \frac{dx}{x^2+2x+2}.$

*The **quadratic formula** says so.*

Solution: Because $b^2 - 4ac < 0$, the denominator is irreducible, so we won't bother trying to factor it.◀ Instead, we'll complete the square:

$$\frac{1}{x^2+2x+2} = \frac{1}{x^2+2x+1+1} = \frac{1}{(x+1)^2+1}.$$

Substituting $u = x+1$ and $du = dx$ completes the problem:

$$\int \frac{dx}{(x+1)^2+1} = \int \frac{du}{u^2+1} = \arctan(x+1) + C. \qquad \square$$

■ **Example 6.**   Find $\int \dfrac{2x+3}{x^2+2x+2}\, dx.$

Solution: Completing the square and substituting $u = x + 1$ helps here, too:[➨]   *Since $u = x+1$, $x = u - 1$.*

$$\int \frac{2x+3}{x^2+2x+2}\, dx \;=\; \int \frac{2x+3}{(x+1)^2+1}\, dx = \int \frac{2(u-1)+3}{u^2+1}\, du$$

$$=\; \int \frac{2u+1}{u^2+1}\, du = \int \frac{2u}{u^2+1}\, du + \int \frac{du}{u^2+1}$$

$$=\; \ln|u^2+1| + \arctan u + C$$

$$=\; \ln|x^2+2x+2| + \arctan(x+1) + C. \qquad \square$$

Reality check: a graphical interlude

We've been throwing plenty of symbols around, but does it all make sense? For some reason, antidifferentiating rational functions seems to spawn logarithms and arctangents. When the dust cleared in Example 2, for instance, we concluded that

$$\int \frac{2+5x+3x^2+3x^3}{x(1+x^2)(x+2)}\, dx = \ln|x| + \arctan x + 2\ln|x+2| + C.$$

A look at graphs helps explain the appearance of these ingredients. Here, together, are graphs of the integrand f and an antiderivative F:

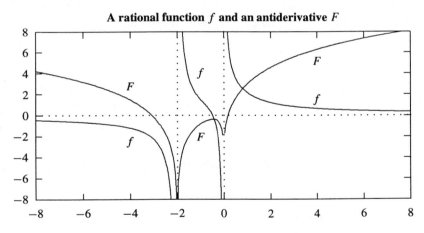

A rational function f and an antiderivative F

The graphs are complicated (like the functions themselves), but they show the right things:

Vertical asymptotes. The integrand f has vertical asymptotes at $x = -2$ and $x = 0$. So does the antiderivative F: near these values of x, the F-graph resembles a logarithm function.

Only two vertical asymptotes. The integrand f has only two vertical asymptotes—the factor $(1 + x^2)$ in the denominator is never zero. For the same reason, the arctangent in F contributes no vertical asymptotes.

In the long run. The general shape of the F-graph looks roughly logarithmic as $x \to \pm\infty$. As $x \to \infty$, for instance, the F graph, like a logarithm, is increasing and concave down.[➨]   *Look closely; do you agree?*

This is as it should be. Here's an intuitive way to see why. Expanding the denominator of f gives

$$f(x) = \frac{2 + 5x + 3x^2 + 3x^3}{x(1 + x^2)(x + 2)} = \frac{2 + 5x + 3x^2 + 3x^3}{2x + x^2 + 2x^3 + x^4}.$$

When $|x|$ is large the highest powers in the numerator and denominator "dominate." Thus, for large $|x|$,

$$f(x) = \frac{2 + 5x + 3x^2 + 3x^3}{2x + x^2 + 2x^3 + x^4} \approx \frac{3x^3}{x^4} \approx \frac{3}{x}.$$

This suggests (and the graph agrees) that for large x, $F(x) \approx 3 \ln|x| + C$.

With our graphical side trip completed, we return to the algebraic mainstream.

Last, not least: how to find the partial fractions

We saved the best for last.

We've seen *why* we'd want to write a rational function as a sum of partial fractions, but not yet *how* to do so.◄ How, for instance, did we arrive at the equation

$$\frac{2 + 5x + 3x^2 + 3x^3}{x(1 + x^2)(x + 2)} = \frac{1}{x} + \frac{1}{1 + x^2} + \frac{2}{x + 2}$$

in Example 2?

Writing a rational function $p(x)/q(x)$ as a sum of partial fractions takes four steps:

Make it proper. If $p(x)/q(x)$ is improper, use long division to rewrite it as the sum of a polynomial and a *proper* rational function.

Often, this step is done for you.

Factor q. Write the denominator, $q(x)$, as a product of linear and irreducible quadratic factors—some may be repeated.◄

Write as a sum. Write $p(x)/q(x)$ as a sum of partial fractions with **undetermined coefficients** (represented by letters) in their numerators.

Find the coefficients. Solve one or more algebraic equations to find numerical values for the coefficients.

The last two steps need elaboration. After brief comments, we'll illustrate everything by example.

The general form of an answer

As earlier examples show, each partial fraction summand comes from some factor of the denominator of the original rational function. For the rational function

$$\frac{2 + 5x + 3x^2 + 3x^3}{x(1 + x^2)(x + 2)},$$

for instance, we expect a partial fraction sum of the general form

$$\frac{2 + 5x + 3x^2 + 3x^3}{x(1 + x^2)(x + 2)} = \frac{A}{x} + \frac{Bx + C}{1 + x^2} + \frac{D}{x + 2};$$

one partial fraction corresponds to each irreducible factor of the denominator. (Since the middle summand has a *quadratic* denominator, its numerator can have degree no more than one. The other summands, which have *linear* denominators, can have only *constant* numerators.)◄

Otherwise a summand would be improper.

Handling repeated factors

If the denominator has a **repeated factor**, that factor contributes *several* partial fractions—one for each power of the factor—to the general form. For example, the partial fraction version of

$$\frac{1}{x^3(x^2+1)^2(x+1)}$$

has the general form

$$\frac{A}{x} + \frac{B}{x^2} + \frac{C}{x^3} + \frac{Dx+E}{x^2+1} + \frac{Fx+G}{(x^2+1)^2} + \frac{H}{x+1}.$$

(Without allowing for these "extra" summands, a partial fraction form might not exist.)

Finding values for the coefficients

Finding values for the unknown constants, as above, is an algebra problem.➤ We'll illustrate several techniques by example.

Not a true calculus problem.

■ **Example 7.** Derive the **partial fraction decomposition**➤

Math jargon—impress your friends.

$$\frac{2}{(1+x)(1-x)} = \frac{1}{1+x} + \frac{1}{1-x}.$$

(We used it in Example 1, page 165.)

Solution: The denominator's factors show that we need constants A and B such that

$$\frac{2}{(1+x)(1-x)} = \frac{A}{1+x} + \frac{B}{1-x}.$$

Multiplying both sides by $(1+x)(1-x)$ clears out the denominators; we get

$$2 = A(1-x) + B(1+x).$$

Now it's easy to find A and B. One way is to expand➤ the right side and collect powers of x:

I.e., multiply out.

$$2 = A(1-x) + B(1+x) = A - Ax + B + Bx = (A+B) + (B-A)x.$$

Equating powers of x on the far left and far right shows that $A+B=2$ and $B-A=0$, so $A=B=1$, as claimed above.

Another route➤ to the same result is to plug judiciously chosen values of x➤ into

Simpler, sometimes.

I.e., values that lead to equations we can solve easily for A and B.

$$2 = A(1-x) + B(1+x).$$

If, say, $x=1$, then $2=2B$, so $B=1$. If $x=-1$, then $2=2A$, so $A=1$. □

■ **Example 8.** Explain why $\dfrac{2+5x+3x^2+3x^3}{x(1+x^2)(x+2)} = \dfrac{1}{x} + \dfrac{1}{1+x^2} + \dfrac{2}{x+2}.$

Solution: We want constants A, B, C, and D such that

$$\frac{2 + 5x + 3x^2 + 3x^3}{x(1 + x^2)(x + 2)} = \frac{A}{x} + \frac{Bx + C}{1 + x^2} + \frac{D}{x + 2}.$$

Some algebra details omitted.

Multiplying both sides by the denominator on the right and collecting powers of x gives:◄

$$
\begin{aligned}
2 + 5x + 3x^2 + 3x^3 &= A(1 + x^2)(x + 2) + (Bx + C)x(x + 2) + Dx(1 + x^2) \\
&= 2A + (A + 2C + D)x + (2A + 2B + C)x^2 + (A + B + D)x^3.
\end{aligned}
$$

Equating the coefficients of various powers of x in the first and last expressions gives *four equations in four unknowns*:

$$2 = 2A; \quad 5 = A + 2C + D; \quad 3 = 2A + 2B + C; \quad 3 = A + B + D.$$

At last!

Solving these equations simultaneously gives $A = 1$, $B = 0$, $C = 1$, $D = 2$. That's why◄

$$\frac{2 + 5x + 3x^2 + 3x^3}{x(1 + x^2)(x + 2)} = \frac{1}{x} + \frac{1}{1 + x^2} + \frac{2}{x + 2},$$

as claimed.   □

Exercises

1. Show algebraically (by adding the terms on the right) that

$$\frac{5x + 7}{(x + 1)(x + 2)} = \frac{2}{x + 1} + \frac{3}{x + 2}.$$

Then use the result to find $\displaystyle\int \frac{5x + 7}{(x + 1)(x + 2)}\, dx$. Finally, check by differentiation that your answer is correct.

2. Show algebraically (by adding the terms on the right) that

$$\frac{2}{x^2 - 1} = (x - 1)^{-1} - (x + 1)^{-1}.$$

Use the result to find $\displaystyle\int \frac{2}{x^2 - 1}\, dx$. Finally, check by differentiation that your answer is correct.

3. Show algebraically (by adding the terms on the right) that

$$\frac{5x + 7}{(x + 1)(x + 2)} = \frac{2}{x + 1} + \frac{3}{x + 2}.$$

Then use the result to find $\displaystyle\int \frac{5x + 7}{(x + 1)(x + 2)}\, dx$. Finally, check by differentiation that your answer is correct.

Evaluate each of the following integrals.

4. $\displaystyle\int \frac{x^2 + 3x - 1}{x(x+1)(x-2)}\, dx$

5. $\displaystyle\int \frac{x^2 - 1}{x\left(x^2 + 4\right)}\, dx$

6. $\displaystyle\int \frac{6}{(x-2)\left(x^2 - 1\right)}\, dx$

7. $\displaystyle\int \frac{x}{x^2 + 2x + 6}\, dx$

8. $\displaystyle\int \frac{4x^2 - 3x + 2}{x\left(2x - 1\right)^2}\, dx$

9. $\displaystyle\int \frac{x^4}{x^4 - 1}\, dx$

10. $\displaystyle\int \frac{5x^2 + 3x - 2}{x^3 + 2x^2}\, dx$

11. $\displaystyle\int \frac{2x + 1}{(x-2)(x+3)}\, dx$

12. $\displaystyle\int \frac{x + 1}{(x-1)(x+2)}\, dx$

13. $\displaystyle\int \frac{x^2 + x}{\left(x^2 + 4\right)^2}\, dx$

14. $\displaystyle\int \frac{x^3}{x^2 + 1}\, dx$

15. $\displaystyle\int \frac{x^3}{x^2 - 1}\, dx$

16. $\displaystyle\int \frac{3x^2 - 1}{(x-1)(x+2)}\, dx$

17. $\displaystyle\int \frac{x^2}{(x^2 + 1)(x+1)^2}\, dx$

18. Use the reduction formula

$$\int \frac{dx}{\left(ax^2 + bx + c\right)^{n+1}} = \frac{2ax + b}{n\left(4ac - b^2\right)\left(ax^2 + bx + c\right)^n} +$$

$$\frac{2(2n-1)a}{n\left(4ac - b^2\right)} \int \frac{dx}{\left(ax^2 + bx + c\right)^n}$$

to evaluate

$$\int \frac{4x^2 + 2x + 1}{\left(4x^2 + 5x + 3\right)^2}\, dx.$$

[HINT: Compute the partial fraction decomposition of the integrand.]

9.3   Trigonometric antiderivatives

In this section we continue our search for antiderivatives of elementary functions. Although the symbolic methods we've developed to date—substitution, integration by parts, and partial fractions—are adequate for many classes of functions, some remain unexplored. We'll consider functions that involve two special ingredients: **powers of trigonometric functions** and **roots of quadratic expressions.** At present we can *check*◄ antiderivatives like these—

By differentiation—try checking one.

$$\int \sin^2 x \cos^3 x \, dx \;=\; \frac{\sin^3 x}{3} - \frac{\sin^5 x}{5} + C;$$

$$\int \sqrt{-x^2 - 2x} \, dx \;=\; \frac{(x+1)\sqrt{-x^2-2x}}{2} + \frac{\arcsin(x+1)}{2} + C,$$

but we can't yet *find* them. After this section we'll be able to do both.

It should be said, too, that in this section we'll meet no truly new or different ideas or methods. Instead we'll explore, mainly through examples, how methods of earlier sections can be combined with properties of trigonometric and algebraic functions to handle new classes of antiderivative problems.

Powers of trigonometric functions

Every antiderivative of the following form—

$$\int \cos^3 x \, dx; \quad \int \cos^3 x \sin^4 x \, dx; \quad \int \cos^2 x \sin^3 x \tan^2 x \, dx; \quad \int \sec^3 x \tan^3 x \, dx$$

(i.e., a product of integer powers of the six trigonometric functions) can be solved in elementary form. As we'll see, all such antiderivatives can be found by combining substitution, integration by parts, and (sometimes) clever applications of trigonometric identities.

Useful properties of trigonometric functions

Before proceeding to antidifferentiation itself, let's recall some properties of and relations among the trigonometric functions.

Two main types.   All six trigonometric functions are defined in terms of sines and cosines. It follows that *every* product of integer powers of trigonometric functions can be written in the form $\sin^n x \cos^m x$, where n and m are (not necessarily positive) integer powers. For instance,

$$\cos^4 x \, \sin^3 x \, \tan^2 x = \cos^4 x \, \sin^3 x \, \frac{\sin^2 x}{\cos^2 x} = \cos^2 x \sin^5 x.$$

See below for some examples.

With enough work,◄ every function of the form $\sin^n x \cos^m x$, *can* be antidifferentiated.

Finding trigonometric antiderivatives tends to be easier when only *non-negative* powers are involved. If negative powers occur, it may help to rewrite a function in terms of positive powers of secants and tangents. For instance,

$$\cos^{-5} x \, \sin^2 x = \cos^{-3} x \, \frac{\sin^2 x}{\cos^2} x = \sec^3 x \tan^2 x.$$

In this section, therefore, we'll mainly consider integrals of two types:

$$\int \sin^n x \, \cos^m x \, dx \quad \text{or} \quad \int \sec^n x \, \tan^m x \, dx,$$

where m and n are non-negative integers.

Reduction formulas. If either $n = 0$ or $m = 0$ in the forms above, then the integral can be handled—with care—using one of these **reduction formulas**:

$$\int \sin^n x \, dx = -\frac{\sin^{n-1} x \cos x}{n} + \frac{n-1}{n} \int \sin^{n-2} x \, dx; \quad (n \neq 0)$$

$$\int \cos^n x \, dx = \frac{\cos^{n-1} x \sin x}{n} + \frac{n-1}{n} \int \cos^{n-2} x \, dx; \quad (n \neq 0)$$

$$\int \tan^n x \, dx = \frac{\tan^{n-1} x}{n-1} - \int \tan^{n-2} x \, dx; \quad (n \neq 1)$$

$$\int \sec^n x \, dx = \frac{\sec^{n-2} x \tan x}{n-1} + \frac{n-2}{n-1} \int \sec^{n-2} x \, dx; \quad (n \neq 1).$$

Large powers may require *repeated* use of the reduction formulas. For an example using tangents, see Example 7, page 80.

Using trigonometric identities. Many integrands can be simplified to one of the forms above, often by using either the **Pythagorean identities**

$$\sin^2 x + \cos^2 x = 1; \qquad \tan^2 x + 1 = \sec^2 x$$

or the **double-angle formulas**➤➤

$$\sin^2 x = \frac{1}{2} - \frac{\cos(2x)}{2}; \qquad \cos^2 x = \frac{1}{2} + \frac{\cos(2x)}{2}.$$

Also known as **half-angle formulas.**

The first two identities permit us, in effect, to trade sines for cosines, secants for tangents, and vice versa. The double-angle formulas can be used to convert even *powers* of sines and cosines into cosines of even *multiples* of x.

We illustrate the use of these facts in finding antiderivatives by concrete examples. We'll usually stop as soon as we've transformed the problem into something we're sure we can handle.

■ **Example 1.**   Find $\int \cos^3 x \, dx$.

Solution: With the substitution $u = \sin x$ in mind, we'll reserve one power of the cosine for du and convert the other two powers to sines:

$$\int \cos^2 x \, \cos x \, dx = \int (1 - \sin^2 x) \cos x \, dx$$

$$= \int (1 - u^2) \, du$$

$$= \sin x - \frac{\sin^3 x}{3} + C.$$

Alternatively, we could have used the reduction formula for cosines—in this case, with $n = 3$:➤➤

$$\int \cos^3 x \, dx = \frac{\cos^2 x \sin x}{3} + \frac{2}{3} \int \cos x \, dx = \frac{\cos^2 x \sin x + 2 \sin x}{3} + C. \quad \square$$

Are the answers comparable? See the exercises.

■ **Example 2.** Find $\int \sin^4 x \, dx$.

Watch the steps.

Solution: We'll use the double-angle formulas to rewrite $\sin^4 x$ in terms of $\cos(2x)$ and $\cos(4x)$, as follows:[◄]

$$\sin^4 x = (\sin^2 x)^2 = \left(\frac{1}{2} - \frac{\cos(2x)}{2}\right)^2 = \frac{1}{4} - \frac{\cos(2x)}{2} + \frac{\cos^2(2x)}{4}$$

The first two summands are now easy to integrate. To the last summand we apply the double-angle formula *again*:

$$\frac{\cos^2(2x)}{4} = \frac{1}{8} + \frac{\cos(4x)}{8}.$$

We've produced another tractable integrand. Here's the result:

$$\int \sin^4 x \, dx = \frac{x}{4} - \frac{\sin(2x)}{4} + \frac{x}{8} + \frac{\sin(4x)}{32} + C.$$

(Other approaches to the same problem may produce an equivalent—but different-appearing—answer.) □

■ **Example 3.** Find $\int \cos^3 x \sin^4 x \, dx$.

Solution: One strategy would be to convert all the sines to cosines. We could attack the result with the cosine reduction formula above. Instead, we'll substitute $u = \sin x$, $du = \cos x \, dx$. We chip off one power of $\cos x$ for du; the remaining two powers of cosine become sines:

$$\int \cos^3 x \sin^4 x \, dx = \int (1 - \sin^2 x)(\sin^4 x) \cos x \, dx = \int (1 - u^2) u^4 \, du.$$

The last integral is simple, so we'll leave it alone. □

■ **Example 4.** Find $\int \sec^3 x \tan^3 x \, dx$.

Solution: If we substitute $u = \sec x$ and $du = \sec x \tan x \, dx$, the remaining tangents can be converted to secants:

$$\begin{aligned}
\int \sec^3 x \tan^3 x \, dx &= \int \sec^2 x \tan^2 x (\sec x \tan x) \, dx \\
&= \int \sec^2 x (\sec^2 x - 1)(\sec x \tan x) \, dx \\
&= \int u^2 (u^2 - 1) \, du.
\end{aligned}$$

The rest is easy. □

Trigonometric substitutions

Trigonometric substitutions help us antidifferentiate certain integrands that involve **roots of quadratic expressions**. The simplest such expressions are these:

$$\sqrt{a^2 - x^2}; \qquad \sqrt{x^2 - a^2}; \qquad \sqrt{a^2 + x^2}.$$

(In each case, a is a positive constant.) That *trigonometric* functions arise at all is at least mildly surprising: so far, there's nothing trigonometric in sight. Yet it turns out that carefully chosen substitutions—with trigonometric ingredients—often reduce integrals of the present type to powers of trigonometric functions. We'll illustrate the idea, and some of the subtleties involved, with a (relatively) simple example.➤

Follow the example carefully; it gives the general idea.

■ **Example 5.** According to our integral table,

$$\int \sqrt{4 - x^2}\, dx = \frac{x\sqrt{4 - x^2}}{2} + 2\,\arcsin\left(\frac{x}{2}\right).$$

How was this found?

Solution: If $x = 2\sin t$, then $dx = 2\cos t\, dt$. These substitutions, and a bit of algebra, produce a simpler-looking integral:➤

Watch carefully.

$$\int \sqrt{4 - x^2}\, dx = 2\int \sqrt{4 - 4\sin^2 t}\,\cos t\, dt = 4\int \cos^2 t\, dt.$$

The last integral is of the type discussed earlier in this section. Using methods developed there (or looking in a table) gives

$$4\int \cos^2 t\, dt = 2t + \sin(2t) + C.$$

So far so good, but we want an answer involving x, not t. To eliminate t in favor of x, we'll need several facts. All follow from having set $x = 2\sin t$:➤

And the formula $\sin(2t) = 2\sin t \cos t$.

$$t = \arcsin\left(\frac{x}{2}\right); \quad 2\cos t = \sqrt{4 - x^2}; \quad \sin(2t) = 2\sin t \cos t = \frac{x\sqrt{4 - x^2}}{2}.$$

Now we can finish our problem:

$$\int \sqrt{4 - x^2}\, dx = 2t + \sin(2t) + C = 2\,\arcsin\left(\frac{x}{2}\right) + \frac{x\sqrt{4 - x^2}}{2} + C. \quad \square$$

Notes on the example. The example needs some amplification:

Inverse substitution. Substituting $x = 2\sin t$ and $dx = 2\cos t\, dt$ represents an "inverse" variant of the more usual u-substitution technique. Such substitutions, though legal, require special care—see below.

Which values of t? The equation $x = 2\sin t$ makes sense for *all* real t. In order to write $t = \arcsin(x/2)$, however, we must (and will, hereafter) assume that t lies in the interval $[-\pi/2, \pi/2]$.➤

I.e., t must lie in the range of the arcsine function.

Absolute values. In the calculation above we used, without comment, the fact that

$$\sqrt{4 - x^2} = \sqrt{4 - 4\sin^2 t} = 2\,|\cos t| = 2\cos t.$$

The last equation assumes, in effect, that $\cos t$ is non-negative. Fortunately that's true: restricting t to the interval $[-\pi/2, \pi/2]$ guarantees so.

The bottom line. Whatever the subtleties of the trigonometric substitution, the bottom line remains: the result *is* a suitable antiderivative, as differentiation can verify.

Trigonometric substitutions: three types

The previous example illustrates one of three types of trigonometric substitutions. Here, telegraphically, are all three:

Trigonometric substitutions			
Radical form	Substitution	t-domain	Result
$\sqrt{a^2 - x^2}$	$x = a\sin t$	$[-\pi/2, \pi/2]$	$\sqrt{a^2 - x^2} = a\cos t$
$\sqrt{a^2 + x^2}$	$x = a\tan t$	$[-\pi/2, \pi/2]$	$\sqrt{a^2 + x^2} = a\sec t$
$\sqrt{x^2 - a^2}$	$x = a\sec t$	$[0, \pi]$, $x \neq \pi/2$	$\sqrt{x^2 - a^2} = \pm a\tan t$

These substitutions, combined (if necessary) with completing the square, produce antiderivatives of many functions that involve square roots of quadratic expressions.

From t back to x: helpful pictures Trigonometric substitutions, if successful, trade a troublesome integral in x for a simpler integral in t. The problem always remains, however, of translating back to an expression in x. As the previous example showed, doing so can be difficult. The following pictures can help jog memory:⬱

Pictures aren't proofs. Careful proofs depend on properties of the trigonometric function and their inverses.

Trigonometric substitutions: pictorial aids

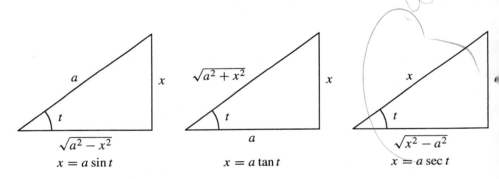

From the middle picture, for instance, one can read such implications as

$$x = a\tan t \implies \sec t = \frac{\sqrt{a^2 + x^2}}{a}.$$

We'll use such facts in the following examples.

■ **Example 6.** Find $\displaystyle\int \frac{dx}{\sqrt{x^2 + 4}}$.

Solution: Let $x = 2\tan t$; then $dx = 2\sec^2 t\,dt$ and $\sqrt{x^2 + 4} = \sqrt{4\tan^2 t + 4} = 2\sec t$.⬱ Therefore,

In the assumed t-domain (see the table above) $\sec t$ is nonnegative.

$$\int \frac{dx}{\sqrt{x^2 + 4}} = \int \frac{2\sec^2 t}{2\sec t}\,dt$$

$$= \int \sec t\,dt = \ln|\sec t + \tan t| + C.$$

To reconvert the result to an expression in x, we'll use the conclusion drawn just before this example:

$$\int \sec t \, dt = \ln \left| \frac{\sqrt{4+x^2}}{2} + \frac{x}{2} \right| + C.$$

$\square$

■ Example 7. Find $\displaystyle\int_0^2 \frac{dx}{\sqrt{x^2+4}}$.

Solution: We just calculated the antiderivative, so there's nothing but algebra left:

$$\int_0^2 \frac{dx}{\sqrt{x^2+4}} = \ln \left| \frac{\sqrt{4+x^2}}{2} + \frac{x}{2} \right| \Big]_0^2 = \ln(\sqrt{2}+1).$$

The previous example also suggests another approach to the same goal: substitution in the *definite integral*. Notice that if $x = 2\tan t$, then $x = 0$ when $t = 0$ and $x = 2$ when $t = \pi/4$. Substituting all this into the original integral gives

$$\int_0^2 \frac{dx}{\sqrt{x^2+4}} = \int_0^{\pi/4} \sec t \, dt = \ln|\sec t + \tan t| \Big]_0^{\pi/4} = \ln(1+\sqrt{2}),$$

just as above.

$\square$

■ Example 8. Find $\displaystyle I = \int \frac{dx}{\sqrt{x^2+2x+5}}$.

Solution: Completing the square in the denominator gives➤➤   *Right?*

$$\int \frac{dx}{\sqrt{x^2+2x+5}} = \int \frac{dx}{\sqrt{(x+1)^2+4}}.$$

Next, we substitute $u = x+1$, $du = dx$; the result looks familiar:

$$I = \int \frac{du}{\sqrt{u^2+4}}.$$

Work already done in the last two examples completes the problem. $\square$

Exercises ——————————————————————

1. We solved Example 1, page 175 in two different ways, and apparently got two different answers. Are the answers really different? Why or why not?

Evaluate each of the following antiderivatives.

2. $\displaystyle\int \cos^3(2x) \sin^2(2x) \, dx$ 4. $\displaystyle\int \cos^2 x \sin^3 x \, dx$

3. $\displaystyle\int \cos^4 x \sin^2 x \, dx$ 5. $\displaystyle\int \sin(2x) \cos^2 x \, dx$

6. $\displaystyle\int \sin^2 x \cos^2 x \, dx$

7. $\displaystyle\int \tan^4 x \, dx$

8. $\displaystyle\int \sec^2 x \tan^2 x \, dx$

9. $\displaystyle\int \sec^3 x \tan^2 x \, dx$

10. $\displaystyle\int \frac{\sin^3 x}{\cos x} \, dx$

11. $\displaystyle\int \sqrt{\cos x} \sin^5 x \, dx$

12. $\displaystyle\int \sqrt{1 + \sin x} \, dx$

13. $\displaystyle\int \frac{dx}{\left(x^2 + 4\right)^2}$

14. $\displaystyle\int x^2 \sqrt{1 - x^2} \, dx$

15. $\displaystyle\int \frac{x^2}{\sqrt{9 - x^2}} \, dx$

16. $\displaystyle\int \frac{dx}{x^2 \sqrt{4 - x^2}}$

17. $\displaystyle\int \frac{dx}{x^2 \sqrt{x^2 - 4}}$

18. $\displaystyle\int \frac{\sqrt{4 - x^2}}{x^2} \, dx$

19. $\displaystyle\int \frac{dx}{x^2 \sqrt{x^2 + 1}}$

20. $\displaystyle\int \frac{dx}{\sqrt{1 + x^2}}$

21. $\displaystyle\int \frac{x + 2}{x \left(x^2 + 1\right)} \, dx$

22. $\displaystyle\int x \arcsin x \, dx$

23. (a) Evaluate $\displaystyle\int_1^2 \frac{\sqrt{x^2 - 1}}{x} \, dx$.

(b) Evaluate $\displaystyle\int_{-2}^{-1} \frac{\sqrt{x^2 - 1}}{x} \, dx$.

Chapter 10

Improper Integrals

10.1 When is an integral improper?

Each expression below is an **improper integral:**➤

$$\int_1^\infty \frac{dx}{x^2} \qquad \int_1^\infty \frac{dx}{x} \qquad \int_{-\infty}^\infty \frac{dx}{1+x^2}$$

$$\int_0^\infty e^{-x^2}\, dx \qquad \int_0^1 \frac{dx}{x^2} \qquad \int_0^1 \frac{dx}{\sqrt{x}}$$

It will pay to keep these standard examples in mind. They illustrate most of what can go right—and wrong—with improper integrals.

"Improper" integrals➤➤ commit no breach of moral or manners.➤➤ The adjective "improper" acts as a warning sticker attached to integrals that differ somehow from the ordinary $\int_a^b f(x)\, dx$ variety, in which $[a, b]$ is a *finite* interval and $f(x)$ is a continuous function. Each integral above, examined carefully, should raise some suspicion.

Like improper fractions.

They're just as useful, moreover, as "proper" integrals.

Integrals can commit two types of "impropriety":

- *The interval of integration may be infinite*, as in the first four examples above. This is technically illegal because the formal definition of definite integral relies on partitions of a *finite* interval.

- *The integrand may be unbounded on the interval of integration,*➤➤ as in the last two examples above. This, too, is technically illegal because the integrand is not even defined at an endpoint of the interval of integration.

The integrand "blows up," in other words.

A few really obstreperous integrals, like

$$\int_0^\infty \frac{dx}{\sqrt{x} + x^2}$$

commit *both* types of impropriety; they need especially strict handling.

Convergence and divergence: basic ideas and examples

Some integrals, despite being improper, have a sensible numerical value. For other integrals, the impropriety is fatal: they have no sensible finite value. Integrals of these two types are called, respectively, **convergent** and **divergent**. First, some concrete examples; formal definitions come later.➤➤

Study the examples carefully; they motivate the formal definitions.

■ **Example 1.** Make sense of $\int_1^\infty \frac{dx}{x^2}$.

181

Solution: What could such an integral mean? Interpreted geometrically, the integral represents the area shown shaded below:

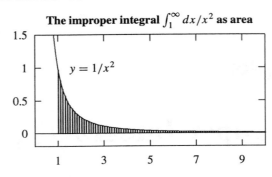

The shaded region extends infinitely far to the right, but could its *area* be finite?
The answer is yes. For any number $t > 1$, consider

$$\int_1^t \frac{dx}{x^2} = \text{area from } x = 1 \text{ to } x = t.$$

Check the easy computation.

We can calculate this area exactly:

$$\int_1^t \frac{dx}{x^2} = \frac{-1}{x}\bigg]_1^t = 1 - \frac{1}{t}.$$

Does this result go against intuition? We'll see many more similar results.

As $t \to \infty$, this quantity tends to 1. In other words, the *total shaded area*, although infinitely long, has *finite* area.
In telegraphic summary:

$$\lim_{t \to \infty} \int_1^t \frac{dx}{x^2} = \lim_{t \to \infty}\left(1 - \frac{1}{t}\right) = 1.$$

We say that our original integral **converges** to 1.   □

■ **Example 2.** Does $\int_1^\infty \frac{dx}{x}$ converge or diverge? Why?

Solution: At first glance, the situation *looks* similar to the previous one:

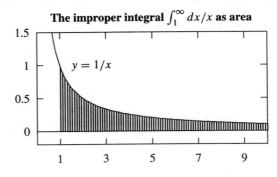

The question, again, is whether the shaded region—unbounded on the right—has finite or infinite area.

The picture alone, being finite, can't tell us the answer. Nor can a comparison

Look carefully; do you see why?

with the previous region: the current region has *larger* area. From what we know so far, the area could be either finite or infinite.

To answer finally, we'll have to *calculate*, much as before. For $t > 1$, the area from $x = 1$ to $x = t$ is

Check details.

$$\int_1^t \frac{dx}{x} = \ln x \Big]_1^t = \ln t.$$

(The area shown above, for instance, is $\ln 10 \approx 2.3026$.) As $t \to \infty$, $\ln t \to \infty$. We conclude, therefore, that the current improper integral **diverges** to infinity. In symbolic shorthand:

$$\lim_{t \to \infty} \int_1^t \frac{dx}{x} = \lim_{t \to \infty} (\ln t) = \infty.$$ □

■ **Example 3.** Does $\displaystyle\int_{-\infty}^{\infty} \frac{dx}{1 + x^2}$ converge? If so, to what?

Solution: The integral is improper at *both* ends, so we'll break it into two pieces—

$$\int_{-\infty}^{\infty} \frac{dx}{1 + x^2} = \int_{-\infty}^{0} \frac{dx}{1 + x^2} + \int_{0}^{\infty} \frac{dx}{1 + x^2}$$

and handle each separately.

One impropriety at a time is plenty!

Consider, first, the last summand. A straightforward calculation—

$$\lim_{t \to \infty} \int_0^t \frac{dx}{1 + x^2} = \lim_{t \to \infty} \arctan x \Big]_0^t = \lim_{t \to \infty} \arctan t = \frac{\pi}{2}$$

shows that the second summand **converges** to $\pi/2$. Because the integrand is *even*, the first summand has the same value as the second. In other words,

Right? How do you know? What does this mean graphically?

$$\int_{-\infty}^{0} \frac{dx}{1 + x^2} = \frac{\pi}{2}$$

also. The conclusion is now clear: the original integral **converges** to π. In symbols:

$$\int_{-\infty}^{\infty} \frac{dx}{1 + x^2} = \int_{-\infty}^{0} \frac{dx}{1 + x^2} + \int_{0}^{\infty} \frac{dx}{1 + x^2} = \pi.$$ □

Caveat grapher

The examples above illustrate one of the subtleties inherent in improper integrals. At first glance the graphs of $1/x^2$ and $1/x$ appear similar: both approach zero asymptotically as $x \to \infty$. Nevertheless, the first graph bounds just *one* unit of area; the second graph bounds *infinite* area.

Another impropriety

In each example above, the improprieties involved infinite *intervals*. Almost the same strategy applies to infinite *integrands*.

■ **Example 4.** Discuss $\displaystyle\int_0^1 \frac{dx}{x^2}$.

Solution: The geometric question, this time, is whether the *vertically* unbounded region shown below has finite or infinite area.

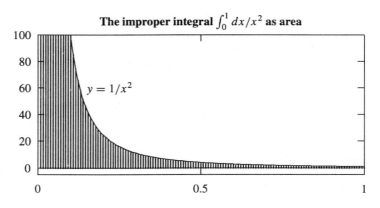

The improper integral $\int_0^1 dx/x^2$ as area

Rather than as $t \to \infty$.

We calculated almost the same integral above.

To settle the question we'll find another limit of areas, this time as t tends to 0 *from the right.*⁤ For any $t > 0$, the area from $x = t$ to $x = 1$ makes good sense and it's easy to find:⁤

$$\int_t^1 \frac{dx}{x^2} = -\frac{1}{x}\Big]_t^1 = \frac{1}{t} - 1.$$

Thus as t tends to 0 from the right, the area in question *blows up to infinity.* Symbolically put:

$$\lim_{t \to 0^+} \int_1^t \frac{dx}{x^2} = \lim_{t \to 0^+} \left(\frac{1}{t} - 1\right) = \infty.$$

To positive infinity.

The original integral, therefore, **diverges.**⁤   □

What the examples say; formal definitions

We applied the same basic idea to every improper integral above—regardless of whether the *interval* or the *integrand* was unbounded. First we *located* the impropriety,⁤ at ∞, at $-\infty$, or at a finite endpoint of the interval of integration. (If an integral is improper at more than one place, we rewrite it as a sum of simpler integrals, each with only *one* impropriety.)

Or improprieties.

Given any integral with one "improper" endpoint, we do the same thing: consider the *limit* as a *variable* endpoint tends, from above or below, to the troublesome value.

In the case of greatest interest to us, the fine print reads like this:

> **Definition:** Consider the integral $I = \int_a^\infty f(x)\,dx$, where f is continuous for $x \geq a$. If the limit
>
> $$L = \lim_{t \to \infty} \int_a^t f(x)\,dx$$
>
> exists and is finite, then I **converges to** L. Otherwise, I **diverges.**

Graphically speaking, the question concerns the long term behavior of the shaded area below—as $t \to \infty$, does the area converge or diverge?

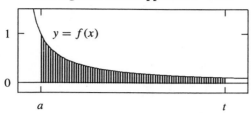

Convergence: what happens as $t \to \infty$?

■ **Example 5.** Does $I = \int_0^\infty \cos x \, dx$ converge or diverge?

Solution: The definition gives a quick answer. Because

$$\int_0^t \cos x \, dx = \sin x \Big]_0^t = \sin t,$$

it follows that

$$\lim_{t \to \infty} \int_0^t \cos x \, dx = \lim_{t \to \infty} \sin t.$$

The last limit *doesn't exist*: as $t \to \infty$, $\cos t$ oscillates endlessly between -1 and 1. Hence I diverges.   □

A similar, but more general, definition of convergence applies to *any* improper integral:

> **Definition:**   Let the integral $I = \int_a^b f(x) \, dx$ be improper either at a or at b. (The cases $a = -\infty$ or $b = \infty$ are allowed.) If either
>
> $$\lim_{t \to a^+} \int_t^b f(x) \, dx \quad \text{or} \quad \lim_{t \to b^-} \int_a^t f(x) \, dx.$$
>
> exists and has finite value L, then I **converges to L**. Otherwise, I **diverges**.

Notice:

A limit of proper integrals. Every convergent *improper* integral is a limit of *proper* integrals. For example,

$$\int_1^\infty f(x) \, dx = \lim_{t \to \infty} \int_1^t f(x) \, dx,$$

if the limit exists.

More than one impropriety. If an integral is improper at *both* ends,➤ it can be broken, in any convenient manner, into two summands; the whole thing converges only if *both* summands converge.

$\int_{-\infty}^\infty dx/(1 + x^2) \, dx$ is doubly improper; we saw above that it converges to π.

■ **Example 6.** Discuss $I = \int_0^\infty \dfrac{dx}{x^2}$.

Solution: The integral is improper at both ends. To separate the improprieties, we'll write

$$I = \int_0^\infty \frac{dx}{x^2} = \int_0^1 \frac{dx}{x^2} + \int_1^\infty \frac{dx}{x^2} = I_1 + I_2.$$

As we've seen in earlier examples, I_2 *converges* to 1, but I_1 *diverges* to ∞. Hence I itself *diverges*. □

Exercises ——————————————————————————

1. Explain why each of the following integrals is improper.

   (a) $\displaystyle\int_0^\infty x^2 e^{-x^2}\, dx$

   (b) $\displaystyle\int_0^1 \frac{x}{\sqrt{x^2 - 3x + 2}}\, dx$

   (c) $\displaystyle\int_1^4 \frac{dx}{x^2 \ln x}$

   (d) $\displaystyle\int_0^3 \frac{dx}{x^2 - x - 2}$

   (e) $\displaystyle\int_0^{\pi/2} \tan x\, dx$

   (f) $\displaystyle\int_0^{2\pi} \frac{\sin x}{\sqrt{1 + \cos x}}\, dx$

2. Explain why the integral $\displaystyle\int_0^{\pi/2} \frac{\cos x}{\sqrt{1 - \sin^2 x}}\, dx$ is *not* improper.

3. Show that the improper integral $\displaystyle\int_0^\infty \frac{dx}{x^2}$ diverges.

Use antiderivatives to evaluate each of the following convergent improper integrals.

4. $\displaystyle\int_0^\infty e^{-x}\, dx$

5. $\displaystyle\int_e^\infty \frac{dx}{x\,(\ln x)^2}$

6. $\displaystyle\int_1^\infty \frac{dx}{x(1 + x)}$

7. $\displaystyle\int_0^4 \frac{dx}{\sqrt{x}}$

8. $\displaystyle\int_{-2}^2 \frac{2x + 1}{\sqrt[3]{x^2 + x - 6}}\, dx$

9. $\displaystyle\int_\pi^\infty e^{-x} \sin x\, dx$

10. (a) Suppose that $f(x) \geq 0$ for all $x \geq 1$ and that $\int_1^\infty f(x)\, dx$ converges. Explain why there is a number a such that $\int_a^\infty f(x)\, dx \leq 10^{-10}$.

    (b) Suppose that $g(x) \geq 0$ for all $x \geq 1$ and that $\int_1^\infty g(x)\, dx$ diverges. Explain why there is a number a such that $\int_1^a g(x)\, dx \geq 10^{10}$.

11. Suppose that $I = \int_0^\infty f(x)\, dx$ converges and that $\left| \int_a^\infty f(x)\, dx \right| \leq 0.0001$. Explain why

$$\left| I - \int_0^a f(x)\, dx \right| \leq 0.0001.$$

For each of the following improper integrals, find a value for the parameter a that makes the value of the integral less than 10^{-5}.

12. $\displaystyle\int_a^\infty e^{-x}\,dx$

13. $\displaystyle\int_a^\infty \frac{dx}{x^2+1}$

14. $\displaystyle\int_a^\infty \frac{dx}{x\,(\ln x)^3}$

15. (a) Show that $\displaystyle\lim_{a\to\infty}\int_{-a}^a x\,dx=0$.

    (b) Show that $\displaystyle\int_{-\infty}^\infty x\,dx$ diverges.

Use antiderivatives to determine whether each of the following improper integrals converges or diverges. Evaluate those that converge.

16. $\displaystyle\int_0^\infty \frac{x}{\sqrt{1+x^2}}\,dx$

17. $\displaystyle\int_0^\infty \frac{\arctan x}{1+x^2}\,dx$

18. $\displaystyle\int_e^\infty \frac{dx}{x\ln x}$

19. $\displaystyle\int_3^\infty \frac{x}{\left(x^2-4\right)^3}\,dx$

20. $\displaystyle\int_{-\infty}^1 e^x\,dx$

21. $\displaystyle\int_0^8 \frac{dx}{\sqrt[3]{x}}$

22. $\displaystyle\int_1^3 \frac{dx}{\sqrt[3]{x-2}}$

23. $\displaystyle\int_2^3 \frac{x}{\sqrt{3-x}}\,dx$

24. $\displaystyle\int_0^2 \frac{dx}{\sqrt{4-x^2}}$

25. $\displaystyle\int_1^\infty \frac{dx}{x(\ln x)^2}$

26. $\displaystyle\int_0^\infty \frac{dx}{(x-1)^2}$

27. $\displaystyle\int_0^\infty \frac{dx}{e^x-1}$

28. $\displaystyle\int_{-\infty}^\infty e^{-x}\,dx$

29. $\displaystyle\int_{-\infty}^\infty \frac{dx}{e^x+e^{-x}}$

30. $\displaystyle\int_0^1 \frac{x}{\sqrt{1-x^2}}\,dx$

31. $\displaystyle\int_0^1 \frac{e^{-\sqrt{x}}}{\sqrt{x}}\,dx$

32. $\displaystyle\int_0^{\pi/2} \frac{\cos x}{\sqrt{\sin x}}\,dx$

33. (a) Show that $\displaystyle\int_1^\infty \frac{dx}{x}$ diverges.

    (b) Show that $\displaystyle\int_1^\infty \frac{dx}{x^p}$ converges if $p>1$.

    (c) Show that $\displaystyle\int_1^\infty \frac{dx}{x^p}$ diverges if $p<1$.

34. (a) Suppose that c is a real number and that $f(x) \geq c > 0$ when $x \geq 0$. Give a graphical argument that explains why $\int_0^\infty f(x)\, dx$ diverges.

 (b) Suppose that $0 \leq g(x) \leq x^{-2}$ for all $x \geq 1$. Give a graphical argument that explains why $\int_1^\infty g(x)\, dx$ converges.

 (c) Suppose that $h(x) \geq x^{-1}$ for all $x \geq 1$. Give a graphical argument that explains why $\int_1^\infty h(x)\, dx$ diverges.

35. For which values of p does $\displaystyle\int_0^1 \frac{dx}{x^p}$ converge?

36. (a) For which values of p does $\displaystyle\int_1^e \frac{dx}{x(\ln x)^p}$ converge?

 (b) For which values of p does $\displaystyle\int_e^\infty \frac{dx}{x(\ln x)^p}$ converge?

Evaluate each of the following improper integrals for all values of the parameter C for which it converges.

37. $\displaystyle\int_0^\infty \left(\frac{2x}{x^2+1} - \frac{C}{2x+1} \right) dx$

40. $\displaystyle\int_0^\infty \left(\frac{1}{\sqrt{x^2+4}} - \frac{C}{x+2} \right) dx$

38. $\displaystyle\int_1^\infty \left(\frac{C}{x+1} - \frac{3x}{2x^2+C} \right) dx$

41. $\displaystyle\int_0^\infty \left(\frac{x}{x^2+1} - \frac{C}{3x+1} \right) dx$

39. $\displaystyle\int_1^\infty \left(\frac{Cx^2}{x^3+1} - \frac{1}{3x+1} \right) dx$

42. $\displaystyle\int_1^\infty \left(\frac{Cx}{x^2+1} - \frac{1}{2x} \right) dx$

Some improper integrals can be transformed into proper integrals by making an appropriate change of variables. For each of the following, use the given substitution to transform the given improper integral into a proper integral with the same value. (Do not evaluate these integrals.)

43. $\displaystyle\int_1^\infty \frac{x}{x^3+1}\, dx; \quad u = x^{-1}$

44. $\displaystyle\int_0^{\pi/2} \frac{\cos x}{\sqrt{\pi - 2x}}\, dx; \quad u = \sqrt{\pi - 2x}$

45. Show that $\displaystyle\int_0^\infty \frac{dx}{1+x^4} = \int_0^\infty \frac{x^2}{1+x^4}\, dx$. [HINT: See exercise #43.]

46. Show that $\displaystyle\int_0^\infty x^3 e^{-x}\, dx = \int_0^1 (-\ln x)^3\, dx$. [HINT: See exercise #43.]

47. Evaluate $\displaystyle\int_0^\infty \frac{x \ln x}{1+x^4}\, dx$.

10.2   Detecting convergence, estimating limits

The last section was about the *idea* of convergence or divergence of an improper integral. In handling an integral of the form $\int_1^\infty f(x)\,dx$, for example, all depends on these questions:

> *Does $\int_1^t f(x)\,dx$ approach a finite limit as t tends to infinity? If so,* what *is this limit?*

Answering these questions was a routine matter in the relatively simple examples above. Each time we found, by antidifferentiation, a simple, explicit formula (in t) for $\int_1^t f(x)\,dx$.➤ Letting t tend to infinity, we could see easily whether a limit existed and, if so, find its value.

E.g., $\int_1^t dx/x^2 = 1 - 1/t$.

Things don't always go so smoothly. Finding a symbolic expression for $\int_1^t f(x)\,dx$ may be difficult (or even impossible). Even if a symbolic expression *is* found, finding a limit—or even deciding whether one exists—may be tricky.

All is not lost. As they do for *proper* definite integrals, numerical methods➤ can help us estimate *improper* integrals. This section tells how. For simplicity, we'll discuss improper integrals of only one type: those with infinite *intervals* of integration.

Such as the trapezoid and midpoint rules.

Important examples

Estimating improper integrals numerically takes special care. One problem is obvious: The improper integral $\int_a^\infty f(x)\,dx$ is over an *infinite* interval, but the error bound formulas for estimating definite integrals involve the *length* of the interval of integration; how can they apply to *infinite* intervals?

The following examples suggest strategies for dealing with (or dodging) such difficulties. They illustrate and motivate the theory that follows.➤

Notice especially how the words **comparison** *and* **tail** *are used. They'll reappear.*

■ **Example 1.** Does $I = \displaystyle\int_1^\infty \frac{dx}{x^5 + 1}$ converge or diverge?

Solution: The question is easy to *state* symbolically:

> *Does the limit* $\displaystyle\lim_{t\to\infty}\left(\int_1^t \frac{dx}{x^5+1}\right)$ *exist?*➤

If so, we'll worry later about its value.

Alas, the question is harder to *answer* symbolically. The given integrand has a very complicated antiderivative;➤ even if we wrote it down, finding a limit (if one exists) symbolically would be very hard.

Even by Maple standards; it fills more than half a computer screen.

It's hard to *calculate* the limit above; perhaps surprisingly, it's easy to see that *some* finite limit does exist. Because the integrand $1/(x^5 + 1)$ is positive for all $x \geq 1$, the question is whether the (positive) area shown below "blows up" or remains bounded as t tends to infinity.

Does I converge? What happens as $t \to \infty$?

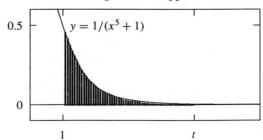

Watch for this word; it will reappear.

The answer is that the area remains bounded, so I converges. A natural **comparison**◂ of the two similar-looking integrals

$$\int_1^\infty \frac{dx}{x^5 + 1} \quad \text{and} \quad \int_1^\infty \frac{dx}{x^5}$$

The computation is easy, but check details.

shows why. Notice first that the *second* integral converges to 1/4:◂

$$\lim_{t \to \infty} \int_1^t \frac{dx}{x^5} = \lim_{t \to \infty} \left(\frac{1}{4} - \frac{1}{4t^4} \right) = \frac{1}{4}.$$

The only x's we care about.

The key idea is that for all $x \geq 1$◂

$$\frac{1}{x^5 + 1} < \frac{1}{x^5}.$$

The graphs agree:

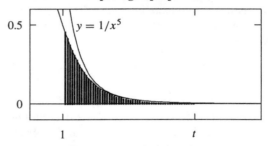

Comparing improper areas

The shaded area, evidently, is *less than* the corresponding area under the graph of $y = 1/x^5$. Our conclusion now follows:

The original integral I converges; its limit is less than 1/4.

Briefly, in symbols:

$$\int_1^\infty \frac{dx}{x^5 + 1} < \int_1^\infty \frac{dx}{x^5} = \frac{1}{4}.$$ □

■ **Example 2.** OK, $I = \displaystyle\int_1^\infty \frac{dx}{x^5 + 1}$ converges. To what value does it converge?

For reasons stated above.

Solution: Symbolic methods aren't promising,◂ so we'll try a numerical approach.

Numerical integration rules apply only to *finite* intervals $[a, b]$. With this restriction in mind, we'll break I into two parts:

$$I = \int_1^\infty \frac{dx}{x^5 + 1} = \int_1^5 \frac{dx}{x^5 + 1} + \int_5^\infty \frac{dx}{x^5 + 1} = I_1 + I_2.$$

The first integral, I_1, is over the *finite* interval $[1, 5]$; we'll estimate its value numerically. (There's nothing sacred about $[1, 5]$; we could have used any other finite interval.) The second integral, I_2, is called, picturesquely, a **tail**◂ of I; it represents the total area—small, we hope—under the graph to the *right* of $x = 5$.

This word, too, will reappear.

Courtesy of Maple.

To start, we'll apply the midpoint with 50 subdivisions to I_1. Here's the result:◂

$$\int_1^5 \frac{dx}{x^5 + 1} \approx M_{50} \approx 0.17991.$$

We expect M_{20} to approximate I_1 closely, but how closely does it approximate I itself? There are two sources of error:

1. The error M_{50} commits in estimating I_1.

2. The error due to ignoring the tail I_2.

We'll bound each error separately.

Bounding the *first* type of error is by now a routine matter. The usual midpoint rule error bound formula guarantees➤ that this error is less than 0.003.

We spare you the details.

Bounding the *second* type of error (aka **bounding the tail**) involves the same comparison as in the previous example. Because the inequality

$$\frac{1}{x^5 + 1} < \frac{1}{x^5}$$

holds for all x of interest, it follows that

$$\int_5^\infty \frac{dx}{x^5 + 1} < \int_5^\infty \frac{dx}{x^5}.$$

Zoologically speaking, the first tail is "skinnier"➤ than the second. The right-hand integral is easy to calculate:

And so has less area.

$$\int_5^\infty \frac{dx}{x^5} = \lim_{t \to \infty} -\frac{1}{4x^4}\Big]_5^t = \frac{1}{2500} = 0.0004.$$

Having bounded *both* types of possible error, we conclude➤ that the estimate $I \approx 0.1781$ commits *total* error less than $0.003 + 0.0004 = 0.0034$. □

With some relief.

Comparing improper integrals: two theorems

The examples above used the **comparisons**

$$\int_1^\infty \frac{dx}{x^5 + 1} < \int_1^\infty \frac{dx}{x^5} \quad \text{and} \quad \int_5^\infty \frac{dx}{x^5 + 1} < \int_5^\infty \frac{dx}{x^5}.$$

From the first inequality we inferred that the "smaller" integral converges; using the second, we bounded a tail. The following theorem guarantees the legitimacy of what we did.➤

Not that it was in much doubt.

Theorem 1 (Comparison for nonnegative improper integrals).
Let f and g be continuous functions. Suppose that for all $x \geq a$,

$$0 \leq f(x) \leq g(x).$$

- If $\int_a^\infty g(x)\,dx$ *converges*, then so does $\int_a^\infty f(x)\,dx$, and

$$\int_a^\infty f(x)\,dx \leq \int_a^\infty g(x)\,dx.$$

- If $\int_a^\infty f(x)\,dx$ *diverges*, then so does $\int_a^\infty g(x)\,dx$.

The theorem's two claims, once understood, are certainly plausible. The first claim says:

> If the area under the g-graph is *finite*, then so is the area under the "lower" f-graph.

The second claim says:

If the area under the f-graph is *infinite*, then so is the area under the "higher" g-graph.

The following picture supports both claims:

What Theorem 1 says, graphically

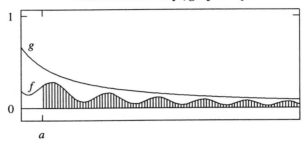

The formal proof would add little, so we omit it.

Using the theorem

We saw both in the previous examples.

The theorem is useful in two ways: for recognizing convergence or divergence, and for estimating limits.

■ **Example 3.** Does $I = \displaystyle\int_0^\infty e^{-x^2}\,dx$ converge? If so, to what limit?

Solution: We'd like to determine the value of the limit

$$\lim_{t\to\infty}\int_0^t e^{-x^2}\,dx,$$

if it exists. Since we can't antidifferentiate the integrand, estimates are called for.

To see that I converges, we'll compare it to a nicer integral, one we *know* converges. Notice first that if $x \ge 1$,

$$e^{-x^2} \le x\,e^{-x^2}.$$

The graphs agree:

Comparing $\exp(-x^2)$ **to** $x\exp(-x^2)$

$$y = \exp(-x^2)$$

$$y = x\exp(-x^2)$$

Now the theorem applies; it says that

$$\int_1^\infty e^{-x^2}\,dx < \int_1^\infty x\,e^{-x^2}\,dx.$$

The right-hand integral can be worked out directly:

$$\int_1^\infty xe^{-x^2}\,dx = \lim_{t\to\infty}\int_1^t xe^{-x^2}\,dx = \lim_{t\to\infty} \left.-\frac{e^{-x^2}}{2}\right]_1^t = \frac{1}{2e}.$$

Therefore I itself➤➤ converges to some (still unknown) limit.

To estimate I numerically, we break it into a convenient finite piece and a tail, and estimate each one separately:

$$I = \int_0^\infty e^{-x^2}\,dx = \int_0^4 e^{-x^2}\,dx + \int_4^\infty e^{-x^2}\,dx = I_1 + I_2.$$

To estimate I_1, we'll use $M_{100} \approx 0.886226$. The usual error bound formula shows➤➤ that M_{100} commits error less than 0.0005 in the process. The tail integral, I_2, turns out to be even smaller. Using the same comparison as above➤➤ gives

$$I_2 = \int_4^\infty e^{-x^2}\,dx < \int_4^\infty xe^{-x^2}\,dx = \frac{1}{2e^{16}} \approx 0.000000056.$$

(The value of the second integral is found as above; we omit the calculation.)

The error from both sources, therefore, is tiny; we conclude that I converges to a limit very close to 0.886226—at least three decimal places are meaningful.➤➤   □

The theorem actually shows that $\int_1^\infty e^{-x^2}\,dx$ (rather than $\int_0^\infty e^{-x^2}\,dx$) converges. That's good enough, since $\int_0^1 e^{-x^2}\,dx$ certainly converges.

We omit the easy computation.

With $a = 4$, this time.

It can be shown by more advanced methods that the exact value of our integral is $\sqrt{\pi}/2 \approx 0.8862269255$.

Compared to what?   Integrals for reference

The *idea* of comparison testing is simple enough. The hardest part, in practice, is to decide *which* known integral to compare to a given unknown integral.

Most of the improper integrals found in elementary calculus courses can be successfully compared to one of these "benchmark" integrals (or a constant multiple thereof):➤➤

Know them.

$$\int_0^\infty e^{-x}\,dx \;=\; 1$$
$$\int_1^\infty \frac{dx}{x^p} \;=\; \frac{1}{p-1} \quad \text{if } p > 1;$$
$$\int_1^\infty \frac{dx}{x^p} \;=\; \infty \quad \text{if } p \le 1.$$

(The last equation means that the integral *diverges to infinity.*)

■ **Example 4.** Does $I = \displaystyle\int_1^\infty \frac{dx}{x+1}$ converge or diverge? Why?

Solution: The obvious comparison is to

$$\int_1^\infty \frac{dx}{x},$$

which we know to *diverge*; we'd expect I, therefore, to *diverge* as well.

Unfortunately, the inequality

$$\frac{1}{x+1} \le \frac{1}{x}$$

runs the wrong way for our purposes. However, the inequality

$$\frac{1}{x+1} \geq \frac{1}{2x}$$

Convince yourself, with graphs if necessary.

is valid for $x \geq 1$.◂ By Theorem 1, therefore,

$$\int_1^\infty \frac{dx}{x+1} \geq \int_1^\infty \frac{dx}{2x} = \frac{1}{2} \int_1^\infty \frac{dx}{x} = \infty,$$

We could also have worked directly, antidifferentiating and taking a limit.

so I diverges, as expected.◂   □

■ **Example 5.** Discuss the convergence of $I = \int_3^\infty \frac{2x^2}{x^4 + 2x + \cos x} \, dx$.

Large values of x determine whether I converges or diverges.

Solution: For large values of x◂ the denominator behaves like x^4.◂ This observation suggests that we compare I with the known-to-converge integral

The other terms are relatively insignificant.

$$\int_3^\infty \frac{2}{x^2} \, dx.$$

The comparison *is* valid. For $x \geq 3$,

$$x^4 + 2x + \cos x > x^4 \implies \frac{2x^2}{x^4 + 2x + \cos x} < \frac{2}{x^2}.$$

By an easy calculation.

Now Theorem 1 guarantees that I converges, and◂ that

$$I < \int_3^\infty \frac{2}{x^2} \, dx = \frac{2}{3}.$$ □

Integrands that change sign; absolute convergence

Does the integral

$$I = \int_1^\infty \frac{\sin x}{x^2} \, dx$$

converge or diverge? Theorem 1 doesn't say—it applies only to *nonnegative* integrands.

Theorem 1 *does*, nevertheless, say something useful. Since

$$\left| \frac{\sin x}{x^2} \right| \leq \frac{1}{x^2}$$

for all $x > 1$, it follows that

$$\int_1^\infty \left| \frac{\sin x}{x^2} \right| \, dx \leq \int_1^\infty \frac{dx}{x^2} = 1.$$

We'd expect, therefore, that I itself converges, and that

$$|I| = \left| \int_1^\infty \frac{\sin x}{x^2} \, dx \right| \leq \int_1^\infty \left| \frac{\sin x}{x^2} \right| \, dx \leq 1.$$

The next theorem justifies all these ruminations.

> **Theorem 2 (Absolute comparison).** Let f and g be continuous functions such that for all $x \geq a$,
>
> $$0 \leq |f(x)| \leq g(x).$$
>
> Suppose further that $\int_a^\infty g(x)\,dx$ converges. Then $\int_a^\infty f(x)\,dx$ also converges and
>
> $$\left| \int_a^\infty f(x)\,dx \right| \leq \int_a^\infty g(x)\,dx.$$

Absolute convergence. Theorem 2 says, among other things, that if $\int_a^\infty |f(x)|\,dx$ converges, then so must $\int_a^\infty f(x)\,dx$. The first condition is called **absolute convergence.**➤

We'll see the same phrase later, for infinite series.

Tips on tails

In any improper integral, say of the form $I = \int_1^\infty f(x)\,dx$, some of the most interesting action happens in the tail. Any such integral can be broken apart, perhaps as follows:

$$\int_1^\infty f(x)\,dx = \int_1^{1000} f(x)\,dx + \int_{1000}^\infty f(x)\,dx.$$

The first term is a *proper* integral; it can be handled either by antidifferentiation or by numerical methods. Whether I converges or diverges depends only on the second term—if the tail converges, then so does I. If, better yet, the tail is *small*,➤ then the first term closely approximates I.

As in the examples above.

Comparison, in the sense of the two theorems above, is the key to keeping tails small. We illustrate with one last example.

■ **Example 6.** Both $I = \int_1^\infty e^{-x}\,dx$ and $J = \int_1^\infty \dfrac{\sin x}{x^2}\,dx$ converge. For each, find a tail with absolute value less than 0.001.

Solution: For each integral, we need an a such that $\int_a^\infty f(x)\,dx < 0.001$.

For I such an a is relatively easy to find. Notice first that for any real number a,

$$\int_a^\infty e^{-x}\,dx = \lim_{t \to \infty} -e^{-x}\Big]_a^t = e^{-a} = \frac{1}{e^a}.$$

Therefore

$$\int_a^\infty e^{-x}\,dx = \frac{1}{e^a} < 0.001 \iff e^a > 1000 \iff a > \ln 1000 \approx 6.9.$$

For J we need Theorem 2. Since

$$\left| \frac{\sin x}{x^2} \right| \leq \frac{1}{x^2},$$

Theorem 2 guarantees that

$$\left| \int_a^\infty \frac{\sin x}{x^2}\,dx \right| \leq \int_a^\infty \frac{1}{x^2}\,dx.$$

An easy calculation shows, finally, that

$$\int_a^\infty \frac{1}{x^2}\,dx = \frac{1}{a} < 0.001 \iff a > 1000. \qquad \square$$

Exercises

1. (a) Explain why the inequalities $x - 1 \le x + \sin x \le x + 1$ are valid for all $x \in \mathbb{R}$.

    (b) Use the comparison test and one of the inequalities in part (a) to show that the improper integral $\int_2^\infty \dfrac{dx}{x + \sin x}$ diverges.

2. (a) Explain why the inequalities $x^2 \le x^2 + \sqrt{x} \le 2x^2$ are valid for all $x \ge 1$.

    (b) Use the comparison test and one of the inequalities in part (a) to show that the improper integral $\int_1^\infty \dfrac{dx}{x^2 + \sqrt{x}}$ converges. (Do not evaluate the integral.)

    (c) Explain why the inequalities $\sqrt{x} \le x^2 + \sqrt{x} \le 2\sqrt{x}$ are valid when $0 \le x \le 1$.

    (d) Use the comparison test and one of the inequalities in part (c) to determine whether the improper integral $\int_0^1 \dfrac{dx}{x^2 + \sqrt{x}}$ converges. (Do not evaluate the integral.)

    (e) Does the improper integral $\int_0^\infty \dfrac{dx}{x^2 + \sqrt{x}}$ converge? Justify your answer.

3. (a) Explain why the inequalities $\frac{1}{2}x^2 \le x^2 - \sqrt{x} \le x^2$ are valid for all $x \ge 2$.

    (b) Use the comparison test and one of the inequalities in part (a) to show that the improper integral $\int_3^\infty \dfrac{dx}{x^2 - \sqrt{x}}$ converges. (Do not evaluate the integral.)

4. (a) Show that the inequalities $\dfrac{1}{2} \le \dfrac{e^x}{1 + e^x} \le 1$ are valid for all $x \ge 0$.

    (b) Use one of the inequalities from part (a) to determine whether the improper integral $\int_0^\infty \dfrac{e^x}{1 + e^x}\, dx$ converges.

5. (a) Explain why the inequalities $\dfrac{1}{\sqrt{2x}} \le \dfrac{x}{\sqrt{1 + x^3}} \le \dfrac{1}{\sqrt{x}}$ are valid for all $x \ge 1$.

    (b) Use the comparison test and one of the inequalities in part (a) to determine whether the improper integral $\int_0^\infty \dfrac{x}{\sqrt{1 + x^3}}\, dx$ converges. (Do not evaluate the integral.)

6. (a) Explain why the inequalities $\sqrt{x} \le \sqrt{x} + 1 \le \sqrt{x^2 + 1}$ are valid for all $x \ge 3$.

    (b) Use the comparison test and one of the inequalities in part (a) to determine whether the improper integral $\int_0^\infty \dfrac{dx}{1 + \sqrt{x}}\, dx$ converges. (Do not evaluate the integral.)

7. (a) Explain why the inequalities $x^4 \le \sqrt{1 + x^9} \le \sqrt{2}\, x^{9/2}$ are valid for all $x \ge 1$.

(b) Use the comparison test and one of the inequalities in part (a) to determine whether the improper integral $\int_0^\infty \dfrac{dx}{\sqrt{1+x^9}}\,dx$ converges. (Do not evaluate the integral.)

8. Consider the integral $I = \int_e^\infty \dfrac{dx}{(\ln x)^2}\,dx$.

   (a) Explain why I is an improper integral.

   (b) Show that $1 \le (\ln x)^2 \le x^2$ for all $x \ge 1$.   [HINT: Use the racetrack principle to show that $1 \le \ln x \le x$ for all $x \ge 1$.]

   (c) Explain why the inequalities in part (b) aren't helpful for determining whether I converges.

   (d) Show that $1 \le \ln x \le \sqrt{x}$ for all $x \ge 1$.

   (e) Show that I diverges.

9. (a) Show that $0 \le \frac{1}{2}x \le \sin x$ when $0 \le x \le 1$.   [HINT: Apply the racetrack principle to the inequality $\frac{1}{2} \le \cos x$ when $0 \le x \le 1$.]

   (b) Use the inequalities in part (a) to show that the improper integral $\int_0^1 \dfrac{dx}{\sqrt{\sin x}}$ converges.

For each of the following (convergent) improper integrals, find a definite integral whose value approximates that of the given improper integral within 10^{-5}. (Do not evaluate these definite integrals.)

10. $\int_0^\infty \dfrac{dx}{x^2 + e^x}$

11. $\int_1^\infty \dfrac{dx}{x^4\sqrt{2x^3+1}}$

12. $\int_0^\infty \dfrac{\arctan x}{(1+x^2)^3}\,dx$

13. $\int_0^\infty \dfrac{e^{-x}}{2+\cos x}\,dx$

Estimate the value of each of the following (convergent) improper integrals to within 0.005 of its exact value. Explain why your estimates have the desired error bound.

14. $\int_1^\infty \dfrac{dx}{x^4 + \sqrt{x}}$

15. $\int_0^\infty e^{-x^2}\sin x\,dx$

16. $\int_0^\infty \dfrac{dx}{e^{x^2} + x}$

17. $\int_1^\infty \dfrac{\arctan x}{(1+x^2)^4}\,dx$

Use the comparison test to determine whether each of the following improper integrals converges or diverges. (Do not evaluate the convergent integrals.)

18. $\int_0^\infty \dfrac{dx}{x^4 + 1}\,dx$

19. $\int_0^\infty \dfrac{dx}{x^4 + x}\,dx$

20. $\int_2^\infty \dfrac{dx}{\sqrt{x} - 1}\,dx$

21. $\int_0^\infty e^{\sin x}\,dx$

22. $\displaystyle\int_1^\infty \frac{dx}{x\sqrt{1+x}}$

23. $\displaystyle\int_0^\infty \frac{dx}{x+e^x}$

24. $\displaystyle\int_0^\infty \frac{dx}{x+e^{-x}}$

25. $\displaystyle\int_1^\infty \frac{dx}{\sqrt[3]{x^6+x}}$

26. $\displaystyle\int_0^\infty \frac{\sqrt{x}}{x+1}\,dx$

27. $\displaystyle\int_2^\infty \frac{\sin x}{x^2\sqrt{x-1}}\,dx$

28. $\displaystyle\int_0^\infty \frac{dx}{1+\sqrt{x}}$

29. $\displaystyle\int_3^\infty \frac{x}{\ln x}\,dx$

30. $\displaystyle\int_0^\infty \frac{dx}{\sqrt{x}(1+x)}$

31. $\displaystyle\int_0^\infty \frac{dx}{\sqrt{x+x^3}}$

32. (a) Show that $\displaystyle\int_1^\infty \frac{\cos x}{x^2}\,dx$ converges.

    (b) Use integration by parts to show that $\displaystyle\int_1^\infty \frac{\sin x}{x}\,dx$ converges.

    (c) Use parts (a) and (b) to show that $\displaystyle\int_0^\infty \sin\left(e^x\right) dx$ converges.
    [HINT: Use the substitution $u = e^x$.]

33. Let $I = \displaystyle\int_0^\infty f(x)\,dx$ where $f(x) = e^{-x^3}\cos^2 x$.

    (a) Explain why I is an improper integral.

    (b) Show that I converges.   [HINT: For every $x > 1$, $x^2 e^{-x^3} > e^{-x^3}$.]

    (c) Compute an estimate of I that is guaranteed to be correct within 0.005 and explain why your estimate has this accuracy.   [HINT: For every $x \geq 0$, $-3 < f''(x) < 3$ and $-57 < f^{(4)}(x) < 46$.]

10.3 Improper integrals and probability

Random behavior

Many real-world phenomena exhibit forms of random behavior: the number of heads in ten throws of a coin; the height of a male college student chosen at random; the actual weight of a nominally 16 oz. cereal package, etc. In these settings, one may want to know➤ such things as the probability that a male student is between 68 and 75 inches tall; the likelihood of getting less than 15 oz. of Cheerios in your package, etc.

Or estimate.

Statisticians call such quantities **random variables**. A random variable that can take only *finitely* many values (e.g., the number of heads in ten coin tosses) is called **discrete**. A random variable that ranges over an *interval* of real numbers is called **continuous**. Adult height is a good example: Any number in some interval—(36, 85), say—might conceivably be someone's height in inches. Here we'll consider only continuous random variables.➤

Even though calculus methods can also be applied in the discrete case.

Basics of continuous probability; the role of calculus

Probability and statistics, although fields of study in their own right, draw heavily on calculus ideas—in particular on the idea of improper integrals. We'll just glimpse one such application here.

By convention, probabilities are measured on a scale from 0 to 1.➤ An event that is *certain* to occur, for instance, has probability 1.

Or, sometimes, using per cent notation.

A continuous random variable X, such as height of college students, has *infinitely many* possible values. It's *not* interesting, therefore, to ask for the probability that X is *precisely* equal to some number.➤ A better question concerns the probability that X lies in some *interval*, such as (68, 75). If this probability is, say, 0.78, then 78% of male students are between 68 and 75 inches tall.

That probability is zero.

Associated to every continuous random variable X is a function f called the **probability density function**, or **pdf**, of X. The connection between f and X is given by an integral:➤

This is all the probability theory we'll need!

$$\int_a^b f(x)\,dx = \text{probability that } X \text{ lies in } (a, b).$$

The shaded region below represents this probability geometrically:

Probability that $a < X < b$

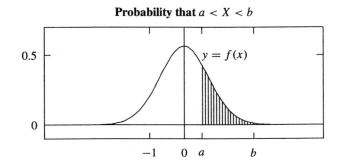

Here are some common-sense facts about f and X:

Domain. The domain of f is the set of possible values of X. (Often, f has an *infinite* domain.)➤

Hence the need for improper integrals.

Near, not exact. For a given x, $f(x)$ measures the comparative likelihood that the value of X falls *near* x. The probability that X takes any *precise* value x is zero.

Certainty. Because X *must* take some real value, $\int_{-\infty}^{\infty} f(x)\,dx = 1$.

A positive function. For all inputs x, $f(x) \geq 0$.

The normal density function

Many real-life random phenomena can be modeled by a **normal** (aka **Gaussian normal**) density function, i.e., one of the form

$$f(x) = \frac{1}{\sqrt{2\pi}s} \exp\left(-\frac{(x-m)^2}{2s^2}\right).$$

(A random variable with this density function is said to be **normally distributed.**)
This function looks messy, but in practice it's not too bad. The letters s and m stand
for constant parameters called, respectively, the **standard deviation** and the **mean**
of the random variable. (Statisticians usually write σ and μ—we'll stick with the
English.) Their numerical values depend on the situation being modeled. The mean
m, as its name suggests, is a kind of *average value* of X. The standard deviation s
is a measure of how "spread out" X's values are.◄ The graphs below illustrate the
geometric effects of varying m and s; all bound the same total (unit) area:

Statisticians use more precise definitions of m and s. For us the general ideas suffice.

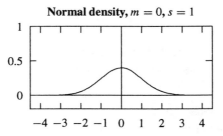

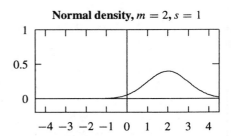

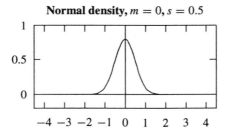

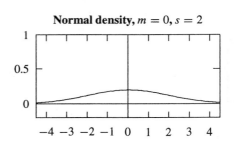

Tips on integrating the normal density—properly and improperly

Using the normal density function

$$f(x) = \frac{1}{\sqrt{2\pi}s} \exp\left(-\frac{(x-m)^2}{2s^2}\right)$$

means, in practice, *integrating* it over various intervals, either finite or infinite.
Because the normal density function has no elementary antiderivative, such integrals
must be estimated numerically. The following properties of f help simplify this
process:

Convergence. For any values of m and s, the improper integral $\int_{-\infty}^{\infty} f(x)\,dx$
converges, and its limit is 1. (Showing that the integral converges is not

too difficult; we showed something very similar in Example 3, page 192. Estimating a limit is easy, but showing rigorously that the limit is *precisely* 1 is considerably harder, and we omit the proof. For more on such matters, see the exercises.)

Symmetry. For any values of m and s, the graph of f is "centered" on the line $x = m$; it bounds total area 0.5 to either side of this line. Using this fact can simplify area computations and improve the accuracy of numerical estimates.

■ **Example 1.** If birth weights (in pounds) of infants born in Northfield, Minnesota, were normally distributed with mean 7.5 pounds and standard deviation 1 pound, what percentage of infants would be expected to weigh between 6.5 and 8.5 pounds? Over 10 pounds?

Solution: Both answers are areas bounded by the normal density function with parameter values $m = 7.5$, $s = 1$. Here they are, shown shaded:➤➤

Look carefully for the tiny shaded area on the far right.

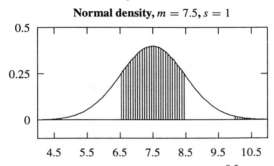

Normal density, $m = 7.5, s = 1$

Written as integrals, the answers are, respectively, $I_1 = \int_{6.5}^{8.5} f(x)\, dx$ and $I_2 = \int_{10}^{\infty} f(x)\, dx$. Applying the midpoint rule to I_1 gives $M_{20} \approx 0.6829.$ ➤➤ Therefore, given our assumptions, about 68% of babies should weigh within a pound of the mean.

We used Maple. We won't bother estimating the (small) error committed.

The picture shows that I_2 is small. To estimate it numerically, we'll use the fact➤➤ that the total area to the right of $x = 7.5$ is➤➤ 0.5. Therefore

cf. the "tips" above.

Exactly!

$$I_2 = \int_{10}^{\infty} f(x)\, dx = 0.5 - \int_{7.5}^{10} f(x)\, dx.$$

Applied to the last integral, the midpoint rule gives $M_{20} \approx 0.4938$, so $I_2 \approx 0.0062$. Therefore only around 0.6% of babies should weigh over 10 pounds. □

Is real life normal? Are real-life phenomena—baby weights, SAT scores, heights of male adults, etc.—*really* normally distributed?

In one sense, certainly not. One obvious problem is that every normal distribution is infinitely "spread out": the variable can take *any* real value—large or small, positive or negative. Few real-life phenomena behave this way. No baby has *negative* weight, for instance; yet every normal distribution assigns some positive (though tiny) probability to this peculiar event.

In another, more practical, sense, normal distributions *do* usefully model many phenomena. "Model" is the key word (here and elsewhere in applied mathematics). In suitable settings, the normal density function provides an effective, albeit imperfect, *approximation* to an underlying reality. Imperfection is no surprise; it comes with the territory of real life. *Effectiveness* is the surprise.

■ **Example 2.** Possible scores on the math and verbal parts of the Scholastic Aptitude Test (SAT) range from 200 to 800. Assuming that test scores are normally distributed, with mean 500 and standard deviation 100,[←] what percentage of students score between 400 and 600? Over 750?

Historically speaking, this assumption, though imprecise, is not far off.

Solution: The analysis—and geometry—is similar to that used in the previous example. Setting $m = 500$ and $s = 100$ gives the picture below, with the areas of interest shaded:

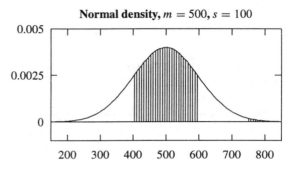

Normal density, $m = 500$, $s = 100$

For reasons we'll see below.

The shaded areas can be estimated numerically, just as above. We omit details, this time, because[←] the shaded areas here have *exactly* the same areas as those in the previous example. We conclude, therefore, that about 68% of SAT scores fall between 400 and 600, i.e., within one standard deviation of the mean; only 0.6% of scores are above 750.   □

Keeping it simple: the standard normal density

Calculations are simplest if $m = 0$ and $s = 1$. With these parameter values, the **standard normal density** function takes the particularly pleasant symbolic form[←]

A graph of this function appears several pages above.

$$n(x) = \frac{1}{\sqrt{2\pi}} \exp\left(\frac{-x^2}{2}\right).$$

Remarkably enough, *any* normal distribution can easily be "standardized"—i.e., reinterpreted as a *standard* normal distribution. In the SAT case, for instance, we could consider how much scores differ from 500, rather than "raw" test scores themselves. Doing so moves the mean to 0; the standard deviation remains at 100. Dividing *these* results by 100 leaves the mean unchanged at 0, but cuts the standard deviation to 1. The results, therefore, have a *standard* normal distribution.

Statisticians call such altered data **Z-scores**. Raw SAT scores of 620 and 450 correspond, for example, to Z-scores of 1.20 and -0.5; i.e., to scores 1.2 standard deviations *above* and 0.5 standard deviations *below* the mean, respectively. Similarly, a 10-pound baby weighs in at 2.5 standard deviations above the mean of 7.5 pounds.

To find the probability of a Z-score between -1 and 1, we integrate the *standard* normal density:

$$\int_{-1}^{1} \frac{1}{\sqrt{2\pi}} \exp\left(\frac{-x^2}{2}\right) dx.$$

The error is small; we'll ignore it.

No *symbolic* antiderivative is available, but numerical integration methods work just fine. The midpoint rule with 20 subdivisions, for instance, gives[←] $M_{20} \approx 0.6829$. This by-now familiar result offers evidence that Z-scores make mathematical sense.

Why Z-scores work: substitution in definite integrals

Also known as u-substitution.

What statisticians call Z-scores, *we* call Z-substitution[←] in a definite integral. We illustrate with an example.

■ **Example 3.** Write integrals, with and without Z-scores, for the probability of scoring between 500 and 700 on the SAT. Why do both integrals give the same result?

Solution: With *raw* scores we use $m = 500$ and $s = 100$; the desired probability is

$$I_1 = \frac{1}{\sqrt{2\pi} \cdot 100} \int_{500}^{700} \exp\left(-\frac{(x-500)^2}{2 \cdot 100^2}\right) dx.$$

With Z-scores we want the probability of a result between 0 and 2,**➤** i.e., the integral

$$I_2 = \frac{1}{\sqrt{2\pi}} \int_0^2 \exp\left(-\frac{x^2}{2}\right) dx.$$

Raw scores of 500 and 700 give Z-scores of 0 and 2.

Numerical evidence that $I_1 = I_2$ is easy to find: applied to *either* integral, $M_{20} \approx 0.47729$. A symbolic calculation confirms the result. The substitution

$$Z = \frac{x-500}{100}; \qquad dZ = \frac{dx}{100}$$

in I_1**➤** yields I_2.**➤** □

Don't forget endpoints:
$x = 500 \implies Z = 0;$
$x = 700 \implies Z = 2.$

Convince yourself; see the exercises.

Z-score tables

In pre-computer days it was difficult to estimate the definite integrals that arise in studying normal distributions. The only recourse was to numerical tables; part of such a table appears below. For a given input x, the table entry is $A(x) = \int_{-\infty}^x n(t)\, dt$, the area shown graphically below:

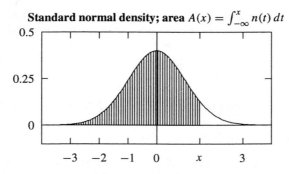

Standard normal density; area $A(x) = \int_{-\infty}^x n(t)\, dt$

Values of $A(x) = \int_{-\infty}^x n(t)\, dt$									
x	0.0	0.1	0.2	0.3	0.4	0.5	0.6	0.7	0.8	0.9
$A(x)$	0.5000	0.5398	0.5793	0.6179	0.6554	0.6915	0.7257	0.7580	0.7881	0.8159
x	1.0	1.1	1.2	1.3	1.4	1.5	1.6	1.7	1.8	1.9
$A(x)$	0.8413	0.8643	0.8849	0.9032	0.9192	0.9333	0.9452	0.9554	0.9641	0.9713
x	2.0	2.1	2.2	2.3	2.4	2.5	2.6	2.7	2.8	2.9
$A(x)$	0.9773	0.9822	0.9861	0.9893	0.9918	0.9938	0.9953	0.9965	0.9974	0.9981

The numbers compare nicely with the graph. As we'd expect, for instance, $A(0) = 0.5$; $A(2.9) \approx 1.00$.

Exercises

1. Sketch a graph of the normal probability density function with the given parameter values.

   (a) $m = 1, s = 1$ (c) $m = 0, s = 1/4$

   (b) $m = 2, s = 1/2$

2. Throughout this section we used the fact that if m is any real number, s is any positive number, and $f(x) = \dfrac{1}{\sqrt{2\pi}s} \exp\left(-\dfrac{(x-m)^2}{2s^2}\right)$, then $\displaystyle\int_{-\infty}^{\infty} f(x)\, dx = 1$. This problem explores this fact.

   (a) We showed in Example 3, page 192, that $\displaystyle\int_{0}^{\infty} e^{-x^2}\, dx$ converges, and we claimed that its limit is $\sqrt{\pi}/2$. Assuming these facts, explain geometrically why $\displaystyle\int_{-\infty}^{\infty} e^{-x^2}\, dx = \sqrt{\pi}$. (Hint: The integrand is even.)

   (b) If $m = 0$ and $s = 1$, then $f(x) = \dfrac{1}{\sqrt{2\pi}} \exp\left(-\dfrac{x^2}{2}\right)$. We'll show that $\displaystyle\int_{-\infty}^{\infty} f(x)\, dx = 1$. Do so by performing the change-of-variable $u = x/\sqrt{2}$ on the integral $\dfrac{1}{\sqrt{2\pi}} \displaystyle\int_{-\infty}^{\infty} \exp\left(-\dfrac{x^2}{2}\right)\, dx$. (Doing so leaves the limits of integration unchanged.)

   (c) The previous part shows that $\dfrac{1}{\sqrt{2\pi}s} \displaystyle\int_{-\infty}^{\infty} \exp\left(-\dfrac{(x-m)^2}{2s^2}\right)\, dx = 1$ in the special case that $m = 0$ and $s = 1$. Assuming this, let's show that the same thing holds for *any* real number m and positive number s. Do so by performing the change-of-variable $u = (x-m)/s$ in the previous integral. (As in the previous part, doing so leaves the limits of integration unchanged.)

3. Estimate the integrals I_1 and I_2 in Example 1 using the trapezoid rule with 20 subintervals and find bounds on the approximation errors. Are these results consistent with those reported in the text? Justify your answer.

4. (a) Let X be a normally distributed random variable with $m = 0$ and $s = 1$. What fraction of the possible values of X lie within three standard deviations of the mean?

   (b) Repeat part (a) assuming $m = 0$ and $s = 2$.

5. Let I_1 and I_2 be the integrals discussed in Example 3. Show that making the substitution $Z = (x - 500)/100$ transforms I_1 into I_2.

6. Using the table giving values of $A(x) = \int_{-\infty}^{x} n(t)\, dt$, compute the probability of getting an SAT score

   (a) between 350 and 500.

   (b) greater than 600.

7. How high must a student score on the SAT to have a score in the top 10%? In the top 5%? [HINT: Use the tabulated values of $A(x)$.]

8. The graph of the probability density function for a random variable X with values in the interval $[0, \infty)$ is shown below. Use this graph to estimate the probability that a random value of X lies in each of the following intervals.

(a) $[0, 0.5]$ (c) $[0, 1.5]$

(b) $[0.5, 1.5]$ (d) $[1.5, \infty)$

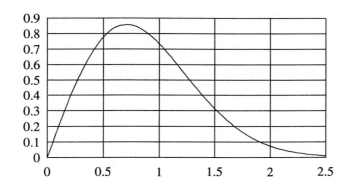

9. Use integration by parts with $u = 1/x$ and $dv = xe^{-x^2}$ to show that

$$\int_a^\infty e^{-x^2}\, dx = \frac{e^{-a^2}}{2a} - \frac{1}{2}\int_a^\infty \frac{e^{-x^2}}{x^2}\, dx.$$

10.4   l'Hôpital's rule: comparing rates

Indeterminate forms: posing the question

Values of simple limits, such as

$$\lim_{x\to\infty}\frac{1}{x^2}=0 \quad \text{and} \quad \lim_{x\to 0}\frac{1}{x^2}=\infty,$$

Mathematicians' jargon.

are easy to guess "by inspection,"** i.e., just by looking at the relative sizes of numerators and denominators.

When is a limit indeterminate?

Other limits are less susceptible to intuition. Consider these:

$$\lim_{x\to\infty}\frac{x^2}{2^x}; \quad \lim_{x\to 0}\frac{\sin(2x)}{x}; \quad \lim_{x\to\infty}xe^{-x}; \quad \lim_{x\to\infty}\frac{x^2+1}{2x^2+3}.$$

More mathematicians' jargon.

Limits like these are called **indeterminate forms**.** They are called "indeterminate" because, in every case, two *conflicting* tendencies operate. In the first limit, for instance, *both* numerator and denominator "blow up":

$$x^2\to\infty \quad \text{and} \quad 2^x\to\infty \quad \text{as} \quad x\to\infty.$$

The ratio, therefore, is **indeterminate of type** ∞/∞. Thus, the real question is *how fast*, compared to each other, the numerator and denominator grow.

The second limit is ambiguous for another reason: both numerator and denominator tend, simultaneously, to zero. Their ratio, therefore, is called **indeterminate of type** $0/0$: again, the question concerns the relative *rates* at which numerator and denominator tend to zero.

It isn't a quotient, this time.

The third limit involves the *product* $x\cdot e^{-x}$.** Because the two factors behave in "opposite" ways—

$$x\to\infty \quad \text{but} \quad e^{-x}\to 0,$$

the limit is called **indeterminate of type** $\infty\cdot 0$.

The last limit, like the first, is indeterminate of type ∞/∞. Much earlier in this course, we saw an algebraic method for detecting "asymptotic behavior" of *rational functions*, like the one here. Dividing top and bottom by x^2 yields a new and simpler** limit—and an answer:

The second limit is not indeterminate!

$$\lim_{x\to\infty}\frac{x^2+1}{2x^2+3}=\lim_{x\to\infty}\frac{1+1/x^2}{2+3/x^2}=\frac{1}{2}.$$

The result means, graphically, that the line $y=1/2$ is a horizontal asymptote for our rational function.

When algebra fails: graphical and numerical evidence

The algebraic limit calculation just above "worked"—but only because both numerator and denominator chanced, by good luck, to be *algebraic* functions.** With the other indeterminate forms above, algebra alone is helpless.

Polynomials, in this case.

Graphical and numerical approaches, although less precise, are more forgiving. The graphs below** strongly suggest values for all three limits as $x\to\infty$:

You decide which is which.

$$\frac{x^2}{2^x}\to 0; \qquad xe^{-x}\to 0; \qquad \frac{x^2+1}{2x^2+3}\to\frac{1}{2}.$$

Three functions' long-run behavior

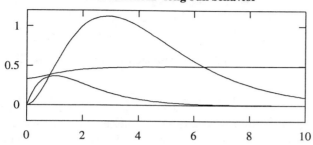

The following table➤ suggests just as convincingly that as $x \to 0$,

We could also have drawn a graph.

$$\frac{\sin(2x)}{x} \to 2.$$

As $x \to 0$, $\sin(2x)/x \to 2$: numerical evidence							
x	-0.1	-0.01	-0.0001	$\ldots$	0.0001	0.01	0.1
$\sin(2x)/x$	1.98670	1.99987	1.99999	$\ldots$	1.99999	1.99987	1.98670

l'Hôpital's rule: finding limits by differentiation

Graphics, numerics, and algebra helped us find➤ plausible values in all four cases above. l'Hôpital's rule offers another, more powerful,➤ approach to indeterminate forms. It handles all four of the indeterminate forms discussed above, and many more besides.

Approximately, anyway.

And slicker.

How it works: simplest examples

l'Hôpital's rule says that, *under appropriate conditions,*➤ an indeterminate form can be evaluated by *differentiating the numerator and the denominator separately.* In symbols:

We'll discuss them below.

(10.4.1) $$\lim_{x \to a} \frac{f(x)}{g(x)} = \lim_{x \to a} \frac{f'(x)}{g'(x)}.$$

Deferring legalities for the moment, we'll illustrate l'Hôpital's method by example.

■ **Example 1.** Use l'Hôpital's rule to show that➤ $\lim_{x \to 0} \dfrac{\sin(2x)}{x} = 2$.

From the numerical data above we expect the answer 2.

Solution: By Equation 10.4.1,

$$\lim_{x \to 0} \frac{\sin(2x)}{x} = \lim_{x \to 0} \frac{2\cos(2x)}{1}.$$

The right-hand limit is no longer indeterminate; simply plugging in $x = 0$ gives 2—the expected answer.   □

■ **Example 2.** Determine $\lim_{x \to \infty} \dfrac{x^2}{2^x}$.

Solution: Assuming for the moment that equation 10.4.1 applies, it says that

$$\lim_{x\to\infty}\frac{x^2}{2^x} = \lim_{x\to\infty}\frac{2x}{2^x\ln 2}.$$

The second limit is *still* indeterminate of type ∞/∞. We'll apply equation 10.4.1 *again*:

$$\lim_{x\to\infty}\frac{x^2}{2^x} = \lim_{x\to\infty}\frac{2x}{2^x\ln 2} = \lim_{x\to\infty}\frac{2}{2^x\ln 2\cdot\ln 2}.$$

The rightmost limit is, at last, not indeterminate. Because the numerator remains constant while the denominator grows without bound, the last limit must be zero. So, therefore, must the first.   □

■ **Example 3.**   A graph suggests[◄] that $\lim_{x\to\infty} xe^{-x} = 0$. Does equation 10.4.1 agree?

But doesn't prove—graphs aren't proofs.

Solution: Yes, but only after a little algebra. In order to use equation 10.4.1, we must first rewrite things in *quotient* form:[◄]

Watch carefully; the step is simple but important.

$$\lim_{x\to\infty} xe^{-x} = \lim_{x\to\infty}\frac{x}{e^x}.$$

Now equation 10.4.1 makes good sense:

$$\lim_{x\to\infty}\frac{x}{e^x} = \lim_{x\to\infty}\frac{1}{e^x} = 0,$$

exactly as expected.   □

When it works, when it doesn't: careful statements

Using l'Hôpital's rule requires a little care; thoughtless applications lead quickly astray. The tempting idea that *every* limit of quotients can be handled *a la* l'Hôpital[◄] is just plain wrong. For instance,

I.e., by differentiating top and bottom.

$$\lim_{x\to 0}\frac{x+42}{x+1} = 42, \qquad\text{but}\qquad \lim_{x\to 0}\frac{(x+42)'}{(x+1)'} = \lim_{x\to 0}\frac{1}{1} = 1.$$

We erred in trying to apply the rule where it doesn't apply—to a limit that wasn't indeterminate to start with.

As a rule, it works only for indeterminate forms.

When, exactly, *does* the rule work?[◄]

> **Theorem 3 (l'Hôpital's rule).** Let f and g be differentiable functions. Suppose further:
>
> (a) As $x\to a$, either (i) $f(x)\to 0$ and $g(x)\to 0$ or
> (ii) $f(x)\to\pm\infty$ and $g(x)\to\pm\infty$;
>
> (b) $\displaystyle\lim_{x\to a}\frac{f'(x)}{g'(x)}$ exists.
>
> Then $\displaystyle\lim_{x\to a}\frac{f(x)}{g(x)} = \lim_{x\to a}\frac{f'(x)}{g'(x)}$.

Note:

Luckily for us—we've already used the result twice with $a=\infty$.

Limits at infinity. The values $a=\pm\infty$ are allowed in the theorem.[◄] *One-sided limits* (e.g., $x\to 0^+$) are also permitted.[◄]

See the next example.

Truly indeterminate. Hypothesis (a) guarantees that $\lim f/g$ is genuinely indeterminate, of type either $0/0$ or ∞/∞.[◄]

Without this proviso, the theorem would fail.

■ **Example 4.** Find $\lim_{x \to 0^+} x \ln x$.

Solution: The theorem applies only to quotients. Let's produce one:

$$\lim_{x \to 0^+} x \ln x = \lim_{x \to 0^+} \frac{\ln x}{1/x}.$$

The right-hand form is indeterminate of type ∞/∞; moreover,

$$\lim_{x \to 0^+} \frac{(\ln x)'}{(1/x)'} = \lim_{x \to 0^+} \frac{1/x}{-1/x^2} = \lim_{x \to 0^+} -x = 0.$$

Hence the theorem applies; the original limit is 0. □

Why it works: comparing rates

The theorem concerns the behavior, as x tends to a, of the ratio $f(x)/g(x)$. If *both* numerator and denominator tend either to zero or to infinity, then the limit—if it exists—depends on the relative *rates* at which f and g tend to their limits. The derivatives $f'(x)$ and $g'(x)$ measure these rates—hence their appearance in l'Hôpital's rule. A formal proof of l'Hôpital's rule➻ makes this general idea rigorous. Among its ingredients is an appeal to the Mean Value Theorem of differential calculus.

We won't give one.

Fine print: pointers toward a proof

To see the general idea of a proof, let's suppose that f and g are differentiable functions, with $f(a) = g(a) = 0$. Then if $x \approx a$, f and g are close to their respective tangent lines at $x = a$. In other words,➻

Convince yourself that the tangent lines have the claimed formulas.

$$\begin{aligned} f(x) &\approx f(a) + f'(a)(x-a) = f'(a)(x-a); \\ g(x) &\approx g(a) + g'(a)(x-a) = g'(a)(x-a). \end{aligned}$$

If $g'(a) \neq 0$, then

$$\frac{f(x)}{g(x)} \approx \frac{f'(a)}{g'(a)}, \quad \text{so} \quad \lim_{x \to a} \frac{f(x)}{g(x)} = \lim_{x \to a} \frac{f'(x)}{g'(x)} = \frac{f'(a)}{g'(a)}.$$

This is what l'Hôpital's rule says, in its simplest form. A fully general proof requires more care, but is based on the same idea.

■ **Example 5.** By l'Hôpital's rule, $\lim_{x \to 0} \frac{\sin(2x)}{\sin x} = \lim_{x \to 0} \frac{2\cos(2x)}{\cos x} = 2$. Interpret the result graphically, using tangent lines.

Solution: Let $f(x) = \sin(2x)$ and $g(x) = \sin x$. Graphs of f, g, and their tangent lines at $x = 0$➻ appear below:

$y = 2x$ and $y = x$, respectively.

Near $x = 0$, $\sin(2x) \approx 2x$, $\sin x \approx x$

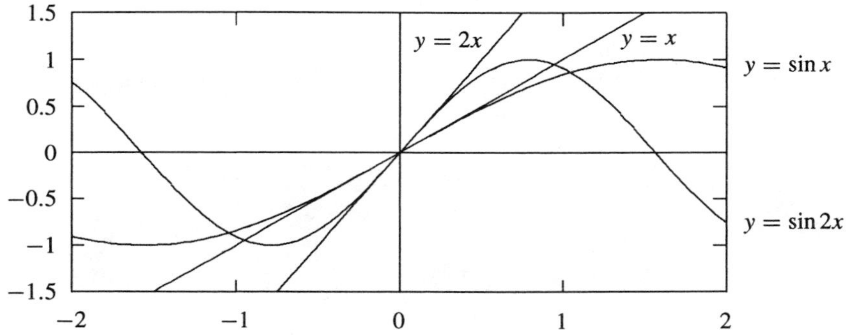

The picture shows that for x near 0, $f(x) \approx 2x$ and $g(x) \approx x$; hence

$$\frac{f(x)}{g(x)} \approx \frac{2x}{x} = 2,$$

as claimed. $\square$

Exercises

1. (a) Using a calculator, fill in the table below.

x	1	10	100	1000
x^2				
2^x				
$(3x + 10)^2$				

(b) Use the entries in the table to guess the limits $\lim\limits_{x\to\infty} x^2/2^x$, $\lim\limits_{x\to\infty} x^2/(3x + 10)^2$, and $\lim\limits_{x\to\infty} 2^x/(3x + 10)^2$.

(c) Use l'Hôpital's rule to verify your guesses in part (b).

Use l'Hôpital's rule to evaluate each of the following limits.

2. $\lim\limits_{x\to\infty} \dfrac{x^2 + 1}{2x^2 + 3}$

3. $\lim\limits_{x\to 0} \dfrac{5x - \sin x}{x}$

4. $\lim\limits_{x\to 0} \dfrac{1 - \cos(5x)}{4x + 3x^2}$

5. $\lim\limits_{x\to\infty} \dfrac{e^x}{x^2 + x}$

6. $\lim\limits_{x\to\infty} e^{-x} \ln x$

7. $\lim\limits_{x\to 0} x \cot x$

8. $\lim\limits_{x\to 8} \dfrac{x - 8}{\sqrt[3]{x} - 2}$

9. $\lim\limits_{x\to 0^+} \dfrac{\sin x}{x + \sqrt{x}}$

10. $\lim\limits_{x\to 0} \dfrac{\sin x}{x - \sin x}$

11. $\lim\limits_{x\to 0} \dfrac{\tan(3x)}{\ln(1 + x)}$

12. $\lim\limits_{x\to 0} \left(e^x + x\right)^{1/x}$

13. $\lim\limits_{x\to\infty} \dfrac{\int_1^x \sqrt{1 + e^{-3t}} \, dt}{x}$

14. Use l'Hôpital's rule to evaluate the (convergent) improper integral $\int_0^\infty x e^{-x} \, dx$.

15. (a) Does the improper integral $\int_0^1 \dfrac{\cos x}{x} \, dx$ converge or diverge? Justify your answer.

(b) Explain why the integral $\int_0^1 \dfrac{\sin x}{x} \, dx$ is *not* improper.
[HINT: Use L'Hôpital's rule.]

(c) Show that $\int_0^\infty \sin\left(e^{-x}\right) dx = \int_0^1 \frac{\sin x}{x} dx$.
[HINT: Use the substitution $u = e^{-x}$.]

16. (a) Guess the value of $\lim_{x \to 0^+} x \ln x$ by looking at a graph.

(b) Evaluate $\lim_{x \to 0^+} x \ln x$ using l'Hôpital's rule. Is the result the same as your answer to part (a)? Why or why not?

(c) Is the integral $\int_0^1 x \ln x \, dx$ improper? Justify your answer.

17. Suppose that f is a differentiable function, f' is continuous on $[1, \infty)$, and $|f(x)| \le e^{-x} \ln x$ when $x \ge 1$. Show that $\int_1^\infty f'(x) \, dx$ converges.

18. Let $\Gamma(x) = \int_0^\infty t^{x-1} e^{-t} \, dt$.

(a) Evaluate $\Gamma(1)$.

(b) Use integration by parts to show that if $x > 1$, $\Gamma(x) = (x - 1)\Gamma(x - 1)$.
[HINT: For every $z \in \mathbb{R}$, $\lim_{t \to \infty} t^z e^{-t} = 0$.]

(c) Show that $0 < \Gamma\left(\frac{2}{3}\right) < 2$.   [HINT: Be careful; the integral has *two* improprieties.]

(d) It is known that $\int_0^\infty e^{-x^2} \, dx = \frac{1}{2}\sqrt{\pi}$. Use this result to evaluate $\Gamma\left(\frac{1}{2}\right)$.
[HINT: Let $u = \sqrt{t}$.]

19. It is a fact that $\int_0^\infty x^n e^{-x} \, dx = n!$ when $n \ge 0$ is an integer. Use this fact to show that

$$\int_0^1 x^m (\ln x)^n \, dx = \frac{(-1)^n \, n!}{(m + 1)^{n+1}}$$

when m and n are positive integers.   [HINT: Make the substitution $x = e^{-u}$ in the second integral.]

10.5   Chapter summary

Both of this chapter's subjects main subjects—improper integrals and l'Hôpital's rule—concern the "long-term" behavior of functions. Both topics lay the groundwork for the later study of infinite series, in which even more delicate questions of long-term behavior arise.

Improper integrals.   Both the integrals

$$\int_1^\infty \frac{1}{x^2}\, dx \quad \text{and} \quad \int_0^1 \frac{1}{x^2}\, dx$$

are improper, for different reasons.   In the first case the interval of integration is infinite; in the second, the integrand itself "blows up" at the left endpoint of integration. Despite such problems, some improper integrals "converge" in a natural sense to finite values.   Whether a given integral converges or diverges depends in a subtle way on how the function behaves on its domain of integration; which alternative occurs for a given integral can be hard to tell at a glance.   The first integral above, for instance, converges (to the value 1), but the second one diverges (to ∞).

Improper integrals, like proper ones, are sometimes best handled numerically. The second section of the chapter treats some of the special problems improper integral raise in this regard.

Probability and improper integrals.   Improper integrals have important applications in probability and statistics. Working with the famous "bell-shaped" or **normal distribution**, for example, requires evaluating—numerically, in most cases—improper integrals like this one:

$$\int_{-1}^1 \frac{1}{\sqrt{2\pi}} \exp\left(\frac{-x^2}{2}\right)\, dx.$$

Doing so enables us to model a large and important variety of random phenomena.

l'Hôpital's rule.   l'Hôpital's rule is a technique for evaluating certain "indeterminate" limits, often of the form

$$\lim_{x\to\infty} \frac{f(x)}{g(x)},$$

where both $f(x)$ and $g(x)$ tend either to 0 or to ∞.   Specialized as such problems may at first seem, they arise often in the context of infinite sequences and infinite series; our study of them here prepares the ground for that topic.

Chapter 11

Infinite Series

11.1   Sequences and their limits

This section, on infinite *sequences*, prepares the ground for the next topic—infinite *series*. Convergent series are defined in terms of the simpler, more basic, idea of convergent *sequences*. This section is a quick introduction to sequences—what they are, what it means for them to converge or diverge, and how to find their limits.

We'll study them in some detail.

The idea and language of sequences; simple examples

A **sequence** is an infinite list of numbers, of the general form

$$a_1, \, a_2, \, a_3, \, a_4, \, \ldots, \, a_k, \, a_{k+1}, \, \ldots.$$

Individual entries are called the **terms** of the sequence; a_3 and a_k, for instance, are called the **third term** and the k-**th term**, respectively. The full sequence is, technically speaking, an **ordered set**; the standard notation

Read "a sub three" and "a sub k."

$$\{a_k\}_{k=1}^{\infty}$$

(or simply $\{a_k\}$) uses set brackets to emphasize this view.

   Our main interest in sequences is in their **limits**. For the simplest sequences, limits are evident at a glance. The first examples below are of this type; we include them mainly to illustrate ideas, notations, and terminology.

■ **Example 1.**   Discuss the sequence $\{a_k\}_{k=1}^{\infty}$ defined by the formula $a_k = \dfrac{1}{k}, \, k = 1, 2, 3, \ldots.$ Does this sequence have a limit?

Solution: Sampling the first few terms—

$$\frac{1}{1}, \, \frac{1}{2}, \, \frac{1}{3}, \, \ldots, \, \frac{1}{10}, \, \frac{1}{11}, \, \ldots, \, \frac{1}{100}, \, \frac{1}{101}, \, \ldots$$

shows that the sequence **converges to zero**: As k increases, the terms approach zero arbitrarily closely. In symbols:

To nobody's surprise.

$$\lim_{k \to \infty} a_k = \lim_{k \to \infty} \frac{1}{k} = 0.$$

□

This time we used j rather than k; the name of the "index variable" doesn't matter.

■ **Example 2.** Suppose that $\{b_j\}_{j=1}^{\infty}$ is defined◀ by $b_j = \dfrac{(-1)^j}{j}$, $\quad j = 1, 2, 3, \ldots$. What's $\displaystyle\lim_{j \to \infty} b_j$?

Solution: Writing out terms shows a similar pattern:

$$-\frac{1}{1}, \frac{1}{2}, -\frac{1}{3}, \ldots, \frac{1}{10}, -\frac{1}{11}, \ldots, \frac{1}{100}, -\frac{1}{101}, \ldots.$$

Although the terms oscillate in *sign*, they approach zero more and more closely. Sooner or later, all the terms—positive or negative—remain within any specified distance from zero.◀ Hence the sequence $\{b_j\}$ also **converges to zero**:

All terms past b_{1000}, for instance, are within 0.001 of zero.

$$\lim_{j \to \infty} b_j = \lim_{j \to \infty} \frac{(-1)^j}{j} = 0.$$

□

■ **Example 3.** Does the sequence $\{c_k\}_{k=0}^{\infty}$ with **general term** $c_k = (-1)^k$ converge?

Solution: No; it **diverges**. Successive terms have this pattern:

$$1, -1, 1, -1, 1, \ldots,$$

never settling on a single limit.

□

Fine points: notes on sequences

Sequences have their own notational quirks and conventions. Here are several to watch for:

Where to start? The sequence in the previous example began with c_0, not c_1. Other starting points, such as a_2 or even b_{-3}, occasionally arise. In practice, fortunately, the difference is usually unimportant. What usually matters for sequences is their *long-run* behavior rather than the presence or absence of a few initial terms.

Index names don't matter. We can define the squaring function by writing either $f(t) = t^2$ or $f(x) = x^2$—the *variable name* makes no difference. In just the same way, a sequence's *index name* is arbitrary: $\{a_k\}_{k=1}^{\infty}$ and $\{a_j\}_{j=1}^{\infty}$ mean exactly the same thing.

Re-indexing. The sequence

$$\frac{1}{1}, \frac{1}{2}, \frac{1}{3}, \ldots, \frac{1}{10}, \frac{1}{11}, \cdots$$

of Example harmseqex can be described symbolically in various different-looking—but still equivalent—ways. Here are two:

$$a_k = \frac{1}{k}, \quad k = 1, 2, \ldots \quad \text{or} \quad a_j = \frac{1}{j+1}, \quad j = 0, 1, 2, \ldots.$$

Depending on the situation, one or the other description may be preferable.

Sequences as functions

Sequences are closely related to functions, as expressions like these—

$$a_k = \frac{\sin k}{k} \quad \text{and} \quad f(x) = \frac{\sin x}{x}$$

attest. The formal definition of sequence makes this connection precise:

> **Definition:** An **infinite sequence** is a real-valued function that is defined for *positive integer inputs.*

Alternative views of sequences. With the definition and the examples above, we have several useful ways to think of a sequence:

As a list. As an *infinite list* of numbers: a_1, a_2, a_3, ...

As a function. As a *function $a(n)$*, where n takes only positive integer values.➤

> Thus $a(1) = a_1, a(2) = a_2, a(3) = a_3$, etc.

The function a might make sense for other inputs, but it doesn't have to.

As a discrete sample. As a *discrete sample* of values of an ordinary function $f(x)$, defined for *real $x \geq 1$.*➤

Most examples above are of this type.

Graphs of sequences, graphs of functions

The graph of a *function f* consists of all points $(x, f(x))$, where x runs through the domain of f. The graph of a *sequence $\{a_k\}$*, therefore, is the set of points (k, a_k), as k runs through *positive integer values.*

■ **Example 4.** Let a function f and a sequence $\{a_k\}$ be defined by

$$f(x) = \frac{\sin x}{x}; \quad a_k = \frac{\sin k}{k}.$$

Plot graphs of both. What do the graphs say about limits?

Solution: Here are the graphs:

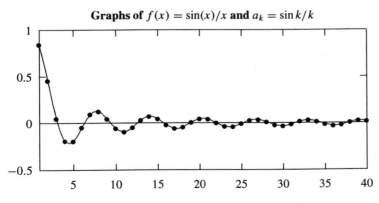

Graphs of $f(x) = \sin(x)/x$ and $a_k = \sin k/k$

They illustrate how the sequence $\{a_k\}$ is a "discrete sample" of the continuous function f. They show, too, that as x (or k) tends to infinity, this sequence and function tend, although oscillating in sign, to zero. □

Not every sequence comes naturally from a familiar function. Still, graphs or tables often suggest limits.

■ **Example 5.** A sequence $\{b_j\}$ has general term

$$b_j = 1 \cdot 2 \cdot 3 \cdot 4 \cdot \ldots \cdot (j-1) \cdot j = j!$$

(In words, b_j is j **factorial**.) Tabulate $\{b_j\}$; find its limit, if any.

Solution: As the table suggests, $\{b_j\}$ diverges—quickly—to infinity:

As $k \to \infty$, $k! \to \infty$: **explosive numerical evidence**						
k	1	2	4	8	16	32	64
$k!$	1	2	24	40320	2.093×10^{13}	2.631×10^{35}	1.269×10^{89}

Graphs don't work well for this sequence. Do you see why not?

Can any doubt remain?◂   □

Limits of sequences, limits of functions

Sequences are special sorts of functions; *limits* of sequences, therefore, are mild variants on *limits at infinity*.◂ An informal definition is will suffice:◂

We used such limits in Chapter 2 to find horizontal asymptotes.

A formal definition describes more precisely what "approaches" means.

> **Definition:** (**Limit of a sequence**). Let $\{a_k\}$ be a sequence and L a real number. If a_k approaches L to within any desired tolerance as k increases without bound, then the sequence **converges to** L. In symbols,
> $$\lim_{k\to\infty} a_k = L.$$
> Otherwise, the sequence **diverges**.

Notice:

Divergence to infinity. If either $a_k \to \infty$ or $a_k \to -\infty$ as k increases without bound, then the sequence **diverges to (positive or negative) infinity**. We write, for instance,
$$\lim_{k\to\infty} k! = \infty.$$

Or no pattern at all!

Other divergence behavior. Example 3 shows that a divergent sequence need not "blow up"; other patterns◂ of wandering behavior are possible.

Asymptotes. A sequence, like a function, converges to a finite limit L if and only if its graph has a **horizontal asymptote** at $y = L$.

Sequences, functions, and limits. Many sequences, as we said above, are "discrete samples" of familiar functions. For such sequences we can often use our knowledge of the underlying function, as the following useful fact attests:

> **Fact:** If f is a function defined for $x \geq 1$, $\lim_{x\to\infty} f(x) = L$, and $a_k = f(k)$ for all $k \geq 1$, then then $\lim_{k\to\infty} a_k = L$.

■ **Example 6.** Does the sequence with general term $a_k = \sin k$ converge or diverge?

Solution: Here's the surprising graph:

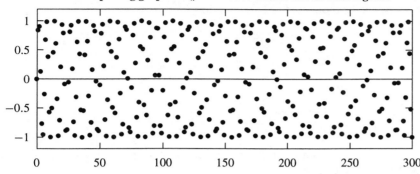

The surprising graph of $a_k = \sin k$: structure but no convergence

Although full of fascinating shapes, the graph never settles on a single limit, so the sequence diverges.➤ □

This example is based on an idea of Gilbert Strang.

Limits of sequences and l'Hôpital's rule

If a sequence happens to be given by a symbolic "formula,"➤ l'Hôpital's rule can often be applied—to the formula—to find the limit.

As many—but not all—are.

■ **Example 7.** Numerical evidence suggests that if $a_k = \dfrac{2^k}{k^2}$, then $\{a_k\}$ diverges to infinity. Use l'Hôpital's rule to show this result symbolically.

Solution: l'Hôpital's rule, applied to the continuous *function* $h(x) = \dfrac{2^x}{x^2}$, says that➤

Differentiate for yourself!

$$\lim_{x \to \infty} \frac{2^x}{x^2} = \lim_{x \to \infty} \frac{2^x \ln 2}{2x} = \lim_{x \to \infty} \frac{2^x \ln 2 \cdot \ln 2}{2} = \infty.$$

(The last equation holds because the numerator blows up as the denominator remains fixed. The other equations follow, therefore, from l'Hôpital's rule.) Because $f(x) \to \infty$, $a_k \to \infty$ too. □

■ **Example 8.** Show that $\displaystyle\lim_{n \to \infty} n^{1/n} = 1$.

Solution: l'Hôpital's rule doesn't apply directly, because no fraction is involved. Applying the natural logarithm helps:

$$a_n = n^{1/n} \implies \ln(a_n) = \frac{\ln n}{n}.$$

Now l'Hôpital's rule *does* apply:

$$\lim_{n \to \infty} \frac{\ln n}{n} = \lim_{n \to \infty} \frac{1/n}{1} = 0.$$

Thus we've shown that as $n \to \infty$, $\ln(a_n) \to 0$; it follows that $a_n \to 1$. □

New sequence limits from old

Limits of sequences, like limits of functions, can be combined in various ways to give *new* limits.

Standard examples

Before combining limits, we'll need some known limits to combine. The limits below can be thought of as basic building blocks.

$$\lim_{n \to \infty} n^{1/n} = 1$$

$$\lim_{n \to \infty} x^{1/n} = 1 \quad \text{(for all } x > 0)$$

$$\lim_{n \to \infty} \frac{1}{n^k} = 0 \quad \text{(for all } k > 0)$$

$$\lim_{n \to \infty} r^n = 0 \quad \text{(if } -1 < r < 1)$$

(We derived the first limit above in an earlier example, using l'Hôpital's rule. The other limits are easier.)

Algebraic combinations

Such plausible calculations as

$$\lim_{k \to \infty} \left(\frac{1}{k} + \frac{3k}{k+1} \right) = \lim_{k \to \infty} \frac{1}{k} + 3 \lim_{k \to \infty} \frac{k}{k+1} = 0 + 3 = 3$$

Already seen, in virtually identical form, for functions.

rely implicitly on the following theorem:◀

Theorem 1 (Algebra with limits). Suppose that as $k \to \infty$,

$$a_k \to L \quad \text{and} \quad b_k \to M,$$

where L and M are finite numbers. Let c be any real constant. Then

$$(ca_k) \to cL, \quad (a_k \pm b_k) \to L \pm M, \quad \text{and} \quad a_k b_k \to LM.$$

If $M \neq 0$, then $\dfrac{a_k}{b_k} \to \dfrac{L}{M}$.

The squeeze principle

For sequences as for functions.

New, unknown, limits can sometimes be found◀ by "squeezing" them between old, known limits.

Theorem 2 (The squeeze principle). Suppose that for all $k > 0$,

$$a_k \leq b_k \leq c_k \quad \text{and} \quad \lim_{k \to \infty} a_k = \lim_{k \to \infty} c_k = L.$$

Then $\lim_{k \to \infty} b_k = L$.

■ **Example 9.** Explain, legalistically, why $\lim_{k \to \infty} \dfrac{\sin k}{k} = 0$.

Convince yourself.

Solution: The "squeeze inequality" $-\dfrac{1}{k} \leq \dfrac{\sin k}{k} \leq \dfrac{1}{k}$ holds◀ for all integers $k > 0$. As $k \to \infty$, both left and right sides tend to 0; so, therefore, must the middle expression.   □

When must a sequence converge? An existence theorem

The ideal way to show that a sequence converges is to *find* a limit. Sometimes, alas, that's difficult.

■ **Example 10.** Does the sequence with general term $a_n = \left(1 + \dfrac{1}{n}\right)^n$ converge or diverge? Why?

Solution: The answer isn't obvious at a glance, so let's tabulate some values:➤➤   *A graph would show the same thing.*

As $n \to \infty$, $(1 + 1/n)^n$ **appears to converge**						
n	16	32	64	128	256	512	1024
$(1 + 1/n)^n$	2.63793	2.67699	2.69734	2.70774	2.71299	2.71563	2.71696

As $n \to \infty$, a_n seems to *increase*, but not to blow up. Therefore, apparently, the sequence converges to some limit.➤➤   □   *The number $e \approx 2.71828$ is a reasonable guess.*

A convergence theorem. The sequence in the previous example seems to converge for two reasons: because it's **increasing** (i.e., $a_1 \le a_2 \le a_3 \dots$) and because it's **bounded above**. Common sense suggests that any increasing sequence either converges to a limit—as the sequence above appears to do—or "blows up" to ∞. For similar reasons, common sense suggests that a **decreasing** sequence (i.e., a sequence for which $a_1 \ge a_2 \ge a_3 \dots$) either converges to a limit or "blows down" to $-\infty$.➤➤   *Think it through; do you agree?*

The theorem below corroborates these common sense impressions. It applies to any **monotone sequence**, i.e., any sequence that is *either* increasing or decreasing.

Theorem 3. Suppose that the sequence $\{a_k\}$ is **increasing** and **bounded above** by a number A. In other words,

$$a_1 \le a_2 \le a_3 \le \dots a_k \le a_{k+1} \le \dots \le A.$$

Then $\{a_k\}$ converges to some finite limit a, with $a \le A$. Similarly, if $\{b_k\}$ is **decreasing** and **bounded below** by a number B, then $\{b_k\}$ converges to a finite limit b, with $b \ge B$.

■ **Example 11.** Consider the increasing sequence $\{a_k\}$ defined by

$$a_1 = 0; \qquad a_{k+1} = \sqrt{6 + a_k} \quad \text{if } k \ge 1.$$

Show that $\{a_k\}$ converges, and the limit is less than or equal to 3.

Solution: The first few terms suggest (and we'll take it for granted) that the sequence is increasing:

$$a_1 = 0; \quad a_2 = \sqrt{6} \approx 2.45; \quad a_3 \approx 2.91; \quad a_4 \approx 2.98; \quad a_5 \approx 2.997.$$

The sequence also seems to be bounded above by 3. This is easy to show. If $a_k < 3$, then by definition

$$a_{k+1} = \sqrt{6 + a_k} < \sqrt{6 + 3} = 3.$$

Thus *no* term can ever exceed 3. It follows from the Theorem, therefore, that the sequence converges to a limit no greater than 3. (The limit *is* 3, of course, but that requires further proof.)   □

Exercises

Find a symbolic expression for the general term a_k of each of the following sequences.

1. $1, -\dfrac{1}{3}, \dfrac{1}{9}, -\dfrac{1}{27}, \dfrac{1}{81}, \ldots$

2. $3, 6, 9, 12, 15, \ldots$

3. $\dfrac{1}{2}, \dfrac{2}{4}, \dfrac{3}{8}, \dfrac{4}{16}, \dfrac{5}{32}, \ldots$

4. $\dfrac{1}{2}, \dfrac{1}{5}, \dfrac{1}{10}, \dfrac{1}{17}, \dfrac{1}{26}, \dfrac{1}{37}, \ldots$

Calculate a_1, a_2, a_5, and a_{10} for each of the following sequences. (Round your answers to four decimal places.) Then determine $\lim\limits_{k \to \infty} a_k$ (if the limit exists).

5. $a_k = (-3/2)^k$

6. $a_k = (-0.8)^k$

7. $a_k = (1.1)^k$

8. $a_k = \left(\sqrt{26}/17\right)^k$

9. $a_k = \left(\dfrac{\pi}{e}\right)^k$

10. $a_k = e^{-k}$

11. $a_k = (\arcsin 1)^k$

12. $a_k = (1/k)^k$

13. $a_k = \sin k$

14. $a_k = \arctan k$

15. $a_k = \cos(1/k)$

16. $a_k = \sin(k\pi)$

17. $a_k = \cos(k\pi)$

18. $a_k = \dfrac{k^2}{k^2 + k + 3}$

19. $a_k = \sqrt{\dfrac{2k}{k+3}}$

20. $a_k = \dfrac{k}{\sqrt{k} + 10}$

21. $a_k = \ln k - \ln(k+1) = \ln\left(\dfrac{k}{k+1}\right)$

22. $a_k = \displaystyle\int_0^k e^{-x}\,dx$

23. $a_k = \displaystyle\int_k^\infty \dfrac{dx}{1+x^2}$

24. $a_k = \dfrac{\cos k}{\ln(k+1)}$

25. $a_k = e^{-k}\sin k$

26. $a_k = 3^{1/k}$

27. (a) For which values of x does $\lim\limits_{n \to \infty} x^n$ diverge?

(b) Find all values of x for which $\lim\limits_{n \to \infty} x^n$ converges to 0.

(c) Are there any values of x for which $\lim\limits_{n \to \infty} x^n$ converges to a number other than 0? For each such x, find the limit.

For which values of x does each of the following sequences converge as $k \to \infty$? Evaluate $\lim\limits_{k \to \infty} a_k$ for these values of x.

28. $a_k = (\ln x)^k$

29. $a_k = e^{kx}$

30. $a_k = (\arcsin x)^k$

31. $a_k = 2^{-k}(\arctan x)^k$

32. Let $a_n = \cos 1 \cdot \cos 2 \cdot \cos 3 \cdot \cos 4 \cdots \cos n$.

(a) Evaluate $a_1, a_2, a_3,$ and a_5. (Round your answers to four decimal places.)

(b) Is the sequence $\{ a_n \}$ bounded? Justify your answer.

(c) Is the sequence $\{ a_n \}$ monotone? Justify your answer.

(d) Is the sequence $\{ |a_n| \}$ bounded? Justify your answer.

(e) Is the sequence $\{ |a_n| \}$ monotone? Justify your answer.

33. Use Theorem 3 to show that the sequence $0.7, 0.77, 0.777, 0.7777, 0.77777,$ $0.777777, 0.7777777, \ldots$ has a limit.

34. Let $a_n = \dfrac{1 \cdot 3 \cdot 5 \cdots (2n - 1)}{2 \cdot 4 \cdot 6 \cdots (2n)}$.

(a) Calculate $a_1, a_2,$ and a_5.

(b) Use Theorem 3 to show that $\lim\limits_{n \to \infty} a_n$ is a finite number.

35. Show that $\lim\limits_{n \to \infty} \sin\left(\dfrac{\pi}{2^2}\right) \cdot \sin\left(\dfrac{\pi}{3^2}\right) \cdots \sin\left(\dfrac{\pi}{n^2}\right) = 0.$ [HINT: $0 < \sin x < x$ when $0 < x < 1$.]

36. Does the sequence defined by $a_1 = 1, a_{n+1} = 1 - a_n$ converge? Justify your answer.

37. Does the sequence defined by $a_1 = 1, a_{n+1} = a_n/2$ converge? Justify your answer.

38. Consider the sequence defined by $a_1 = 1, a_{n+1} = \left(\dfrac{n}{n + 1}\right) a_n$.

(a) Show that the sequence converges.

(b) Find the limit.   [HINT: Write out the first few terms and look for a pattern.]

39. Suppose that $\lim\limits_{k \to \infty} a_k = L$, where L is a finite number, and that the terms of the sequence $\{ b_k \}$ are defined by $b_k = L - a_k$. Explain why $\lim\limits_{k \to \infty} b_k = 0$.

40. Give an example of a sequence that is

(a) convergent but not monotone.

(b) bounded but not monotone.

(c) monotone but not convergent.

(d) monotone decreasing and unbounded.

(e) monotone decreasing and convergent.

(f) unbounded but not monotone.

41. Show that $\lim_{n \to \infty} x^{1/n} = 1$ for all $x > 0$.

42. For which values of $x \geq 0$ does the sequence defined by $a_1 = x$, $a_{n+1} = \sqrt{a_n}$ converge? Justify your answer.

For each sequence below, calculate a_{100} (to four decimal places), then find the limit of each sequence.   [HINT: L'Hôpital's rule may be useful.]

43. $a_n = \dfrac{n+2}{n^3+4}$

46. $c_k = \dfrac{\ln\left(1+k^2\right)}{\ln\left(4+3k\right)}$

44. $a_m = m^2/e^m$

47. $d_n = n\sin(1/n)$

45. $b_j = \ln j/\sqrt[3]{j}$

48. $a_k = \left(2^k + 3^k\right)^{1/k}$

49. (a) Suppose that f is a differentiable function and that $f(0) = 0$. Show that
$\lim_{n \to \infty} nf(1/n) = f'(0)$.

(b) Use the result in part (a), not l'Hôpital's rule, to evaluate $\lim_{n \to \infty} n \arctan(1/n)$.

50. Let $a_k = \left(1 + \dfrac{x}{k}\right)^k$, where x is a real number.

(a) Show that $\lim_{k \to \infty} \ln(a_k) = x$.

(b) Use part (a) to evaluate $\lim_{k \to \infty} a_k$.

51. Let $a_n = \displaystyle\sum_{k=1}^{n} k/n^2$.

(a) Evaluate a_{10}.

(b) Explain why $\lim_{n \to \infty} a_n = \displaystyle\int_0^1 x\,dx$.   [HINT: $\frac{k}{n^2} = \left(\frac{k}{n}\right)\cdot\frac{1}{n}$.]

52. Let $a_n = \displaystyle\sum_{k=1}^{n} k^2/n^3$. Evaluate $\lim_{n \to \infty} a_n$.   [HINT: Think of a_n as a Riemann sum approximation to an integral.]

53. Let $a_n = \displaystyle\sum_{k=1}^{n} \dfrac{1}{n+k}$. Evaluate $\lim_{n \to \infty} a_n$.   [HINT: $\frac{1}{n+k} = \frac{1}{1+k/n}\cdot\frac{1}{n}$.]

54. Suppose that $\{\,a_n\,\}$ is a sequence with the property $a_{n+1}/a_n \leq (n+3)/(2n+1)$ for all $n \geq 1$. Show that $\lim_{n \to \infty} a_n = 0$.   [HINT: Start by showing that $a_{n+1}/a_3 \leq (6/7)^{n-2}$ for all $n \geq 3$.]

11.2 Infinite series, convergence, and divergence

Introduction

An **infinite series** is an **infinite sum**, i.e., an expression of the form

"Series" for short.

*On the left side we used **sigma notation** for brevity.*

$$\sum_{k=1}^{\infty} a_k = a_1 + a_2 + a_3 + a_4 + \cdots + a_k + a_{k+1} + \ldots .$$

A **series**, in other words, results from *adding* the terms of a **sequence** $a_1, a_2, a_3, \ldots$. If, say, $a_k = 1/k^2$, then

$$\sum_{k=1}^{\infty} a_k = \sum_{k=1}^{\infty} \frac{1}{k^2} = \frac{1}{1} + \frac{1}{4} + \frac{1}{9} + \frac{1}{16} + \ldots .$$

If $a_k = k$, then

$$\sum_{k=1}^{\infty} a_k = \sum_{k=1}^{\infty} k = 1 + 2 + 3 + \ldots .$$

The idea of an infinite sum raises immediate questions:

- What does it *mean* to add infinitely many numbers?

- Which series "add up" to a *finite* number? Which "blow up"?

- If a series *has* a finite sum, how can we find (or estimate) it?

The rest of this chapter addresses just these questions.

Improper sums, improper integrals

Standard examples of infinite series include

$$\sum_{k=1}^{\infty} \frac{1}{k^2}, \quad \sum_{k=1}^{\infty} \frac{1}{k}, \quad \sum_{k=1}^{\infty} \frac{1}{\sqrt{k}}, \quad \text{and} \quad \sum_{k=1}^{\infty} \frac{1}{2^k}.$$

The striking resemblance to the improper integrals

Typographically, even.

$$\int_{1}^{\infty} \frac{dx}{x^2}, \quad \int_{1}^{\infty} \frac{dx}{x}, \quad \int_{1}^{\infty} \frac{dx}{\sqrt{x}}, \quad \text{and} \quad \int_{1}^{\infty} \frac{dx}{2^x}$$

is no accident; infinite series and improper integrals are similar in almost every respect. We'll see, in fact, that the first and fourth series above converge, while the second and third diverge—exactly as for the corresponding integrals.

We'll return repeatedly to this important analogy.

Why series matter: a preview

Understanding series thoroughly takes some legwork. As a first glimpse of why the work is worth doing, consider the fact that for any real number x,

We'll see at the end of this chapter why it's true.

(11.2.1) $$\cos x = 1 - \frac{x^2}{2!} + \frac{x^4}{4!} - \frac{x^6}{6!} + \frac{x^8}{8!} - \ldots .$$

If, say, $x = 1$, then

The "dots" in each equation mean that the pattern continues.

(11.2.2) $$\cos 1 = 1 - \frac{1}{2!} + \frac{1}{4!} - \frac{1}{6!} + \frac{1}{8!} - \ldots .$$

Transcendental values

Why should we care about such equations? Why write something familiar—the cosine function—in terms of something exotic—a series?

One good answer is that the cosine function, although crucial in applications,[*] has no *algebraic* formula (i.e., it's *transcendental*). Without a concrete formula, finding accurate *numerical values* of cos x for given inputs x is a genuine problem.[*]

The equations above help solve this problem. Equation 11.2.1, although not a formula in the ordinary sense,[*] gives a concrete, computable recipe for *approximating* cos x, for any input x: Given an input x, calculate as far out as practically possible in the "infinite polynomial"

$$1 - \frac{x^2}{2!} + \frac{x^4}{4!} - \frac{x^6}{6!} + \frac{x^8}{8!} - \dots .$$

With luck the result should closely approximate the "true" value of cos x.

Equation 11.2.2 shows how to approximate cos 1. It's easy to check[*] that to 7 decimals,

$$1 - \frac{1}{2!} = 0.5000000;$$

$$1 - \frac{1}{2!} + \frac{1}{4!} = 0.5416667;$$

$$1 - \frac{1}{2!} + \frac{1}{4!} - \frac{1}{6!} = 0.5402778;$$

$$1 - \frac{1}{2!} + \frac{1}{4!} - \frac{1}{6!} + \frac{1}{8!} = 0.5403026.$$

The results converge, with gratifying speed, to the "right" answer—the true value of cos 1 (≈ 0.5403023).

Open questions. These calculations raise at least as many questions as they answer:

> *Where did Equation 11.2.1 come from? What do the all the "dots" really mean? Are similar equations available for* other *functions—sine, arctangent, logarithm, etc.? How many terms are needed to guarantee accuracy to, say, 5 decimals?*

Questions like these guide our study of series.

Convergence: definitions and terminology

Working successfully with series requires some up-front investment in definitions and technical language.[*] After stating terms and definitions, we'll show by example why they're natural.

Series language: terms, partial sums, tails, convergence, limit

Let $\sum_{k=1}^{\infty} a_k = a_1 + a_2 + a_3 + \dots + a_k + a_{k+1} + \dots$ be an infinite series.[*] The summand a_k is called the k**th term** of the series. The n**th partial sum**, denoted S_n, is the (finite) sum of all terms *through index n*:

$$S_n = a_1 + a_2 + a_3 + \dots + a_{n-1} + a_n = \sum_{k=1}^{n} a_k .$$

The *n*th tail, denoted R_n, is the (infinite) sum of all terms *beyond index n*:

$$R_n = a_{n+1} + a_{n+2} + a_{n+3} + \cdots = \sum_{k=n+1}^{\infty} a_k.$$

As the notation R_n suggests, the *n*th tail is a **remainder**—what remains after adding the terms through index n. In symbols:

$$\sum_{k=1}^{\infty} a_k = (a_1 + a_2 + \cdots + a_n) + (a_{n+1} + a_{n+2} + a_{n+3} + \ldots) = S_n + R_n.$$

The crucial definition➤ of convergence involves partial sums:

With notation as above.

> **Definition:** If $\lim\limits_{n \to \infty} S_n = S$, for some finite number S, then the series $\sum_{k=1}^{\infty} a_k$ **converges to the limit** S. Otherwise, the series **diverges**.

Notice:

Divergent series. A **divergent** series is one for which the sequence of partial sums does *not* converge to a finite limit S. One possibility—but not the only one—is that the partial sums S_n blow up to infinity. Another possibility is that the partial sums remain bounded, but never settle on a specific limit.

Improper integrals, improper sums. The definition above says, in symbols, that

$$\sum_{k=1}^{\infty} a_k = \lim_{n \to \infty} \sum_{k=1}^{n} a_k,$$

if the limit exists. Convergence for improper integrals means much the same thing:

$$\int_{x=1}^{\infty} f(x)\, dx = \lim_{n \to \infty} \int_{x=1}^{n} f(x)\, dx,$$

if *this* limit exists.➤ The analogy means that an infinite series is an "improper sum" in *exactly* the sense that an integral may be "improper." Convergence, too, is defined similarly.

In each case we take the limit of something proper.

What convergence means for partial sums and tails. To say that $\sum a_k$ converges to the limit (or **sum**) S means that $S_n \to S$ as $n \to \infty$. Since for all n, $S = S_n + R_n$, it follows that $R_n \to 0$ as $n \to \infty$.➤ In particular: To decide whether a *series* converges, we need to examine the *sequence* of partial sums.

As for improper integrals!

Two sequences: keep them straight. Every series $\sum a_k$ involves *two* sequences: the sequence $\{a_k\}$ of *terms* and the sequence $\{S_n\}$ of *partial sums*. Keeping these related but different sequences separate is essential.➤ We'll pay special attention to that in the next examples.

Many, maybe most, student difficulties arise from confusing them.

■ **Example 1.** (A geometric series.) Does the series $\sum_{k=0}^{\infty} \dfrac{1}{2^k} = 1 + \dfrac{1}{2} + \dfrac{1}{4} + \dfrac{1}{8} + \ldots$ converge? If so, to what limit?

Solution: In this series the index k began at 0, not 1. So, therefore, does the sequence of partial sums. Direct calculation yields:

$$S_0 = 1 \quad = \quad 2 - 1;$$
$$S_1 = 1 + \frac{1}{2} = \frac{3}{2} \quad = \quad 2 - \frac{1}{2};$$
$$S_2 = 1 + \frac{1}{2} + \frac{1}{4} = \frac{7}{4} \quad = \quad 2 - \frac{1}{4};$$
$$S_3 = 1 + \frac{1}{2} + \frac{1}{4} + \frac{1}{8} = \frac{15}{8} \quad = \quad 2 - \frac{1}{8};$$
$$S_4 = 1 + \frac{1}{2} + \frac{1}{4} + \frac{1}{8} + \frac{1}{16} = \frac{31}{16} \quad = \quad 2 - \frac{1}{16}$$

The pattern is easy to see: For any $n \geq 1$,

$$S_n = 2 - \frac{1}{2^n}.$$

From this explicit formula for S_n, the conclusion follows: Because $S_n \to 2$ as $n \to \infty$, the series *converges* to 2.

A table of numerical values gives the same impression—and also illustrates the behavior of tails. Note that for each n,

$$S_n = \sum_{k=0}^{n} \frac{1}{2^k}; \quad R_n = \sum_{k=n+1}^{\infty} \frac{1}{2^k} = 2 - \sum_{k=0}^{n} \frac{1}{2^k}.$$

Check some for yourself.

Here are the numbers:◂◂

Partial sums and tails of $1 + 1/2 + 1/4 + \ldots$										
n	0	1	2	3	4	5	6	7	8	9	10
S_n	1	1.5	1.75	1.875	1.938	1.969	1.984	1.992	1.996	1.998	1.999
R_n	1	0.5	0.25	0.125	0.063	0.031	0.016	0.008	0.004	0.002	0.001

■ **Example 2.** Does $\displaystyle\sum_{k=1}^{\infty} (-1)^k = -1 + 1 - 1 + 1 - 1 + \ldots$ converge?

Solution: No. Successive partial sums are of the form $-1, 0, -1, 0, \ldots$. Since the sequence $\{S_n\}$ *diverges*, so does the series. □

■ **Example 3.** Does the **harmonic series** $\displaystyle\sum_{k=1}^{\infty} \frac{1}{k}$ converge?

Solution: The answer depends, as always, on the partial sums S_n. By definition,

$$S_n = 1 + \frac{1}{2} + \frac{1}{3} + \frac{1}{4} + \frac{1}{5} + \frac{1}{6} + \cdots + \frac{1}{n}.$$

Unlike in the last two examples.

Results to 3 decimal places.

This time,◂◂ unfortunately, no simple formula for S_n in terms of n comes to mind, so we'll investigate the sequence $\{S_n\}$ numerically. A computer makes quick work of the calculations: Here are some results:◂◂

$$S_1 = 1; \quad S_{20} = 3.598; \quad S_{50} = 4.499; \quad S_{100} = 6.793; \quad S_{1000} = 7.485.$$

The numerical evidence is ambiguous; the S_n's seem to keep growing, though slowly. Whether the S_n's converge to some finite number or diverge to infinity is not yet clear.◂◂ □

In fact, the series diverges. We'll show fact this below by comparing it to the integral $\int_1^{\infty} 1/x \, dx$.

When does a series (or an integral) converge?

Deciding whether a given infinite series converges or diverges is a delicate question; so was the analogous question for improper integrals. Such series as

$$\sum_{k=1}^{\infty} \frac{1}{k} \quad \text{and} \quad \sum_{k=1}^{\infty} \frac{1}{k^2}$$

pose the typical dilemma: Although successive *terms*➤ tend to zero, the *number* of terms is infinite. Convergence or divergence hinges on which of these conflicting tendencies "wins" in the long run.

Of both series.

The same dilemma arose➤ for the improper integrals

$$\int_{1}^{\infty} \frac{1}{x} \, dx \quad \text{and} \quad \int_{1}^{\infty} \frac{1}{x^2} \, dx.$$

Remember? See Chapter 10, Section 1.

Although the integrands tend to zero as x grows without bound, the total area bounded by their graphs may or may not converge to a finite limit. Indeed, we discovered earlier that these two integrals have opposite outcomes: the first integral *diverges to infinity*; the second *converges to one.*➤

Remember why? If not, recalculate.

Convergence vs. divergence, graphically

Deciding whether the two *series* above converge or diverge is a bit harder than for the corresponding integrals, because (as we saw in Example 3) no convenient *formulas* for S_n are available.

For a series $\sum a_k$, plotting *both* $\{a_k\}$ and $\{S_n\}$➤ on the same axes illustrates the connection between the two—and sometimes suggests whether the series converge or diverges. The **harmonic series** $\sum 1/k$ generates this picture:

The terms and the partial sums, respectively.

Terms and partial sums for $\sum 1/k$

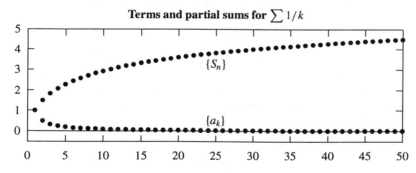

The picture for the series $\sum 1/k^2$ gives a different impression:

Terms and partial sums for $\sum 1/k^2$

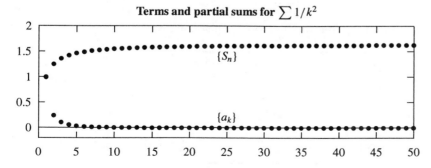

The question—for any series—is whether the sequence $\{S_n\}$ converges to a limit. The pictures suggest➤ that the second series above converges and the first diverges. (From the second picture, what would you estimate as the limit of the series $\sum_{k=1}^{\infty} 1/k^2$?)

But don't prove!

Geometric series

Geometric series form the simplest and most important class of infinite series. A **geometric series** is one of the form

$$a + ar + ar^2 + ar^3 + ar^4 + \cdots = \sum_{k=0}^{\infty} ar^k \, ;$$

Each term is r times the previous term.

a is called the **leading term**; r is the **ratio**. The series

$$1 + \frac{1}{2} + \frac{1}{4} + \frac{1}{8} + \cdots = \sum_{k=0}^{\infty} \left(\frac{1}{2}\right)^k$$

of Example 1 is geometric, with $a = 1$ and $r = 1/2$.

Partial sums of geometric series

Unlike for many other series.

Geometric series have one great advantage: it's *easy* to decide whether they converge or diverge and, if they converge, to find their limits. The reason it's easy is that there's a convenient, explicit formula for an arbitrary partial sum S_n. The formula depends on the fact that if $r \neq 1$ and $n \geq 1$, then

$$1 + r + r^2 + r^3 + \cdots + r^n = \frac{1 - r^{n+1}}{1 - r}.$$

(Multiply both sides of the formula by $(1 - r)$ to convince yourself that it holds.)

Here, then, is our formula for S_n. If $r \neq 1$ and $n \geq 1$, then

$$(11.2.3) \qquad S_n = a + ar + ar^2 + ar^3 + \cdots + ar^n = a\frac{1 - r^{n+1}}{1 - r}.$$

From this formula follows the whole story of convergence and divergence for geometric series. The question, as for any series, is how the sequence $\{S_n\}$ of partial sums behaves as n tends to infinity. From the facts that

$$\lim_{n \to \infty} r^n = 0 \quad \text{if} \quad |r| < 1;$$
$$\lim_{n \to \infty} r^n = \infty \quad \text{if} \quad r \geq 1;$$
$$\lim_{n \to \infty} r^n \text{ does not exist} \quad \text{if} \quad r \leq -1;$$

the conclusion follows. It's worth emphasizing:

Fact: If $|r| < 1$, the **geometric series**

$$\sum_{k=0}^{\infty} ar^k = a + ar + ar^2 + ar^3 + \ldots$$

converges to $\dfrac{a}{1 - r}$. If $a \neq 0$ and $|r| \geq 1$, the series diverges.

■ **Example 4.** The series $\dfrac{1}{3} - \dfrac{1}{6} + \dfrac{1}{12} - \dfrac{1}{24} + \ldots$ converges. To what?

Solution: The series is geometric with $a = 1/3$ and $r = -1/2$. It converges, therefore, to

$$\frac{a}{1 - r} = \frac{1/3}{1 + 1/2} = \frac{2}{9} \approx 0.2222222.$$

A look at partial sums and tails supports the computation.

\multicolumn{12}{c}{**Partial sums and tails of** $\frac{1}{3} - \frac{1}{6} + \frac{1}{12} - \frac{1}{24} + \cdots$}
n	0	1	2	3	4	5	6	7	8	9	10
S_n	0.3333	0.1667	0.2499	0.2083	0.2291	0.2187	0.2239	0.2213	0.2226	0.2220	0.2223
R_n	−0.1111	0.0556	−0.0277	0.0140	−0.0068	0.0036	−0.0016	0.0010	−0.0003	0.0004	−0.0001

$\square$

Telescoping series

Geometric series are the most important variety for which partial sums can be found explicitly. Telescoping series offer the same possibility. The next example illustrates how they work—and the reason for the name.

■ **Example 5.** Show that $\displaystyle\sum_{k=1}^{\infty} \frac{1}{k(k+1)}$ converges, and find its limit.

Solution: A bit of algebra lets us rewrite the series in a more helpful form:

$$\sum_{k=1}^{\infty} \frac{1}{k(k+1)} = \sum_{k=1}^{\infty} \left(\frac{1}{k} - \frac{1}{k+1} \right).$$

Writing out a few terms shows the "telescoping" pattern:

$$\begin{aligned} S_n &= \sum_{k=1}^{n} \left(\frac{1}{k} - \frac{1}{k+1} \right) \\ &= \left(\frac{1}{1} - \frac{1}{2} \right) + \left(\frac{1}{2} - \frac{1}{3} \right) + \left(\frac{1}{3} - \frac{1}{4} \right) + \cdots + \left(\frac{1}{n} - \frac{1}{n+1} \right) \\ &= 1 - \frac{1}{n+1}. \end{aligned}$$

Now it's clear that as $n \to \infty$, $S_n \to 1$. That's the limit.

Again the numbers agree:➤➤

Watch the tails go to zero.

\multicolumn{11}{c}{**Partial sums and tails of** $\sum_{k=1}^{\infty} 1/(k^2 + k)$}
n	1	2	3	4	5	6	7	8	9	10	11
S_n	0.5000	0.6667	0.7500	0.8000	0.8333	0.8571	0.8750	0.8889	0.9000	0.9091	0.9167
R_n	0.5000	0.3333	0.2500	0.2000	0.1667	0.1429	0.1250	0.1111	0.1000	0.0909	0.0833

$\square$

Algebra with series

Combining series➤➤ algebraically produces new series. Combining *convergent* series produces new *convergent* series, with limits related in the expected way:

As with functions and sequences.

> **Theorem 4.** Suppose that $\sum_{k=1}^{\infty} a_k$ converges to S and that $\sum_{k=1}^{\infty} b_k$ converges to T. Let c be any constant. Then
>
> $$\sum_{k=1}^{\infty} (a_k \pm b_k) \qquad \text{converges to } S \pm T;$$
>
> $$\sum_{k=1}^{\infty} ca_k \qquad \text{converges to } cS.$$

After all, the limit of a series is defined as the limit of the sequence of partial sums.

These plausible properties of *convergent series* follow directly from the analogous properties of *convergent sequences.*◄

A little series algebra, cleverly applied, can immensely simplify finding limits of certain series.

■ **Example 6.** Evaluate $\sum_{k=0}^{\infty} \dfrac{4 + 2^k}{3^k}$.

We are able to find sums of geometric series.

Solution: The given series is the sum of two convergent geometric series.◄ The theorem says, therefore, that

$$\sum_{k=0}^{\infty} \frac{4 + 2^k}{3^k} = \sum_{k=0}^{\infty} \frac{4}{3^k} + \sum_{k=0}^{\infty} \left(\frac{2}{3}\right)^k = 6 + 3 = 9. \qquad \square$$

■ **Example 7.** For the geometric series $\sum_{k=0}^{\infty} \dfrac{3}{2^k}$, calculate the tail $R_{10} = \sum_{k=11}^{\infty} \dfrac{3}{2^k}$.

Solution: A little algebra is all that's needed:

$$
\begin{aligned}
R_{10} = \sum_{k=11}^{\infty} \frac{3}{2^k} &= \frac{3}{2^{11}} + \frac{3}{2^{12}} + \frac{3}{2^{13}} + \cdots \\
&= \frac{3}{2^{11}} \left(1 + \frac{1}{2} + \frac{1}{2^2} + \frac{1}{2^3} + \cdots\right) \\
&= \frac{3}{2^{11}} \cdot 2 = \frac{3}{2^{10}} = \frac{3}{1024}.
\end{aligned}
$$

(The last series above sums to 2; hence the final answer.) $\qquad \square$

Detecting divergent series

And their limits.

For us, *convergent* series◄ are of interest. The previous theorem, for instance, applies safely *only* to convergent series. It's important, therefore, to recognize divergence when it occurs. The *nth term test* sometimes helps.

A series $\sum a_k$ converges if and only if its partial sums S_n converge to a limit. For this to occur, the *difference* $S_n - S_{n-1}$ between successive partial sums must tend to zero.◄ This difference is simply the *n*th term:

Otherwise the partial sum sequence wouldn't "level off."

$$S_n - S_{n-1} = (a_1 + a_2 + \ldots a_{n-1} + a_n) - (a_1 + a_2 + \ldots a_{n-1}) = a_n.$$

Thus, for a series to converge, it's necessary for its *terms* to tend to zero. We'll restate this fact in its most useful form:

> **Theorem 5 (The *n*th term test for divergence).** If $\lim_{n \to \infty} a_n \neq 0$, then $\sum a_n$ diverges.

What the theorem doesn't say.   It's important to notice that the theorem does *not* guarantee that $\sum a_n$ converges if $a_n \to 0$. As the harmonic series $\sum 1/k$ shows, the terms of a *divergent* series may tend to zero. The rather blunt nth term test may detect *divergence*, but never convergence. Sharper instruments are needed to detect convergence. We'll develop several in the next section.

■ **Example 8.**   Does $\displaystyle\sum_{k=1}^{\infty} \frac{k}{k+1000}$ converge?

Solution: The nth term test says no. Because

$$a_n = \frac{n}{n+1000} \to 1 \quad \text{as} \quad n \to \infty,$$

the series diverges.   □

■ **Example 9.**   Given that $\displaystyle\sum_{k=0}^{\infty} \frac{3^k}{k!}$ converges, find $\displaystyle\lim_{k\to\infty} \frac{3^k}{k!}$.

Solution: The limit is zero. By the previous theorem, the terms of *any* convergent series must tend to zero. (Tabulating values➤ of $3^k/k!$ supports this conclusion.) □   *Try some.*

Exercises ────────────────────────────

1. This exercise is about partial sums of the geometric series $\sum_{k=0}^{\infty} ar^k$. By definition, a partial sum S_n of this series is $S_n = \sum_{k=0}^{n} ar^k = a + ar + ar^2 + ar^3 + \cdots + ar^n$. According to Equation 11.2.3, page 228, there's a simpler, more explicit formula for S_n:

   $$S_n = a\,\frac{1 - r^{n+1}}{1 - r}.$$

   (a) Let $a = 1$, $r = 2$, and $n = 5$. Calculate both sides of the expression above. Are they equal?

   (b) Repeat the previous part, using $a = 1$, $r = 0.01$, and $n = 5$.

   (c) Suppose that $r = 1$. Explain why the equation above does *not* hold in this case. Then find another (even simpler) formula for S_n that does hold when $r = 1$.

   (d) Calculate and simplify the quantity $S_n - rS_n$.

   (e) Use your result from the previous part to show that the formula above holds if $r \neq 1$. [HINT: $S_n - rS_n = (1-r)S_n$.]

   (f) Find $3 + 6 + 12 + 24 + 48 + 96 + \cdots + 3072$. [HINT: Use the formula for S_n.]

2. The series $\displaystyle\sum_{k=0}^{\infty} a_k = \sum_{k=0}^{\infty} \frac{1}{k!}$ converges to $e \approx 2.718282$.

(a) Evaluate a_1, a_2, a_5, and a_{10}.

(b) Evaluate S_1, S_2, S_5, and S_{10}.

(c) Explain why $\{ S_n \}$ is an increasing sequence.

(d) Evaluate R_1, R_2, R_5, and R_{10}.

(e) Show that $R_k > 0$ for all $k \geq 0$.

(f) Explain why $\{ R_n \}$ is a decreasing sequence.

(g) Using a calculator or a computer, determine a value of n for which S_n differs from e by less than 0.001.

(h) Using a calculator or a computer, determine a value of n for which S_n differs from e by less than 10^{-5}.

(i) Use parts (f) and (h) to show that $R_{50} < 10^{-5}$.

(j) Explain why $\displaystyle\lim_{n \to \infty} R_n = 0$.

3. The series $\displaystyle\sum_{k=0}^{\infty} a_k = \sum_{k=0}^{\infty} \frac{1}{k^2}$ converges to $\dfrac{\pi^2}{6}$.

(a) Evaluate a_1, a_2, a_5, and a_{10}.

(b) Evaluate S_1, S_2, S_5, and S_{10}.

(c) Is $\{ S_n \}$ an increasing sequence? Justify your answer.

(d) Evaluate R_1, R_2, R_5, and R_{10}.

(e) Is $\{ R_n \}$ a decreasing sequence? Justify your answer.

(f) Show that $0 < R_n < 0.05$ for all $n \geq 20$.

(g) Evaluate $\displaystyle\lim_{n \to \infty} R_n = 0$.

4. Consider the series $\displaystyle\sum_{k=0}^{\infty} a_k = \sum_{k=0}^{\infty} \frac{1}{5^k}$.

(a) Evaluate a_1, a_2, a_5, and a_{10}.

(b) Evaluate S_1, S_2, S_5, and S_{10}.

(c) Show that the sequence $\{ S_n \}$ is increasing and bounded above. What does this imply about the sequence of partial sums?

(d) Find the sum of the series (i.e., $\displaystyle\lim_{n \to \infty} S_n$).

(e) Evaluate R_1, R_2, R_5, and R_{10}.

(f) Show that $\{ R_n \}$ is decreasing and bounded below.

(g) Evaluate $\displaystyle\lim_{n \to \infty} R_n$.

5. Consider the series $\sum_{k=0}^{\infty} a_k = \sum_{k=0}^{\infty} (-0.8)^k$.

   (a) Evaluate a_1, a_2, a_5, and a_{10}.

   (b) Evaluate S_1, S_2, S_5, and S_{10}.

   (c) Find the sum of the series (i.e., $\lim_{n \to \infty} S_n$).

   (d) Evaluate R_1, R_2, R_5, and R_{10}.

   (e) Is the sequence $\{ S_n \}$ increasing? Justify your answer.

   (f) Show that the sequence $\{ R_n \}$ is neither increasing nor decreasing.

   (g) Show that the sequence $\{ |R_n| \}$ is decreasing.

   (h) Evaluate $\lim_{n \to \infty} R_n$.

6. Consider the series $\sum_{k=0}^{\infty} a_k = \sum_{k=0}^{\infty} \frac{1}{k + 2^k}$.

   (a) Evaluate S_1, S_2, S_5, and S_{10}.

   (b) Show that the sequence $\{ S_n \}$ is increasing.

   (c) Explain why $a_k \le 2^{-k}$ when $k \ge 0$.

   (d) Use part (c) to show that $S_n \le 2 - 2^{-n} < 2$.
       [HINT: $\sum_{k=0}^{n} 2^{-k}$ is a geometric series.]

   (e) Use parts (b) and (d) to show that the series $\sum_{k=0}^{\infty} a_k$ converges (i.e., that
       $\lim_{n \to \infty} S_n$ exists).

   (f) Explain why $\lim_{n \to \infty} R_n = 0$.

7. Consider the series $\sum_{j=0}^{\infty} a_j = \sum_{j=0}^{\infty} \frac{1}{2 + 3^j}$.

   (a) Evaluate S_1, S_2, S_5, and S_{10}.

   (b) Show that the sequence $\{ S_n \}$ is increasing and bounded above.

   (c) Does $\sum_{j=0}^{\infty} a_j$ converge? Justify your answer.

8. It is known that $\sum_{m=1}^{\infty} \frac{1}{m^4} = \frac{\pi^4}{90}$. Use this fact to evaluate

   (a) $\sum_{i=0}^{\infty} \frac{1}{(i + 1)^4}$

   (b) $\sum_{k=3}^{\infty} \frac{1}{k^4}$

9. It is known that $\sum_{i=1}^{\infty} \frac{1}{i^2} = 1 + \frac{1}{4} + \frac{1}{9} + \frac{1}{16} + \frac{1}{25} + \cdots = \frac{\pi^2}{6}$.

   (a) Use this fact to evaluate $\sum_{j=1}^{\infty} \frac{1}{(2j)^2} = \frac{1}{4} + \frac{1}{16} + \frac{1}{36} + \frac{1}{64} + \frac{1}{100} + \cdots$.

(b) Use part (a) to show that $\displaystyle\sum_{k=0}^{\infty} \frac{1}{(2k+1)^2} = 1 + \frac{1}{9} + \frac{1}{25} + \frac{1}{49} + \cdots = \frac{\pi^2}{8}$.

(c) Evaluate $\displaystyle\sum_{m=1}^{\infty} \frac{(-1)^{m+1}}{m^2} = 1 - \frac{1}{4} + \frac{1}{9} - \frac{1}{16} + \frac{1}{25} - \frac{1}{36} + \cdots$.

Find the limit of each of the following convergent series.

10. $\displaystyle\frac{1}{16} + \frac{1}{32} + \frac{1}{64} + \frac{1}{128} + \cdots + \frac{1}{2^{i+4}} + \cdots$

11. $\displaystyle 2 - 5 + 9 + \frac{1}{3} + \frac{1}{9} + \frac{1}{27} + \frac{1}{81} + \cdots + \frac{1}{3^n} + \cdots$

12. $\displaystyle\sum_{n=0}^{\infty} e^{-n}$

13. $\displaystyle\sum_{k=3}^{\infty} \left(\frac{e}{\pi}\right)^k$

14. $\displaystyle\sum_{m=1}^{\infty} (\arctan 1)^m$

15. $\displaystyle\sum_{i=10}^{\infty} \left(\frac{2}{3}\right)^i$

16. $\displaystyle\sum_{j=5}^{\infty} \left(-\frac{1}{2}\right)^j$

17. $\displaystyle\sum_{j=0}^{\infty} \frac{3^j + 4^j}{5^j}$

For each of the following geometric series, find all values of x for which the series converges. Then state the limit as a simple expression involving x. (Assume $x^0 = 1$ for all x.)

18. $\displaystyle\sum_{k=0}^{\infty} x^k$

19. $\displaystyle\sum_{m=2}^{\infty} \left(\frac{x}{5}\right)^m$

20. $\displaystyle\sum_{j=5}^{\infty} x^{2j}$

21. $\displaystyle\sum_{k=1}^{\infty} x^{-k}$

22. $\displaystyle\sum_{n=3}^{\infty} (1 + x)^n$

23. $\displaystyle\sum_{j=4}^{\infty} \frac{1}{(1-x)^j}$

24. Find the limit of the sequence defined by $S_1 = 1$, $S_{n+1} = S_n + 1/3^n$.
    [HINT: Write out the first few terms to see the pattern.]

25. Find the limit of the sequence defined by $a_1 = 4$, $a_{n+1} = a_n - 1/2^n$.

For each of the following series, find an expression for the partial sum S_n. Then use the expression for S_n to decide whether the series converges and, if so, to find its limit.

26. $\displaystyle\sum_{k=0}^{\infty}(\arctan(k+1) - \arctan k)$

29. $\displaystyle\sum_{j=1}^{\infty}\ln\left(1+\frac{1}{j}\right) = \sum_{j=1}^{\infty}(\ln(j+1) - \ln j)$

27. $\displaystyle\sum_{i=2}^{\infty}\frac{1}{i(i-1)} = \sum_{i=2}^{\infty}\left(\frac{1}{i-1} - \frac{1}{i}\right)$

30. $\displaystyle\sum_{k=0}^{\infty}\cos(k\pi)$

28. $\displaystyle\sum_{m=1}^{\infty}\left(\frac{1}{\sqrt{k}} - \frac{1}{\sqrt{k+2}}\right)$

31. $\displaystyle\sum_{j=2}^{\infty}(-1)^k k$

Determine whether each of the following series converges or diverges. If a series converges, find its limit. Justify your answers.

32. $2 - 2 + 2 - 2 + 2 - 2 + \cdots$

33. $\dfrac{3}{10} - \dfrac{3}{20} + \dfrac{3}{40} - \dfrac{3}{80} + \dfrac{3}{160} - \dfrac{3}{320} + \cdots$

34. $1 - \dfrac{1}{2} + \dfrac{1}{2} - \dfrac{1}{3} + \dfrac{1}{3} + \cdots$

35. $1 - 1 + 2 - 1 - 1 + 3 - 1 - 1 - 1 + 4 - 1 - 1 - 1 - 1 + \cdots$

36. $\dfrac{4}{7^{10}} + \dfrac{4}{7^{12}} + \dfrac{4}{7^{14}} + \dfrac{4}{7^{16}} + \dfrac{4}{7^{18}} + \cdots$

37. $\displaystyle\sum_{n=0}^{\infty}\frac{n+1}{2n+1}$

41. $\displaystyle\sum_{k=1}^{\infty}\left(\int_{k}^{k+1}\frac{dx}{x^2}\right)$

38. $\displaystyle\sum_{j=0}^{\infty}(\ln 2)^j$

42. $\displaystyle\sum_{n=1}^{\infty}\left(1+\frac{1}{n}\right)^n$

39. $\displaystyle\sum_{k=1}^{\infty}\frac{k^{\pi}}{k^e}$

43. $\displaystyle\sum_{j=1}^{\infty}\sqrt[j]{\pi}$

40. $\displaystyle\sum_{m=2}^{\infty}\frac{1}{(\ln 3)^m}$

44. $\displaystyle\sum_{n=1}^{\infty}\frac{\ln n}{\ln(3+n^2)}$

45. Let $S_n = \displaystyle\sum_{k=1}^{n} \frac{1}{\sqrt{k}}$.

  (a) Evaluate $\displaystyle\lim_{k\to\infty} \frac{1}{\sqrt{k}}$.

  (b) Show that $S_n \geq \dfrac{n}{\sqrt{n}} = \sqrt{n}$ for all $n \geq 1$.

  (c) Use part (b) to show that $\displaystyle\sum_{k=1}^{\infty} \frac{1}{\sqrt{k}}$ diverges.

46. A certain series $\displaystyle\sum_{k=1}^{\infty} a_k$ has partial sums $S_n = \displaystyle\sum_{k=1}^{n} a_k = 5 - \frac{3}{n}$.

  (a) Evaluate $S_{100} = \displaystyle\sum_{k=1}^{100} a_k$.

  (b) Evaluate $\displaystyle\sum_{k=1}^{\infty} a_k$.

  (c) Evaluate $\displaystyle\lim_{k\to\infty} a_k$.

47. A certain series $\displaystyle\sum_{j=1}^{\infty} b_j$ has partial sums $S_n = \displaystyle\sum_{j=1}^{n} b_j = \ln\left(\frac{2n+3}{n+1}\right)$.

  (a) Evaluate $\displaystyle\lim_{n\to\infty} S_n$.

  (b) Does the series converge? Justify your answer.

48. Suppose that the partial sums of the series $\displaystyle\sum_{k=1}^{\infty} a_k$ satisfy the inequality

$$\frac{6\ln n}{\ln\left(n^2+1\right)} < S_n < 3 + ne^{-n}$$

for all $n \geq 100$.

  (a) Does the series converge? If so, to what limit? Justify your answers.

  (b) What, if anything, can be said about $\displaystyle\lim_{k\to\infty} a_k$? Explain.

49. Let $S_n = \displaystyle\sum_{k=1}^{n} a_k$ and suppose that $0 \leq S_n \leq 100$ for all $n \geq 1$.

  (a) Give an example of a sequence $\{a_k\}$ satisfying the conditions above but for which $\displaystyle\sum_{k=1}^{\infty} a_k$ diverges.

  (b) Show that if $a_k > 0$ for all $k \geq 1$, then $\displaystyle\sum_{k=1}^{\infty} a_k$ converges.

  (c) Show that if $a_k > 0$ for all $k \geq 10^6$, then $\displaystyle\sum_{k=1}^{\infty} a_k$ converges.

50. Suppose that $\sum_{k=1}^{\infty} a_k$ diverges.

   (a) Explain why $a_k > 0$ for all $k \geq 1$ implies that $\lim_{n \to \infty} S_n = \infty$.

   (b) Give an example of a divergent series for which $\lim_{n \to \infty} S_n$ does not exist.

51. Let S_n denote the nth partial sum of the **harmonic series** $\sum_{k=1}^{\infty} \frac{1}{k} = 1 + \frac{1}{2} + \frac{1}{3} + \frac{1}{4} + \cdots$
   (i.e., $S_n = \sum_{k=1}^{n} \frac{1}{k}$).

   (a) Complete the table below; report answers to 3 decimal places. In the third row, enter *differences* between successive entries in the second row. A few entries are given.

n	10	20	30	40	50	60	70	80	90	100
S_n	2.929	3.598								5.187
ΔS_n	—	0.671								

   (b) Do the numbers above suggest clearly whether the harmonic series converges or diverges? Why or why not?

   (c) Complete the table below; report answers to 3 decimal places. In the third row, enter *differences* between successive entries in the second row. A few entries are given.

n	2	4	8	16	32	64	128	256	512	1024
S_n	1.500	2.083								7.509
ΔS_n	—	0.583								0.693

   (d) Do the numbers above suggest clearly whether the harmonic series converges or diverges? Why or why not?

   (e) The bottom row of the table in (c) shows that doubling n causes S_n to increase by about 0.693. Use this fact to guess values for S_{2048}, S_{4096}, and S_{8192}. (Don't try to calculate these numbers directly!)

52. What is wrong with the following argument?

   Let $S = 1 + 2 + 4 + 8 + \cdots$. Then $2S = 2 + 4 + 8 + \cdots = S - 1$, so $S = -1$.

53. Let $H_n = \sum_{k=1}^{n} \frac{1}{k}$ and $I_n = \int_{1}^{n+1} \frac{dx}{x}$.

   (a) Let L_n be the left Riemann sum approximation, with n equal subdivisions, to I_n. Show that $L_n = H_n$.

   (b) Use part (a) to show that the harmonic series diverges. [HINT: Start by comparing L_n and I_n.]

54. Let $H_n = \sum_{k=1}^{n} \dfrac{1}{k}$ and $a_m = \sum_{j=1}^{2^{m-1}} \dfrac{1}{2^{m-1}+j}$. Then $H_{2^n} = 1 + \sum_{m=1}^{n} a_m$.

[NOTE: a_m is the sum of a "block" of 2^{m-1} consecutive terms of the harmonic series—those from $n = 2^{m-1} + 1$ through $n = 2^m$.]

(a) Show that $a_1 = \dfrac{1}{2}$, $a_2 = \dfrac{7}{12}$, and $a_3 = \dfrac{533}{840}$.

(b) Show that $H_8 = 1 + a_1 + a_2 + a_3 = \dfrac{761}{280}$.

(c) Show that $a_k \geq \dfrac{1}{2}$ for all $k \geq 1$.

[HINT: $\dfrac{1}{2^{m-1}+j} \leq \dfrac{1}{2^{m-1}+2^{m-1}}$ when $1 \leq j \leq 2^{m-1}$.]

(d) Use part (b) to show that $\lim\limits_{n \to \infty} H_n = \infty$ (i.e., the harmonic series diverges).

55. Let $H_n = \sum_{k=1}^{n} \dfrac{1}{k}$ and $I = \int_{1}^{2} \dfrac{dx}{x}$.

(a) Let L_n be the left Riemann sum approximation to I with n equal subdivisions. Show that $L_n = \sum_{k=0}^{n-1} \dfrac{1}{n+k} = \dfrac{1}{n} + \dfrac{1}{n+1} + \dfrac{1}{n+2} + \cdots + \dfrac{1}{2n-1}$.

(b) Explain why $L_n > \ln 2$ for all $n \geq 1$.

(c) Show that $H_{2n} - H_n = L_n - \dfrac{1}{2n}$ for all $n \geq 1$.

(d) Show that $\lim\limits_{n \to \infty} H_{2n} - H_n = \ln 2$.

(e) Explain why part (d) implies that the harmonic series diverges.

(f) Explain why $H_{2n} - H_n > \ln 2 - \dfrac{1}{2}$ for all $n \geq 1$.

[HINT: Use parts (b) and (c).]

(g) Show that $H_2 > 1 + \ln 2 - \frac{1}{2}$.

(h) It follows from parts (d) and (e) that $H_{2^m} > 1 + m \left(\ln 2 - \frac{1}{2} \right)$ when $m \geq 1$.

56. Consider the series $\sum_{k=1}^{\infty} \dfrac{1}{k^p}$ with $p > 1$. This exercise outlines a proof that this series converges.

(a) Let S_n denote the partial sum $\sum_{k=1}^{n} \dfrac{1}{k^p}$. Show that the sequence of partial sums $\{ S_n \}$ is increasing.

(b) Show that $S_{2m+1} = 1 + \sum_{k=1}^{m} \dfrac{1}{(2k)^p} + \sum_{k=1}^{m} \dfrac{1}{(2k+1)^p}$.

(c) Explain why $S_{2m+1} < 1 + 2 \sum_{k=1}^{m} \dfrac{1}{(2k)^p}$.

[HINT: $1/(x+1) < 1/x$ for all $x > 0$.]

(d) Show that $S_{2m+1} < 1 + 2^{1-p} S_{2m+1}$.

[HINT: First show that $S_{2m+1} < 1 + 2^{1-p} S_m$.]

(e) Show that $\{ S_n \}$ is bounded above.

11.3   Testing for convergence; estimating limits

Converge or diverge: what's the question?

In theory, the question of convergence is simple:

> *The series $\sum a_k$ converges to S if the sequence $\{S_n\}$ of partial sums tends to S.*

The trouble, in practice, is that a simple, explicit formula for S_n is often unavailable.➤

The harmonic series poses this problem.

It might seem, then, that testing for convergence—let alone finding a limit—would be difficult or impossible. Surprisingly, this isn't so. All the convergence tests of this section and the next (comparison test, integral test, ratio test, etc.) offer clever, indirect, ways of testing whether $\{S_n\}$ converges—even without knowing each S_n exactly.

Nonnegative series

This apparent sleight-of-hand depends on a surprisingly simple observation:➤

Convince yourself of this simple but important fact.

> *If $a_k \geq 0$ for all k, then the sequence $\{S_n\}$ of partial sums of $\sum a_k$ is* nondecreasing:
>
> $$S_1 \leq S_2 \leq S_3 \leq \ldots S_n \leq S_{n+1} \ldots .$$

(A series with this convenient property is called **nonnegative**.) By Theorem 3 of Section 11.1, any nondecreasing sequence must either converge or *diverge to infinity*. To decide whether a nonnegative series $\sum a_k$ converges, therefore, it's enough to answer this key question:

> *As $n \to \infty$, do the partial sums $\{S_n\}$ blow up or remain bounded?*

If a simple formula for S_n is available (as, e.g., for *geometric* series) the answer may be obvious. If not, our best recourse is to try to **compare**➤ the given series to something (another series, an integral—anything that works) better understood. Sometimes an obvious comparison suggests itself.

Watch for more uses of this word.

■ **Example 1.**   Does $\displaystyle\sum_{k=0}^{\infty} \frac{1}{2^k + 1}$ converge?

Solution: We saw in the previous section➤ that the similar series $\sum_{k=0}^{\infty} 1/2^k$ converges to 2. Because for all k,

cf. Example 1, page 225.

$$\frac{1}{2^k + 1} < \frac{1}{2^k},$$

each partial sum of $\sum_{k=0}^{\infty} 1/(2^k + 1)$ is *less than* the corresponding partial sum of $\sum_{k=0}^{\infty} 1/2^k$. Because the latter sums tend to 2, the former must tend to a limit less than 2. Briefly, in symbols:

$$\frac{1}{2^k + 1} < \frac{1}{2^k} \implies \sum_{k=0}^{\infty} \frac{1}{2^k + 1} < \sum_{k=0}^{\infty} \frac{1}{2^k} = 2.$$

Numerical evidence➤ agrees. For the original series,

Courtesy of Maple.

$$S_{10} \approx 1.263523535; \quad S_{20} \approx 1.264498826; \quad S_{100} \approx 1.264499781. \quad \square$$

The comparison test: one series *vs.* another

In the previous example we showed that one series converges by comparing[←] it to another series that is *known* to converge. The following theorem makes this idea precise, duly noting all necessary hypotheses.

Theorem 6 (Comparison test for nonnegative series). Suppose that for all $k \geq 1$, $0 \leq a_k \leq b_k$. Consider the series $\sum_{k=1}^{\infty} a_k$ and $\sum_{k=1}^{\infty} b_k$.

- If $\sum_{k=1}^{\infty} b_k$ *converges*, so does $\sum_{k=1}^{\infty} a_k$, and

$$\sum_{k=1}^{\infty} a_k \leq \sum_{k=1}^{\infty} b_k.$$

- If $\sum_{k=1}^{\infty} a_k$ *diverges*, so does $\sum_{k=1}^{\infty} b_k$.

Notice:

What we'd expect. The theorem's assertion is simple and plausible, and not difficult to prove. A formal proof depends on the fact that like the terms themselves, the *partial sums* of $\sum a_k$ are *all* less than the corresponding partial sums of $\sum b_k$.

Successful comparisons. In a "successful" comparison, either both series converge or both diverge. To use the comparison test, therefore, it's necessary first to *guess* whether the series in question converges or diverges.[←]

Comparing tails. We assumed above that the comparison inequality

$$0 \leq a_k \leq b_k$$

holds for *all* possible k. This assumption isn't really necessary; if N is *any* positive integer, and $0 \leq a_k \leq b_k$ for all $k \geq N$, then

$$\sum_{k=N}^{\infty} a_k \leq \sum_{k=N}^{\infty} b_k.$$

We'll use this fact below to estimate a limit numerically.

Comparing integrals, comparing series. An almost identical comparison theorem applies to *improper integrals*.[←] Theorem 1, Section 10.2, says that if $0 \leq a(x) \leq b(x)$ for all $x \geq 1$, then

$$\int_1^{\infty} a(x)\, dx \leq \int_1^{\infty} b(x)\, dx.$$

■ **Example 2.** As we saw above, $\displaystyle\sum_{k=0}^{\infty} \frac{1}{2^k + 1}$ converges. How closely does $S_{100} \approx 1.264499781$ approximate the true limit S?

Solution: The error in using S_{100} to estimate the limit of *any* series comes from ignoring the "tail" R_{100}. In symbols:

$$S = \sum_{k=0}^{\infty} a_k = \sum_{k=0}^{100} a_k + \sum_{k=101}^{\infty} a_k = S_{100} + R_{100}.$$

Because for our series

$$\frac{1}{2^k + 1} < \frac{1}{2^k}$$

for all $k \geq 101$,➤ it follows that

The inequality actually holds for all k; here we care only about $k \geq 101$.

$$\sum_{k=101}^{\infty} \frac{1}{2^k + 1} < \sum_{k=101}^{\infty} \frac{1}{2^k}.$$

We can calculate the right side directly:

$$\sum_{k=101}^{\infty} \frac{1}{2^k} = \frac{1}{2^{101}} + \frac{1}{2^{102}} + \frac{1}{2^{103}} + \cdots = \frac{1}{2^{101}}\left(1 + \frac{1}{2} + \frac{1}{2^2} + \cdots\right) = \frac{1}{2^{100}}.$$

Thus, as expected, we commit very little➤ error by ignoring the tail R_{100}; the estimate S_{100} differs from S by less than $1/2^{100} \approx 8 \times 10^{-32}$.   □

Very, very little.

Compared to what?

The idea of comparison is easy. Harder, in practice, is deciding what to compare a series *to*. Our only reliable "benchmark" series, so far, are geometric series. The **integral test**➤ will enlarge considerably our stock of "known" series.

Just below.

The integral test: series *vs.* integrals

Since

$$\int_1^{\infty} \frac{1}{x}\,dx \quad \text{diverges, but} \quad \int_1^{\infty} \frac{1}{x^2}\,dx \quad \text{converges,}$$

one might guess, by analogy, that

$$\sum_{k=1}^{\infty} \frac{1}{k} \quad \text{diverges, but} \quad \sum_{k=1}^{\infty} \frac{1}{k^2} \quad \text{converges.}$$

These guesses are correct, as graphical evidence has already suggested.

Integrals and series: comparing areas

Thinking of the terms of a positive series $\sum_{k=1}^{\infty} a_k$ as rectangular *areas*➤ clarifies the connection with the integral $\int_1^{\infty} a(x)\,dx$.

Each has base 1.

The first picture shows one way of doing so:

If $\int_1^{\infty} a(x)\,dx$ diverges, so must $\sum_{k=1}^{\infty} a_k$

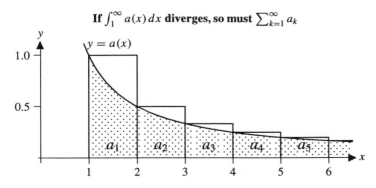

Look closely:

Total areas. The successive rectangles have *heights* $a(1) = a_1, a(2) = a_2, a(3) = a_3$, etc.; each *base* is 1. Their respective *areas*, therefore, are a_1, a_2, a_3, etc. Here's the key:

> *The series* $a_1 + a_2 + a_3 + a_4 + \ldots$ *represents the total "left-rule" rectangular area from 1 to* ∞.

An important inequality. The shaded area—*less than the rectangular area*—represents the *integral* $\int_1^\infty a(x)\,dx$. Thus if the *integral* diverges, so must the series. ◄ Here's the picture's message, in inequality form:

We'll collect our results in the next theorem.

$$a_1 + a_2 + a_3 + \cdots \geq \int_1^\infty a(x)\,dx.$$

If the right side diverges to infinity, so must the left.

A decreasing integrand. The reasoning that led to the inequality above requires that the integrand a be *decreasing* for $x \geq 1$, as shown in the picture. We'll collect such hypotheses carefully in the next theorem.

In the next picture an integral bounds a series *from above*:

If $\int_1^\infty a(x)\,dx$ converges, so must $\sum_{k=1}^\infty a_k$

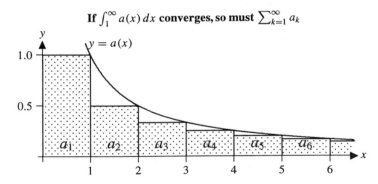

Again, successive rectangles have areas a_1, a_2, a_3, etc. This time, comparing areas shows that ◄

Convince yourself; notice how the first term is "broken off."

$$a_1 + a_2 + a_3 + \cdots \leq a_1 + \int_1^\infty a(x)\,dx.$$

If the right side converges, so must the left.

Combining both inequalities above gives upper *and* lower bounds for the series:

$$\int_1^\infty a(x)\,dx \leq \sum_{k=1}^\infty a_k \leq a_1 + \int_1^\infty a(x)\,dx.$$

In particular:

> *The integral* $\int_1^\infty a(x)\,dx$ *and the series* $\sum_{k=1}^\infty a_k$ *either both converge or both diverge.*

The final picture relates the tails of an integral and of a series:

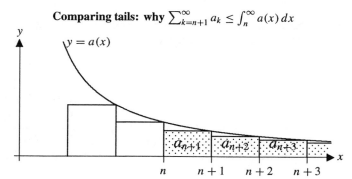

Comparing tails: why $\sum_{k=n+1}^{\infty} a_k \leq \int_n^{\infty} a(x)\, dx$

It shows—again, by comparing areas:

For any n, the tail R_n satisfies

$$R_n = \sum_{k=n+1}^{\infty} a_k \leq \int_n^{\infty} a(x)\, dx.$$

The pictures above show a particular function $a(x)$, but the conclusions we drew are "generic": they hold for *any* function $a(x)$ which is both *positive and decreasing.*➧

It's time to collect all the observations above in a theorem. As above, $a_k = a(k)$, for all integers $k \geq 1$.

What could go wrong if $a(x)$ isn't decreasing?

> **Theorem 7 (Integral test for positive series).** Suppose that for all $x \geq 1$, the function $a(x)$ is continuous, positive, and decreasing. Consider the series $\sum_{k=1}^{\infty} a_k$ and the integral $\int_1^{\infty} a(x)\, dx$.
>
> - If either diverges, so does the other.
>
> - If either converges, so does the other. In this case,
>
> $$\int_1^{\infty} a(x)\, dx \leq \sum_{k=1}^{\infty} a_k \leq a_1 + \int_1^{\infty} a(x)\, dx.$$
>
> - If the series converges, then
>
> $$R_n = \sum_{k=n+1}^{\infty} a_k \leq \int_n^{\infty} a(x)\, dx.$$

P-series, convergent and divergent

Series of the form $\displaystyle\sum_{k=1}^{\infty} \frac{1}{k^p}$ are called *p*-**series**.

■ **Example 3.**   Which *p*-series converge?

Solution: We found in Chapter 10 that the integral $\int_1^{\infty} dx/x^p$ converges only if $p > 1$. Therefore, using the integral test:➧

Are all hypotheses really satisfied?

The p-series $\displaystyle\sum_{k=n+1}^{\infty} \frac{1}{k^p}$ *converges if and only if $p > 1$.*

In particular, the harmonic series $\sum 1/k$ diverges.   □

(**Harmonic divergence.**) The fact that the **harmonic series**

$$1 + \frac{1}{2} + \frac{1}{3} + \frac{1}{4} + \frac{1}{5} + \dots$$

diverges to infinity—even though the terms themselves tend to zero—has fascinated mathematicians for many centuries. One early proof (not the one given here!) is attributed to Nicole Oresme, a 14th-century French bishop, scientist, and mathematician.

■ **Example 4.**   The p-series $\sum_{k=1}^{\infty} 1/k^3$ converges, by the integral test, to some limit S. How large must n be to ensure that S_n differs from S by less than 0.0001?

See the "Comparing tails" picture.

Solution: We'll choose n so that the tail R_n is less than 0.0001. The theorem shows how. By the last part,◄

$$R_n \le \int_n^{\infty} \frac{1}{x^3}\, dx.$$

What do you get?

The right side is easily calculated; we get $1/2n^2$.◄ This quantity, and hence also R_n, is less than 0.0001 if $n \ge 71$. Hence $S_{71} \approx 1.20196$ differs from the true limit by less than 0.0001.   □

■ **Example 5.**   Does $\displaystyle\sum_{k=1}^{\infty} \frac{1}{10k+1}$ converge?

Solution: The integral test says no.

$$\int_1^{\infty} \frac{1}{10x+1}\, dx = \lim_{n\to\infty} \int_1^n \frac{1}{10x+1}\, dx = \lim_{n\to\infty} \left.\frac{\ln(10x+1)}{10}\right]_1^n = \infty.$$

The comparison test *also* says no. Because for all $k \ge 1$,

$$\frac{1}{10k+1} \ge \frac{1}{11k},$$

it follows that

$$\sum_{k=1}^{\infty} \frac{1}{10k+1} \ge \sum_{k=1}^{\infty} \frac{1}{11k} = \frac{1}{11} \sum_{k=1}^{\infty} \frac{1}{k}.$$

Since the last series diverges, so must the first.   □

The ratio test: comparison with a geometric series

In a geometric series

$$a + ar + ar^2 + ar^3 + ar^4 + \dots$$

Why? See the Fact on page 228.

the **ratio of successive terms** is r; the series converges if and only if $|r| < 1$.◄

The **ratio test** is based on the same principle. It amounts, in the end, to a slightly disguised form of comparison with a geometric series.

> **Theorem 8 (Ratio test for positive series).** Suppose that $\sum_{k=1}^{\infty} a_k$ is a positive series, and that
>
> $$\lim_{k \to \infty} \frac{a_{k+1}}{a_k} = L.$$
>
> - If $L < 1$, then $\sum a_k$ converges.
>
> - If $L > 1$, then $\sum a_k$ diverges.
>
> - If $L = 1$, either convergence or divergence is possible. The test is inconclusive.

Notice:

When the ratio tends to 1. For many series, unfortunately, $a_{k+1}/a_k \to 1$ as $k \to \infty$. In such cases, the ratio test says *nothing*. This happens for *every* p-series $\sum 1/k^p$, for instance.

When the ratio test works best. The ratio test works best for such series as $\sum 1/k!$, $\sum r^k$, and $\sum 1/(2^k + 3)$, in which the index k appears in an exponent or a factorial.

Fine print: the idea of a proof

To illustrate the connection between the ratio test and geometric series, and to give the idea of a proof, let's suppose that

$$\lim_{k \to \infty} \frac{a_{k+1}}{a_k} = \frac{1}{2}.$$

Why must $\sum a_k$ converge?

The idea is that for large k, $a_{k+1} \approx a_k/2$, so the series behaves similarly to a geometric series. Suppose, for instance, that $a_{k+1} < 0.6a_k$ for all $k \geq 1000$. ➤ Then

The inequality must hold for all large k because of the limit above.

$$a_{1001} < (0.6)a_{1000}, \quad a_{1002} < (0.6)a_{1001} < (0.6)^2 a_{1000}, \quad a_{1003} < (0.6)^3 a_{1000}, \quad \dots ;$$

so

$$\sum_{k=1000}^{\infty} a_k = a_{1000} + a_{1001} + a_{1002} + \dots < a_{1000} \left(1 + (0.6) + (0.6)^2 + (0.6)^3 + \dots \right).$$

The last inequality is the point: It shows that $\sum a_k$ converges *by comparison with the geometric series* $\sum a_{1000}(0.6)^k$.

The *divergence* statement can be proved in a similar way.

■ **Example 6.** Show that $\sum_{k=0}^{\infty} \frac{1}{k!}$ converges. Guess a limit.

Solution: The ratio test works nicely. ➤ Since

Check details carefully.

$$\lim_{k \to \infty} \frac{a_{k+1}}{a_k} = \lim_{k \to \infty} \frac{k!}{(k+1)!} = \lim_{k \to \infty} \frac{1}{k+1} = 0,$$

the series *converges*. In fact, it converges very, very fast. Here are some representative partial sums:

$S_5 \approx 2.716666667;$ $S_{10} \approx 2.718281803;$ $S_{30} \approx 2.718281828459045235360287477133.$

Maple says so.

Is e somehow involved? To 30 decimals, ◂

$$e = 2.718281828459045235360287471135.$$

In fact, this series *does* converge to e. We'll explore this phenomenon further in later sections.   ☐

■ **Example 7.** Does $\displaystyle\sum_{k=0}^{\infty} \frac{100^k}{k!}$ converge?

Solution: Yes, by the ratio test:

$$\lim_{k\to\infty} \frac{a_{k+1}}{a_k} = \lim_{k\to\infty} \frac{100^{k+1}}{(k+1)!} \cdot \frac{k!}{100^k} = \lim_{k\to\infty} \frac{100}{k+1} = 0.$$

Notice what the result means: even though 100^k grows very fast, $k!$ grows faster still.   ☐

Exercises

1. Consider the series $S_n = \displaystyle\sum_{k=0}^{n} a_k$ where $a_k = \dfrac{1}{k + 2^k}$.

   (a) Use the comparison test to show that the series converges.
   [HINT: $a_k \le 2^{-k}$ for all $k \ge 0$.]

   (b) Show that $0 \le R_{10} \le 2^{-10}$.

   (c) Compute an estimate of the limit of the series that is guaranteed to be within 0.001 of the exact value.

   (d) Is your estimate in part (c) an overestimate or an underestimate? Justify your answer.

2. Consider the series $S_n = \displaystyle\sum_{j=0}^{n} \dfrac{1}{2 + 3^j}$.

   (a) Show that the series converges.

   (b) Estimate the value of the limit of the series within 0.01.

   (c) Is your estimate in part (b) an overestimate or an underestimate? Justify your answer.

3. Suppose that for all $k \ge 1$, $0 \le a_k \le b_k$. Let $S_n = \displaystyle\sum_{k=1}^{n} a_k$ and $T_n = \displaystyle\sum_{k=1}^{n} b_k$.

   (a) Suppose that $\displaystyle\sum_{k=1}^{\infty} b_k$ converges. Explain why there is a number M such that $S_n \le T_n \le M$ for all $n \ge 1$.

   (b) Explain why $\{\, S_n \,\}$ is an increasing sequence.

   (c) Explain why parts (a) and (b) together imply that $\displaystyle\sum_{k=1}^{\infty} a_k$ converges.

(d) Suppose that $\sum_{k=1}^{\infty} a_k$ diverges. Explain why $\lim_{n \to \infty} S_n = \infty$.

(e) Suppose that $\sum_{k=1}^{\infty} a_k$ diverges. Use part (d) to show that $\sum_{k=1}^{\infty} b_k$ diverges.

Suppose that $a(x)$ is continuous, positive, and decreasing for all $x \geq 1$ and that $a_k = a(k)$ for all integers $k \geq 1$.

4. Rank the values $\int_1^n a(x)\,dx$, $\sum_{k=1}^{n-1} a_k$, and $\sum_{k=2}^{n} a_k$ in increasing order.

   [HINT: Draw a picture.]

5. Rank the values $\int_n^{\infty} a(x)\,dx$, $\sum_{k=n+1}^{\infty} a_k$, and $\int_{n+1}^{\infty} a(x)\,dx$ in increasing order.

6. Draw a carefully annotated picture that shows why $\int_1^{n+1} a(x)\,dx \leq \sum_{k=1}^{n} a_k$.

7. Draw a carefully annotated picture that shows why $\sum_{k=2}^{n} a_k \leq \int_1^n a(x)\,dx$.

8. Draw a carefully annotated picture that shows why

$$\sum_{k=n+1}^{\infty} a_k \leq a_{n+1} + \int_{n+1}^{\infty} a(x)\,dx \leq \int_n^{\infty} a(x)\,dx.$$

9. Suppose that $a(x)$ is continuous, positive, and decreasing for all $x \geq 1$, that $a_k = a(k)$ for all integers $k \geq 1$, and that $\int_1^{\infty} a(x)\,dx$ converges.

   (a) Explain why the sequence of partial sums $\{ S_n \}$ is an increasing sequence.

   (b) Explain why $\int_1^n a(x)\,dx \leq \int_1^{\infty} a(x)\,dx$.

   (c) Use parts (a) and (b) to show that the sequence of partial sums $\{ S_n \}$ converges.

Use the integral test to find upper and lower bounds on the limit of each of the following series.

10. $\sum_{k=0}^{\infty} \dfrac{1}{k^2 + 1}$

12. $\sum_{j=1}^{\infty} je^{-j}$

11. $\sum_{k=1}^{\infty} \dfrac{1}{k\sqrt{k}}$

13. Use the ratio test to show that each of the following series converges.

(a) $\displaystyle\sum_{j=0}^{\infty} \frac{j^2}{j!}$

(b) $\displaystyle\sum_{k=1}^{\infty} \frac{2^k}{k!}$

14. Let $H_n = \displaystyle\sum_{k=1}^{n} \frac{1}{k}$ and $S_n = \displaystyle\sum_{k=0}^{n} \frac{1}{2k+1}$.

(a) Explain why $\displaystyle\lim_{n\to\infty} H_n = \infty$.

(b) Show that $S_n \geq \frac{1}{2} H_n$.

(c) What do the results in parts (a) and (b) imply about $\displaystyle\sum_{k=0}^{\infty} \frac{1}{2k+1}$? Explain.

15. Let $H_n = \displaystyle\sum_{k=1}^{n} \frac{1}{k}$ be the nth partial sum of the harmonic series.

(a) Show that $\ln(n+1) < H_n < 1 + \ln n$.   [HINT: Use the integral test.]

(b) Use part (a) to show that $H_N > 10$ implies that $N > 8000$.

(c) Show that $\displaystyle\lim_{n\to\infty} \frac{H_n}{\ln n} = 1$

(d) Show that the sequence defined by $a_n = H_n - \ln n$ is decreasing.
[HINT: Explain why $\ln(n+1) - \ln n = \int_n^{n+1} x^{-1}\, dx > (n+1)^{-1}$.]

(e) Use part (d) to show that $\displaystyle\lim_{n\to\infty} a_n$ exists. (This limit, denoted γ, is called **Euler's constant**; $\gamma \approx 0.577$.)

16. For which values of p does $\displaystyle\sum_{n=2}^{\infty} \frac{1}{n(\ln n)^p}$ converge? Justify your answer.

17. Consider the series $\displaystyle\sum_{k=1}^{\infty} \frac{2 + \sin k}{k^2}$.

(a) Explain why Theorem 7 can't be used to prove that this series converges.

(b) Show that this series converges.

18. Consider the series $\displaystyle\sum_{k=1}^{\infty} a_k = \sum_{k=1}^{\infty} \frac{\ln k}{k}$.

(a) Use the integral test to show that the series diverges.   [HINT: The function $\ln x / x$ is monotone on $[3, \infty)$. Start by showing that $\sum_{k=3}^{\infty} a_k$ diverges.]

(b) Use the comparison test to show that the series diverges.

(c) Can the ratio test be used to show that the series diverges? Explain.

19. Consider the series $\displaystyle\sum_{k=1}^{\infty} a_k = \frac{1}{2} + \frac{1}{3} + \frac{1}{2^2} + \frac{1}{3^2} + \frac{1}{2^3} + \frac{1}{3^3} + \cdots$.

(a) Explain why $\displaystyle\lim_{k\to\infty} a_{k+1}/a_k$ does not exist.

(b) What does the ratio test say about the convergence of the series $\displaystyle\sum_{k=1}^{\infty} a_k$?

(c) Show that the series converges and evaluate its limit.   [HINT: Rewrite the given series as the sum of two series.]

20.   (a) What does the ratio test say about the convergence of the series

$$\frac{1}{2} + \frac{1}{2} + \frac{1}{4} + \frac{1}{4} + \frac{1}{8} + \frac{1}{8} + \cdots ?$$

(b) Does the series in part (a) converge or diverge? Explain.

Use the comparison test to show that each of the following series converges. Then find an upper bound on the limit of each series.

21. $\displaystyle\sum_{n=1}^{\infty} \frac{1}{n^2 + \sqrt{n}}$

23. $\displaystyle\sum_{m=1}^{\infty} \frac{1}{m\sqrt{1 + m^2}}$

22. $\displaystyle\sum_{j=0}^{\infty} \frac{1}{j + e^j}$

24. $\displaystyle\sum_{k=1}^{\infty} \frac{k}{\left(k^2 + 1\right)^2}$

Determine whether each of the following series converges or diverges. If a series converges, find an upper bound on its limit. Justify your answers.

25. $\dfrac{1}{100} + \dfrac{1}{200} + \dfrac{1}{300} + \cdots$

26. $\displaystyle\sum_{n=1}^{\infty} \frac{\arctan n}{1 + n^2}$

27. $\displaystyle\sum_{m=1}^{\infty} \frac{m^3}{m^5 + 3}$

28. $\displaystyle\sum_{k=2}^{\infty} \frac{1}{k \ln k}$

29. $\displaystyle\sum_{k=1}^{\infty} \frac{1}{\ln\left(10^k\right)}$

30. $\displaystyle\sum_{j=1}^{\infty} \frac{j}{2^j}$

31. $\displaystyle\sum_{k=0}^{\infty} \frac{k^2}{5k^2 + 3}$

32. $1 - \dfrac{1}{2} - \dfrac{1}{3} - \dfrac{1}{4} - \dfrac{1}{5} - \cdots$

33. $\displaystyle\sum_{n=2}^{\infty} \frac{1}{\sqrt[3]{n^2 - 1}}$

34. $\displaystyle\sum_{k=1}^{\infty} \frac{\sqrt{k}}{k^2 + k + 1}$

35. $\displaystyle\sum_{m=0}^{\infty} e^{-m^2}$

36. $\displaystyle\sum_{n=0}^{\infty} \frac{n!}{(2n)!}$

37. $\displaystyle\sum_{k=0}^{\infty} \frac{k!}{(k + 1)! - 1}$

38. $\displaystyle\sum_{j=2}^{\infty} \frac{\ln j}{j^2}$

39. $\displaystyle\sum_{n=1}^{\infty} \left(\sum_{k=1}^{n} k^{-1} \right)$

40. $1 - \dfrac{1}{2} + \dfrac{1}{2} - \dfrac{1}{4} + \dfrac{1}{3} - \dfrac{1}{6} + \dfrac{1}{4} - \dfrac{1}{8} + \dfrac{1}{5} - \dfrac{1}{10} + \cdots$

For each series below, compute an estimate of its limit that is guaranteed to be in error by no more than 0.005. Justify your answer.

41. $\displaystyle\sum_{k=0}^{\infty} \frac{k}{k^6 + 17}$

42. $\displaystyle\sum_{k=2}^{\infty} \frac{1}{k\,(\ln k)^5}$

43. (a) Where in the proof of the integral test (Theorem 7) is the assumption that $a(x)$ is a decreasing function used?

 (b) Suppose that the requirement that $a(x)$ be decreasing for all $x \geq 1$ is replaced by the "weaker" requirement that $a(x)$ be decreasing for all $x \geq 10$. How does this change in assumptions affect the conclusions of Theorem 7?

44. Give an example of a divergent series $\sum a_k$ such that $a_k > 0$ and $a_{k+1}/a_k < 1$ for all $k \geq 1$.

45. Consider the series $\displaystyle\sum_{k=1}^{\infty} \frac{1}{k!}$.

 (a) Explain why $\dfrac{1}{k!} \leq \dfrac{1}{2^{k-1}}$ for all $k \geq 1$. [HINT: $n \geq 2 \implies 1/n \leq 1/2$.]

 (b) Show that $\dfrac{1}{k!} \leq \dfrac{1}{10!\,10^{k-10}}$ for all $k \geq 10$.

 (c) Explain why S_{10} underestimates the limit of the series. Find a bound on the magnitude of the approximation error. [HINT: Use part (b) to bound R_{10}.]

46. Let r be a positive number less than 1. Suppose that $\{a_k\}$ is a sequence of positive terms and that $a_{k+1}/a_k \leq r$ for all $k \geq 1$.

 (a) Show that $a_2 \leq a_1 r$ and that $a_3 \leq a_1 r^2$. (A similar argument shows that $a_{k+1} \leq a_1 r^k$.)

 (b) Use the result mentioned parenthetically in part (a) to show that $\displaystyle\sum_{k=1}^{\infty} a_k$ converges. [HINT: Use the comparison test and the formula for the sum of geometric series.]

 (c) Show that $R_n = \displaystyle\sum_{k=n+1}^{\infty} a_k \leq \dfrac{a_{n+1}}{1-r}$.

47. Let $a_k = 1/k$. Then $a_{k+1}/a_k < 1$ for all $k \geq 1$. Why can't the ideas outlined in the previous exercise be used to "prove" that the harmonic series converges?

48. Let x be a positive real number and N an integer such that $N \geq x$.

 (a) Show that $\dfrac{x^k}{k!} \leq \dfrac{x^N}{N!} \left(\dfrac{x}{N+1}\right)^{k-N}$ for all $k \geq N$.

 (b) Use part (a) to show that $\displaystyle\sum_{k=N}^{\infty} \frac{x^k}{k!} \leq \frac{x^N}{N!} \cdot \frac{1}{1 - \frac{x}{N+1}}$.

11.4 Absolute convergence; alternating series

Not-necessarily-positive series

The integral, comparison, and ratio tests, as stated in the last section, apply only to *nonnegative* series.➤ Not every interesting series, however, is nonnegative.

I.e., series without negative terms.

■ **Example 1.** From numerical and graphical evidence, does the **alternating harmonic series**

$$\sum_{k=1}^{\infty} \frac{(-1)^{k+1}}{k} = 1 - \frac{1}{2} + \frac{1}{3} - \frac{1}{4} + \frac{1}{5} - \dots$$

converge or diverge? To what limit?

Solution: As for any series, the question is how partial sums behave. Tabulating some of them➤ shows a pattern:

Computed by machine, of course.

Partial sums of $1 - \frac{1}{2} + \frac{1}{3} - \frac{1}{4} + \frac{1}{5} - \dots$										
n	1	2	3	4	5	6	7	8	9	10	...
S_n	1	0.500	0.833	0.583	0.783	0.616	0.759	0.634	0.745	0.645	...
n	51	52	53	54	55	56	57	58	59	60	...
S_n	0.703	0.683	0.702	0.684	0.702	0.684	0.702	0.684	0.701	0.685	...

Successive partial sums seem to hop back and forth *across* some limiting value. *Plotting* partial sums (together with terms) shows the same pattern:

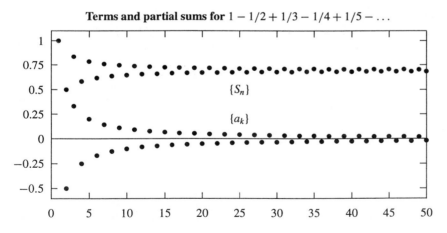

Terms and partial sums for $1 - 1/2 + 1/3 - 1/4 + 1/5 - \dots$

Because the *terms* alternate in sign, the *partial sums* successively rise and fall, alternately overshooting and undershooting the limiting value—apparently around 0.69.➤ □

It can be shown (with considerable effort) that the exact limit is $\ln 2 \approx 0.69315$.

In this section we develop tools for handling such not-necessarily-positive series.

Absolute *vs.* conditional convergence

The **alternating harmonic series** illustrates the phenomenon of **conditional convergence**. Although

$$1 - \frac{1}{2} + \frac{1}{3} - \frac{1}{4} + \frac{1}{5} - \dots$$

Obtained from the previous series by taking the absolute value of each term.

itself *converges*, as the previous example leads one to expect, the *ordinary* harmonic series◂

$$1 + \frac{1}{2} + \frac{1}{3} + \frac{1}{4} + \frac{1}{5} + \cdots$$

By the integral test.

diverges.◂

■ **Example 2.** Does $\displaystyle\sum_{k=1}^{\infty} \frac{\sin k}{k^2}$ converge? Does $\displaystyle\sum_{k=1}^{\infty} \frac{|\sin k|}{k^2}$? Estimate limits.

Solution: The first series, like the alternating harmonic series, has both negative and positive terms—though in no regular order, this time. Plotting terms and partial *But doesn't prove.* sums suggests◂ that this series, too, converges:

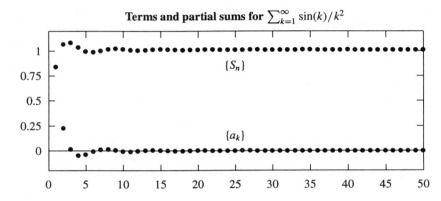

The partial sums wander slightly up and down, but still appear to approach a horizontal asymptote, perhaps near $y = 1$. We'll show below that this impression is correct.

The second series *also* seems to converge, this time to some limit near 1.25:

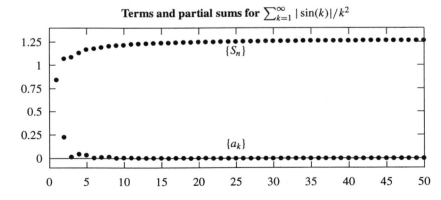

If $\sum a_k$ is the first series, $\sum |a_k|$ is the second.

The second series comes from the first by taking the *absolute* value of each term.◂ The new series is *nonnegative*, so the *comparison test* applies. Since

$$0 \le \frac{|\sin k|}{k^2} \le \frac{1}{k^2}$$

for all $k \ge 1$, and $\sum_{k=1}^{\infty} 1/k^2$ converges, so must $\sum_{k=1}^{\infty} |\sin k|/k^2$. □

This example illustrates the phenomenon of **absolute convergence**: not only does the original series $\sum_{k=1}^{\infty} a_k$ converge, but so does its "absolute version" $\sum_{k=1}^{\infty} |a_k|$.

Definition: Let $\sum_{k=1}^{\infty} a_k$ be any series.

- If $\sum_{k=1}^{\infty} |a_k|$ diverges but $\sum_{k=1}^{\infty} a_k$ converges, then $\sum_{k=1}^{\infty} a_k$ **converges conditionally.**

- If $\sum_{k=1}^{\infty} |a_k|$ converges, then $\sum_{k=1}^{\infty} a_k$ **converges absolutely.**

The wacky world of conditional convergence. Conditionally convergent series have some surprising properties. Here is one of the oddest:

Let $\sum a_k$ be conditionally convergent, and let L be *any* real number. Then the terms of $\sum a_k$ can be reordered in such a way that the resulting series converges to L.

(For more details, see your instructor.) Notice how drastically this peculiar property of conditionally convergent series upsets the naive hope that addition is commutative.

Plusses and minuses of plusses and minuses

Let $\sum_{k=1}^{\infty} a_k$ be any series. If, perchance, $a_k \geq 0$ for all k, the *advantage* is in simplicity: the partial sums are *nondecreasing*. The *disadvantage*—as the harmonic series shows—is that the partial sums may tend to infinity.

Mixing positive and negative terms may cost something in simplicity, but it's an *advantage* for convergence. As the *alternating* harmonic series shows, positive and negative terms can offset each other, thus helping in the cause of convergence.

Absolute convergence implies ordinary convergence

We saw in Chapter 10 that for *integrals* that if $\int_1^{\infty} |f(x)|\, dx$ converges, then so must $\int_1^{\infty} f(x)\, dx$, and

$$\left| \int_1^{\infty} f(x)\, dx \right| \leq \int_1^{\infty} |f(x)|\, dx.$$

Exactly the same principle applies to infinite series:

Theorem 9. If $\sum_{k=1}^{\infty} |a_k|$ converges, then so does $\sum_{k=1}^{\infty} a_k$, and

$$\left| \sum_{k=1}^{\infty} a_k \right| \leq \sum_{k=1}^{\infty} |a_k|.$$

Notice:

- The idea of a rigorous proof is to write the original series as a sum of two new series, one entirely positive and the other entirely negative. Using the comparison test, one can show that each of the new series converges.

- The theorem shows that the series $\sum_{k=1}^{\infty} \sin(k)/k^2$ of Example 2 does indeed converge, because $\sum_{k=1}^{\infty} |\sin k|/k^2$ does. Our limit estimates are also consistent with the theorem:

As the picture suggested.

Look back at the figures.

$$1 \approx \left| \sum_{k=1}^{\infty} \frac{\sin k}{k^2} \right| \leq \sum_{k=1}^{\infty} \frac{|\sin k|}{k^2} \approx 1.25.$$

■ **Example 3.** For which values of x does the **power series**

$$\sum_{k=1}^{\infty} kx^k = x + 2x^2 + 3x^3 + 4x^4 + \dots$$

A power series is something like an "infinite polynomial." More on power series in the next section.

converge?◂

Solution: We'll use the ratio test to check, first, for *absolute convergence.*◂

Watch the algebra.

$$\lim_{k \to \infty} \left| \frac{(k+1)x^{k+1}}{kx^k} \right| = \lim_{k \to \infty} \left| \frac{k+1}{k} \right| \cdot |x| = 1 \cdot |x| = |x|.$$

If $|x| < 1$, the original series converges absolutely. (Therefore, by Theorem 9 it also converges *without* absolute value signs.)

See for yourself.

It's easy to see◂ that for $|x| \geq 1$, the series *diverges*, by the nth term test. □

Using the theorem to estimate limits

Estimating a limit for any series—nonnegative or not—depends upon keeping the upper tail small. Theorem 9, combined with earlier estimates, can help.

We used Maple.

■ **Example 4.** For the series $\sum_{k=1}^{\infty} (-1)^{k+1} k^{-3}$, we find◂ that $S_{100} \approx 0.901542$. How closely does S_{100} approximate S, the true limit of the series?

Solution: Since

$$S = \sum_{k=1}^{\infty} \frac{(-1)^{k+1}}{k^3} = \sum_{k=1}^{100} \frac{(-1)^{k+1}}{k^3} + \sum_{k=101}^{\infty} \frac{(-1)^{k+1}}{k^3} = S_{100} + R_{100},$$

The first inequality uses Theorem 9; the second uses the integral test.

we need only estimate R_{100}.◂

$$|R_{100}| = \left| \sum_{k=101}^{\infty} \frac{(-1)^{k+1}}{k^3} \right| \leq \sum_{k=101}^{\infty} \frac{1}{k^3} < \int_{100}^{\infty} \frac{1}{x^3} \, dx.$$

Its lower limit is 100, not 101. Theorem 7, page 243, says why.

The last integral◂ is easy to calculate:

$$\int_{100}^{\infty} \frac{1}{x^3} \, dx = \frac{-1}{2x^2} \Bigg]_{100}^{\infty} = \frac{1}{20000} = 0.00005.$$

The estimate $S \approx S_{100} \approx 0.901542$, in other words, is good to at least four decimal places. □

Alternating series: convergence and estimation

For most series with both positive and negative terms, testing for absolute convergence is usually the best option. In the special◂ case that the terms alternate *strictly* in sign, we can sometimes do better.

But surprisingly useful.

> **Definition:** An **alternating series** is one whose terms alternate strictly in sign, i.e., a series of the form
>
> $$c_1 - c_2 + c_3 - c_4 + c_5 - c_6 + \dots ,$$
>
> where each c_i is positive.

The alternating harmonic series$^{➤}$

See Example 1—especially the picture.

$$1 - \frac{1}{2} + \frac{1}{3} - \frac{1}{4} + \frac{1}{5} - \ldots$$

represents the best possibility. Because successive terms alternate in sign and decrease in size, successive partial sums *straddle* smaller and smaller intervals. If the terms also tend to zero, then the partial sums narrow down on a limit.$^{➤}$

Here, more than ever, the picture is key.

The following theorem makes these observations formal, and gives a simple, convenient error bound.

Theorem 10 (Alternating series test). Consider the series

$$\sum_{k=1}^{\infty} (-1)^{k+1} c_k = c_1 - c_2 + c_3 - c_4 + \ldots,$$

where

- $c_1 \geq c_2 \geq c_3 \geq \cdots \geq 0$;

- $\lim_{k \to \infty} c_k = 0$.

Then the series converges, and its limit S lies between any two successive partial sums—i.e., for each $n \geq 1$, either $S_n \leq S \leq S_{n+1}$ or $S_{n+1} \leq S \leq S_n$. In particular,

$$|S - S_n| < c_{n+1}.$$

A formal proof is slightly tricky, but the underlying idea is simple. Because the limit S lies *between* successive partial sums, adding another term to any partial sum always "overshoots" the limit—hence the final inequality.

Although the hypotheses of this theorem seem restrictive, surprisingly many interesting series turn out to satisfy them.

Using the theorems: miscellaneous examples

Combining Theorems 9 and 10 with results from earlier sections, we can handle many not-necessarily-positive series, detecting convergence or divergence and, when possible, estimating limits. The following examples illustrate some useful tricks of this trade.

■ **Example 5.** (**An alternating p-series: another look.**) What does Theorem 10 say about $\sum_{k=1}^{\infty} (-1)^{k+1}/k^3$ and its 100th partial sum, $S_{100} \approx 0.9015422$?

Solution: In this context, $c_k = 1/k^3$. Now Theorem 10 says$^{➤}$ not only that the series converges—which we already knew—but also that

As it should.

$$|S - S_{100}| < c_{101} = \frac{1}{101^3} \approx 0.0000009.$$

Thus $S_{100} \approx 0.9015422^{➤}$ lies within 0.0000009 of the true limit S. Equivalently, S lies *between* $S_{100} \approx 0.9015422$ and $S_{101} \approx 0.9015431$.   □

Does S_{100} overshoot or undershoot? Why?

■ **Example 6.** (**The n-th term test: always available.**) Does $\sum_{j=1}^{\infty} (-1)^j \dfrac{j}{j+1}$ converge or diverge? Why?

Solution: The alternating series test looks tempting at first glance, but it doesn't apply. The given series *is* alternating, and its terms *do* decrease in magnitude, but the remaining hypothesis isn't satisfied:

$$\lim_{j \to \infty} \frac{j}{j+1} = \lim_{j \to \infty} \frac{1}{1 + 1/j} = 1,$$

Maybe we should call it the "j-th term test," here. In any case, the index name is immaterial.

Successive terms are alternately near 1 and −1.

not zero as Theorem 10 requires. The *n*-th term test◄ *does* apply, however. Since $j/(j+1) \to 1$ as $j \to \infty$, it follows that $(-1)^j j/(j+1)$ has *no limit* as $j \to \infty$.◄ It follows that the given series diverges.   □

■ **Example 7.**   Does the series

$$1 + 2 + 3 + 4 + 5 - \frac{1}{6} + \frac{1}{7} - \frac{1}{8} + \frac{1}{9} - \cdots$$

converge? If so, find or estimate the limit.

We saw basics of series algebra in Section 11.2.

Solution: The alternating series test doesn't apply right out of the box, because the first five terms break the desired pattern. The problem isn't fatal, however. Basic algebra◄ lets us group our terms into two blocks, as follows—

$$(1 + 2 + 3 + 4 + 5) - \left(\frac{1}{6} - \frac{1}{7} + \frac{1}{8} - \frac{1}{9} + \cdots \right).$$

The first block is finite, so convergence isn't an issue; its sum is 15. The second block, clearly, satisfies all hypotheses of the alternating series test, and so converges to some limit L. Any partial sum of the second block, moreover, differs from L by less than the magnitude of the *next* term.◄

By the the last line of Theorem 10.

The entire series, therefore, converges to $S = 15 + L$, and any partial sum differs from S by no more than the next term. The partial sum $S_9 = 1 + 2 + \cdots + 1/9 \approx 14.962$, for instance, overshoots the true limit by less than 1/10. In other words, the exact limit S satisfies $14.862 \le S \le 14.962$.◄   □

Adding more terms would approximate S more closely.

■ **Example 8.**   Does $\displaystyle\sum_{n=1}^{\infty} \frac{\sin n}{n^3 + n^2 + n + 1 + \cos n}$ converge or diverge? Why?

Solution: The problem is easier than it looks. Hoping for absolute convergence, we'll start by taking absolute values:

$$\left| \frac{\sin n}{n^3 + n^2 + n + 1 + \cos n} \right| = \frac{\cdot\, |\sin n|}{n^3 + n^2 + n + 1 + \cos n}.$$

From the general appearance of numerator and denominator, comparison suggests itself. Some simple inequalities make the job much easier:◄

Convince yourself of each.

$$\frac{|\sin n|}{n^3 + n^2 + n + 1 + \cos n} \le \frac{1}{n^3 + n^2 + n} \le \frac{1}{n^3}.$$

It's a p-series with $p = 3$.

We know that $\sum 1/n^3$ converges;◄ so, by comparison, must the absolute value version of the given series. By Theorem 9, the original series must converge, too. □

Exercises ——————————————————————————

1. We showed in Example 7, page 256, that the series $1 + 2 + 3 + 4 + 5 - \dfrac{1}{6} + \dfrac{1}{7} - \dfrac{1}{8} + \dfrac{1}{9} - \cdots$ converges to some limit S.

   (a) Does the series converge conditionally or absolutely? Why?

   (b) Calculate S_{15} for the series above. Does S_{15} overestimate or underestimate S? How do you know?

   (c) According to *Maple*, $S_{60} \approx 14.902$. Use this result to find good upper and lower bounds for S. Explain your answer.

   (d) We said in this section that the alternating harmonic series can be shown to converge to $\ln 2$. Use this fact to find the limit of the series above exactly.

2. In Example 8, page 256, we showed that $\displaystyle\sum_{n=1}^{\infty} \frac{\sin n}{n^3 + n^2 + n + 1 + \cos n}$ converges to some limit S, but we didn't find or estimate S.

   (a) Compute (not by hand!) S_{50}.

   (b) Explain why $|R_{50}| \le \displaystyle\int_{50}^{\infty} \frac{dx}{x^3}$.

   (c) Use the previous part to give good upper and lower bounds for S.

3. Consider the series $\displaystyle\sum_{k=1}^{\infty} (-1)^{k+1} a_k$. Suppose that the terms of the sequence $\{ a_k \}$ are positive and decreasing for all $k \ge 10^9$ and $\displaystyle\lim_{k \to \infty} a_k = 0$, but that $a_{10^9} > a_1$. Explain why the series converges. [HINT: Theorem 10 doesn't apply directly.]

Determine whether each of the following series converges absolutely, converges conditionally, or diverges. Find upper and lower bounds on the limit of each convergent series. Justify your answers.

4. $\displaystyle\sum_{k=1}^{\infty} \frac{(-1)^k}{\sqrt{k}}$

5. $\displaystyle\sum_{j=1}^{\infty} \frac{(-1)^{j+1}}{j^2}$

6. $\displaystyle\sum_{n=1}^{\infty} \frac{(-3)^n}{n^3}$

7. $\displaystyle\sum_{k=4}^{\infty} (-1)^k \frac{\ln k}{k}$

8. $\displaystyle\sum_{m=8}^{\infty} \frac{\sin m}{m^3}$

9. $\displaystyle\sum_{n=1}^{\infty} \frac{\cos(n\pi)}{n}$

10. $\displaystyle\sum_{k=0}^{\infty} (-1)^k \frac{k}{2k + 1}$

11. $\displaystyle\sum_{m=0}^{\infty} (-1)^m \frac{4m^3}{2^m}$

12. $\displaystyle\sum_{k=0}^{\infty} \frac{(-2)^k}{3^k + k}$

13. $\displaystyle\sum_{j=0}^{\infty} (-1)^j \frac{j!}{(j^2)!}$

14. $\displaystyle\sum_{n=1}^{\infty} (-1)^{n+1} \frac{\arctan n}{n}$

For each series below, show that the series converges. Then compute an estimate of the limit that is guaranteed to be in error by no more than 0.005.

15. $\displaystyle\sum_{k=1}^{\infty}(-1)^k\frac{3}{k^4}$

16. $\displaystyle\sum_{k=1}^{\infty}\frac{(-1)^k}{k^2+2^k}$

17. $\displaystyle\sum_{k=0}^{\infty}\frac{(-2)^k}{7^k+k}$

18. $\displaystyle\sum_{k=0}^{\infty}\frac{(-3)^k}{(k^2)!}$

19. $\displaystyle\sum_{k=5}^{\infty}(-1)^k\frac{k^{10}}{10^k}$

20. Does the infinite series

$$1-\frac{1}{2^3}+\frac{1}{3^2}-\frac{1}{4^3}+\frac{1}{5^2}-\frac{1}{6^3}+\frac{1}{7^2}-\frac{1}{8^3}+\cdots$$

converge or diverge? Justify your answer.

21. Give an example of a convergent series $\displaystyle\sum_{k=1}^{\infty}a_k$ with the property that $\displaystyle\sum_{k=1}^{\infty}(a_k)^2$ diverges.

22. Suppose that $a_k\geq 0$ for all $k\geq 1$. Is it possible that $\displaystyle\sum_{k=1}^{\infty}a_k$ converges conditionally? Explain.

23. Suppose that $\displaystyle\sum_{k=1}^{\infty}a_k$ diverges. Is it possible that $\displaystyle\sum_{k=1}^{\infty}|a_k|$ converges? Justify your answer.

24. Let $a_k=\dfrac{|\sin k|}{k^2}$.

(a) Show that $\dfrac{a_{k+1}}{a_k}=|\cos 1+\sin 1\cot k|\cdot\dfrac{n^2}{(n+1)^2}$.

(b) Explain why the result in part (a) implies that $\{a_k\}$ never becomes a decreasing sequence.

(c) Show that $\displaystyle\sum_{k=1}^{\infty}(-1)^{k+1}a_k$ converges.

25. (a) Show that $\displaystyle\sum_{k=n+1}^{\infty}\frac{(-1)^{k+1}}{k^2}\leq\sum_{k=n+1}^{\infty}\frac{1}{k^2}$ for any $n\geq 1$.

(b) Explain why the result in part (a) implies that $\displaystyle\sum_{k=1}^{\infty}\frac{(-1)^{k+1}}{k^2}$ converges faster than $\displaystyle\sum_{k=1}^{\infty}\frac{1}{k^2}$.

26. This exercise outlines a proof of the alternating series test.

Let $S_n = \sum_{k=1}^{n} (-1)^{k+1} c_k$ denote the partial sum of the first n terms of a series satisfying the hypotheses of the alternating series test (Theorem 10).

(a) Show that the sequence of even partial sums, $S_2, S_4, S_6, S_8, \ldots$, is monotone increasing.

(b) Show that the sequence of odd partial sums, $S_1, S_3, S_5, S_7, \ldots$, is monotone decreasing.

(c) Show that $S_{2m} \leq S_{2m-1}$ for any integer $m \geq 1$.

(d) Use part (c) to show that the sequence of even partial sums and the sequence of odd partial sums both converge. [NOTE: Although both sequences converge, we must still show that they converge to the *same* limit.]

(e) Show that $\lim_{m \to \infty} (S_{2m+1} - S_{2m}) = 0$. From this it follows that there is a real number S such that $\lim_{n \to \infty} S_n = S$.

(f) Explain why $0 < S - S_{2m} < c_{2m+1}$ and $0 < S_{2m+1} - S < c_{2m+2}$.

11.5   Power series

Basic ideas and examples

A **power series** is a series of the form

$$a_0 + a_1 x + a_2 x^2 + a_3 x^3 + a_4 x^4 + \cdots + a_n x^n + \cdots = \sum_{k=0}^{\infty} a_k x^k.$$

Each a_k is a *fixed constant*, called the **coefficient of** x^k. The symbol x denotes a *variable*.

■ **Example 1.   (A geometric power series.)** Among the simplest and most useful power series is

$$S(x) = 1 + x + x^2 + x^3 + x^4 + x^5 + \cdots = \sum_{k=0}^{\infty} x^k.$$

(For all $k \geq 0$, $a_k = 1$.) For which real numbers x does $S(x)$ converge?

Why?

Solution: Setting $x = 1$ gives the *divergent*[◄◄] series

$$S(1) = 1 + 1 + 1^2 + 1^3 + 1^4 + \cdots = 1 + 1 + 1 + 1 + 1 + \ldots.$$

Check the arithmetic.

If, say, $x = 1/2$, the series *converges to 2*:[◄◄]

$$S(1/2) = 1 + \frac{1}{2} + \frac{1}{4} + \frac{1}{8} + \cdots + \frac{1}{2^n} + \cdots = 2.$$

Recall these properties of geometric series? cf. Section 11.2.

Indeed, for *any* value of x, $S(x)$ is a *geometric series* in x, so it converges if and only if $|x| < 1$. We even know the limit:[◄◄]

If $|x| < 1$, then $1 + x + x^2 + x^3 + x^4 + \ldots$ converges to $\dfrac{1}{1 - x}$.

□

In the next section we'll see why.

■ **Example 2.**   It's a fact[◄◄] that for any number x,

$$e^x = 1 + x + \frac{x^2}{2!} + \frac{x^3}{3!} + \frac{x^4}{4!} + \cdots.$$

Interpret the right side in the language of power series.

Solution: Writing the series in the form

$$1 + x + \frac{x^2}{2!} + \frac{x^3}{3!} + \cdots = \sum_{k=0}^{\infty} \frac{1}{k!} x^k$$

shows the pattern of coefficients.

For what values of x does the series *converge*?

We'll use the *ratio test* to decide.   Since the ratio test works only for *positive series*, we'll use it to check for *absolute* convergence.   For any input x,

$$\lim_{k \to \infty} \frac{|a_{k+1} x^{k+1}|}{|a_k x^k|} = \lim_{k \to \infty} \frac{|x|^{k+1}}{(k+1)!} \cdot \frac{k!}{|x|^k} = \lim_{k \to \infty} \frac{|x|}{k+1} = 0.$$

Because $0 < 1$, the ratio test guarantees that this series➤ converges absolutely—and therefore in the ordinary sense—for *all* values of x.➤ We haven't yet shown *why* it's so, but the series converges to e^x for any x. If, say, $x = 1$, then the series converges to e, as partial sums attest numerically:

Unlike the previous one.

Recall: If a series converges with absolute value signs, then it converges without. cf. Theorem 9, page 253.

$$S_{10} \approx 2.718281803; \quad S_{20} \approx 2.718281830; \quad S_{30} \approx 2.71828182845905.$$

(The last number agrees with e in all 15 decimal places.)   □

Power series and polynomials

Power series are, roughly speaking, "unending" or "infinite-degree" polynomials. More precisely:

Terms are power functions. For both polynomials and power series, each summand is of the form $a_k x^k$, with k a nonnegative integer.

Partial sums are ordinary polynomials. Any partial sum S_n of the power series $\sum_{k=0}^{\infty} a_k x^k$ is

$$S_n = a_0 + a_1 x + a_2 x^2 + + a_3 x^3 + \cdots + a_n x^n,$$

an ordinary polynomial of degree n.

Easy to use. Polynomials are easy to differentiate and integrate, term by term; so—with due care➤ taken for convergence—are power series. We'll return soon to this theme and to its practical importance.

An important proviso!

Choosing base points

The polynomial expressions

$$p(x) = x^2 - 2x + 2, \quad q(x) = (x-1)^2 + 1, \quad \text{and} \quad r(x) = (x-2)^2 + 2(x-2) + 2$$

all represent the same function.➤ The differences have to do with different choices of **base point**. Version q, for instance, is said to be **expanded about the base point** $x = 1$, because q is written in powers of $(x - 1)$. Versions p and r are expanded about the base points $x = 0$ and $x = 2$, respectively.

Are you convinced? If not, multiply out q and r.

   Which version is "best" depends on the problem at hand. Version q, for instance, focuses attention most clearly on the graph's vertex, which occurs at the base point $x = 1$. Finding values and derivatives of the function at the base point is especially easy.

   The same choice of base point applies to power series. For example, the two power series➤

They're different functions, this time.

$$\sum_{k=0}^{\infty} 2^k x^k \quad \text{and} \quad \sum_{k=0}^{\infty} 2^k (x-1)^k$$

are written, respectively, in powers of x and powers of $(x - 1)$; their respective base points are $x = 0$ and $x = 1$. Mathematically, the difference isn't great: The second form amounts only to a "shift" of one unit to the right. We'll usually, but not always, treat power series based at $x = 0$.

Power series as functions

Any power series

$$S(x) = \sum_{k=0}^{\infty} a_k x^k = a_0 + a_1 x + a_2 x^2 + a_3 x^3 + \dots$$

defines, in a natural way, a *function* of x. For a given input x, $S(x)$ is the *limit*—if one exists—of the power series above.

Domains

For example, the natural domain of $\sqrt{x}$ is the set of nonnegative numbers.

Any function given by a "formula" in x has a *natural domain*: the set of x for which the formula makes sense.◄ Power series are no different. The domain of a function $S(x)$ given by a power series is the set of inputs x for which the series converges—also known as the **interval of convergence**. We saw above, for instance, that $\sum_{k=0}^{\infty} x^k$ converges for x in $(-1, 1)$; $\sum_{k=0}^{\infty} x^k/k!$ converges for x in $(-\infty, \infty)$.

■ **Example 3.** A function $S(x)$ is defined by the power series

$$S(x) = \sum_{k=0}^{\infty} 2^k x^k = 1 + 2x + 4x^2 + 8x^3 + \dots.$$

What's the domain of S? Is a simpler formula available?

In other words, $S(x)$ is "geometric in $2x$."

Solution: Think of $S(x)$ as a *geometric series* $1 + r + r^2 + r^3 + \dots$, with $r = 2x$:◄

$$S(x) = 1 + 2x + (2x)^2 + (2x)^3 + \dots.$$

A geometric series converges if $|r| < 1$; *this* one converges, therefore, if $|2x| < 1$, i.e., if $|x| < 1/2$. In that case the *limit* is $1/(1-r) = 1/(1-2x)$. To summarize: The power series $S(x)$ converges for x in $(-1/2, 1/2)$; on that domain,

$$S(x) = 1 + 2x + (2x)^2 + (2x)^3 + \dots = \frac{1}{1-2x}.$$

□

Finding the interval of convergence

The first task, given a power series, is to find the interval of convergence. In principle, finding this interval might be difficult; it practice, it's often simple. In many cases of interest the *ratio test* is all that's needed. We illustrate with several important examples.

■ **Example 4.** Show that the series

$$1 + 2x + 3x^3 + 4x^3 + 5x^4 + \dots = \sum_{k=1}^{\infty} k x^{k-1}$$

converges only for x in $(-1, 1)$. Guess a limit.

Solution: We'll use the ratio test to check for *absolute convergence*. The ratio of successive terms is

$$\frac{\left|a_{k+1}x^{k+1}\right|}{\left|a_k x^k\right|} = \frac{(k+1)|x|^k}{k|x|^{k-1}} = |x|\frac{k+1}{k}.$$

As $k \to \infty$, this ratio tends to $|x|$. Therefore the series converges—with or without absolute value signs—if $|x| < 1$.

If $|x| = 1$, the ratio test is inconclusive. It's easy to see, though, that if $x = \pm 1$,

$$\left|kx^{k-1}\right| = |k| \to \infty \quad \text{as} \quad k \to \infty.$$

Thus, by the nth term test, the series diverges. For the same reason, the series diverges if $|x| > 1$. The convergence interval, therefore, is $(-1, 1)$, as claimed.

Let's guess a limit. We saw above➤ that the equation

cf. Example 1, above.

$$S(x) = 1 + x + x^2 + x^3 + x^4 + \cdots = \frac{1}{1-x}$$

holds for $|x| < 1$. *Differentiating* all three quantities suggests a limit. If $|x| < 1$,

$$S'(x) = 1 + 2x + 3x^2 + 4x^3 + \cdots = \frac{1}{(1-x)^2}.$$

If, say, $x = 1/2$, we'd expect that

$$\sum_{k=1}^{\infty} \frac{k}{2^{k-1}} = 1 + \frac{2}{2} + \frac{3}{4} + \frac{4}{8} + \cdots = \frac{1}{(1-1/2)^2} = 4.$$

Numerical evidence suggests we're right. For the series above, $S_{20} \approx 3.999958037$.

Our guess is reasonable—and correct—but it raises important questions:

- Is it legitimate to differentiate a series term by term?

- On what interval does the resulting series converge?

We'll address these questions in the next section. $\square$

■ **Example 5.** Antidifferentiating the geometric series $S(x) = 1 + x + x^2 + x^3 + \ldots$ term-by-term gives the new series

$$T(x) = x + \frac{x^2}{2} + \frac{x^3}{3} + \frac{x^4}{4} + \frac{x^5}{5} + \cdots = \sum_{k=1}^{\infty} \frac{x^k}{k}.$$

Where does $T(x)$ converge? Guess a limit.

Solution: Because $S(x)$ converges➤ for $|x| < 1$, it's reasonable to expect the same of $T(x)$. The ratio test agrees:➤

Absolutely.

Check algebraic details.

$$\lim_{k\to\infty} \frac{\left|a_{k+1}x^{k+1}\right|}{\left|a_k x^k\right|} = \lim_{k\to\infty} |x| \cdot \frac{k+1}{k} = |x|.$$

Therefore, as expected, $T(x)$ converges absolutely on the interval $(-1, 1)$.

What happens at the endpoints, $x = \pm 1$? Setting $x = \pm 1$ in $T(x)$ produces two by-now familiar series:➤

The harmonic series and the alternating harmonic series.

$$T(1) = \sum_{k=1}^{\infty} \frac{1}{k}; \qquad T(-1) = \sum_{k=1}^{\infty} \frac{(-1)^k}{k}.$$

By the alternating series theorem. As we've seen earlier, the first diverges; the second converges conditionally.[←] Thus T converges for x in $[-1, 1)$.

Since $T(x)$ came from $S(x)$ by *antidifferentiation*, it's reasonable to *guess* a similar relationship for limits:

$$1 + x + x^2 + x^3 + \cdots = \frac{1}{1-x} \implies x + \frac{x^2}{2} + \frac{x^3}{3} + \cdots = -\ln(1-x).$$

Numerical evidence suggests, as before, that we're right. If $x = 1/2$, the series gives $S_{20} \approx 0.69314714$—not far from $-\ln 1/2 \approx 0.69314718$.

Similar questions to those above arise again:

- Is it legitimate to *antidifferentiate* a series term by term?

- On what interval does the resulting series converge?

We'll address these questions in the next section.   □

■ **Example 6.**   Where does the power series $\sum_{k=0}^{\infty} k! \, x^k$ converge?

Solution: Like *any* power series, this one converges at its base point, $x = 0$. But if $x \neq 0$,

$$\lim_{k \to \infty} \frac{|a_{k+1} x^{k+1}|}{|a_k x^k|} = \lim_{k \to \infty} \frac{(k+1)! \, |x|^{k+1}}{k! \, |x|^k} = \lim_{k \to \infty} (k+1) \cdot |x| = \infty.$$

This result amounts to a dramatic violation of the nth term test—far from tending to zero, successive terms grow larger and larger. Hence this power series has, in a sense, the smallest possible domain: it converges *only* if $x = 0$.   □

Power series convergence: lessons from the examples

The examples above illustrate several useful properties of power series and their convergence sets.

In the last example the power series converged only for $x = 0$. Stretching a point slightly, we'll call the set $\{0\}$ an interval of radius 0.

In particular, the convergence set is symmetric about the "base point."

The radius of convergence.   Every power series $\sum_{k=0}^{\infty} a_k x^k$ converges for $x = 0$. The real question is this:

How far from zero can x be without destroying convergence?

In every example above the convergence set turned out to be an *interval, centered at* 0.[←] This was no accident. The convergence domain for *any* power series is an interval centered at zero;[←] its radius is called the **radius of convergence** of the power series.

The following theorem guarantees all these claims:

> **Theorem 11.** Let $S(x) = \sum_{k=0}^{\infty} a_k x^k$ be a power series and let C be any real number. If $S(x)$ converges for $x = C$, then $S(x)$ also converges for $|x| < |C|$.

The idea of proof for $C = 1$: Suppose that $\sum_{k=0}^{\infty} a_k x^k$ converges for $x = 1$. Then $\sum_{k=0}^{\infty} a_k$ converges. By the nth term test, $a_k \to 0$ as $k \to \infty$. Since the a_k's can't blow up, they must be bounded in absolute value.➹ In other words, there's a number $M > 0$ such that for all k,

$$|a_k| \leq M .$$

If $|x| < 1$, the inequality

$$\left| a_k x^k \right| \leq M |x|^k$$

holds for all k. Now the comparison test applies: because the geometric series $\sum_{k=0}^{\infty} M|x|^k$ converges,➹ so must $\sum_{k=0}^{\infty} \left| a_k x^k \right|$. That's what we wanted to show.➹ $\square$

If the a_k's weren't bounded, they couldn't tend to zero.

Because $|x| < 1$.

A general proof, for any value of C, isn't much harder.

At endpoints, anything can happen. In several examples above the series converged on the *open* interval $(-1, 1)$; in another example, it converged on $[-1, 1)$.➹ A series' interval of convergence may include either, both, or neither of its endpoints—any combination is possible.➹ In practice, however, what really counts is a series' **radius of convergence**: what happens at the endpoints, though sometimes interesting, is usually less important.

In both cases, the radius of convergence is 1.

See the exercises.

Any radius of convergence is possible. In several examples above, power series turned out to converge on $(-1, 1)$. Actually, *any* (positive) radius of convergence is possible. Indeed, for any positive constant R the series $\sum_{k=0}^{\infty} x^k / R^k$ has radius of convergence precisely R.

Power series convergence, graphically

For any $n \geq 0$, the nth partial sum of the power series $S(x) = \sum_{k=0}^{\infty} a_k x^k$ is the *polynomial*

$$p_n(x) = a_0 + a_1 x + a_2 x^2 + a_3 x^3 + \cdots + a_n x^n.$$

To say that the power series converges for x in $(-R, R)$ means that for any x in that interval, there's a number $S(x)$ such that $p_n(x) \to S(x)$ as $n \to \infty$.

Sorting out precisely what this means is a worthy challenge; the following picture gives a graphical sense of the situation for the geometric power series $S(x) = \sum_{k=0}^{\infty} x^k$:

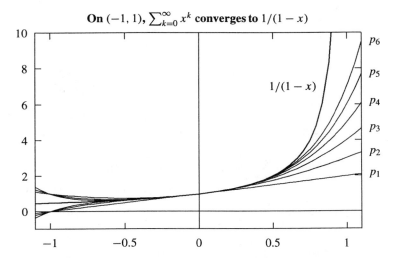

On $(-1, 1)$, $\sum_{k=0}^{\infty} x^k$ converges to $1/(1-x)$

In the picture the polynomial graphs➹ represent the first six partial sums. Over the interval $(-1, 1)$, they appear to approach the graph of the limiting function more

Labeled p_1 through p_6.

and more closely. *Outside* that interval, the polynomial graphs appear—literally—to diverge, rather than to approach any common limiting function.

Exercises

1. The power series $\sum\limits_{k=0}^{\infty} \dfrac{x^k}{k}$ has radius of convergence 1. Plot the partial sum polynomials of degree 1, 2, 4, 6, 8, and 10 over the interval $[-2, 2]$. Is the interval of convergence of the series apparent? Explain.

Find the radius of convergence of each of the following power series.

2. $\sum\limits_{j=1}^{\infty} \left(\dfrac{x}{2}\right)^j$

3. $\sum\limits_{k=1}^{\infty} \dfrac{x^k}{\sqrt{k}}$

4. $\sum\limits_{m=1}^{\infty} \dfrac{x^m}{m^2 + 1}$

5. $\sum\limits_{n=0}^{\infty} \dfrac{x^n}{n! + n}$

6. $\sum\limits_{k=1}^{\infty} \dfrac{x^k}{k 2^k}$

7. $\sum\limits_{n=1}^{\infty} n^n x^n$

8. $\sum\limits_{n=0}^{\infty} (x - 2)^n$

9. $\sum\limits_{n=2}^{\infty} \dfrac{(x - 3)^{2n}}{n^4}$

10. $\sum\limits_{n=2}^{\infty} \dfrac{(x - 5)^n}{n \ln n}$

11. $\sum\limits_{n=1}^{\infty} \dfrac{(x + 1)^n}{n}$

12. Find the radius and the interval of convergence for each of the following power series.

(a) $\sum\limits_{k=1}^{\infty} (3x)^k$

(b) $\sum\limits_{m=0}^{\infty} \dfrac{(3x)^m}{m!}$

(c) $\sum\limits_{n=1}^{\infty} \dfrac{(3x)^n}{n}$

(d) $\sum\limits_{j=1}^{\infty} \dfrac{(3x)^j}{j^2}$

13. Let $R > 0$ be any positive constant.

(a) Show that $\sum\limits_{k=0}^{\infty} \dfrac{x^k}{R^k}$ converges on $(-R, R)$.

(b) Show that $\sum\limits_{k=1}^{\infty} \dfrac{x^k}{k R^k}$ converges on $[-R, R)$.

(c) Show that $\sum\limits_{k=1}^{\infty} \dfrac{x^k}{k^2 R^k}$ converges on $[-R, R]$.

(d) Concoct a power series that converges on $(-R, R]$.

14. For each of the following intervals, give an example of a power series that has the given interval as its interval of convergence. [HINT: See the previous exercise.]

(a) $[-4, 4)$

(b) $[-1, 5]$

(c) $(-4, 0)$

(d) $(8, 16]$

(e) $[-11, -3)$

15. Suppose that the power series $\sum_{k=0}^{\infty} a_k x^k$ converges only when $-2 < x \leq 2$.

(a) Explain why the radius of convergence of this power series is 2.

(b) Explain why the power series $\sum_{k=0}^{\infty} a_k (x - 1)^k$ has radius of convergence 2.

(c) Show that the interval of convergence of the power series $\sum_{k=0}^{\infty} a_k (x - 3)^k$ is $(1, 5]$.

(d) Find the interval of convergence of the power series $\sum_{k=0}^{\infty} a_k (x + 1)^k$.

16. Suppose that the power series $\sum_{k=0}^{\infty} a_k (x - b)^k$ converges when $-11 \leq x < 17$.

(a) What is the radius of convergence of the power series?

(b) Determine the value of b.

Find the interval of convergence (endpoint behavior too!) of each of the following power series.

17. $\sum_{m=0}^{\infty} \left(\frac{x - 3}{2} \right)^m$

18. $\sum_{j=0}^{\infty} \frac{(x - 2)^j}{j!}$

19. $\sum_{k=1}^{\infty} \frac{(x - 1)^k}{k4^k}$

20. $\sum_{n=1}^{\infty} \frac{(x - 1)^n}{\sqrt{n}}$

21. $\sum_{i=1}^{\infty} \frac{(x + 5)^i}{i(i + 1)}$

22. $\sum_{m=1}^{\infty} \frac{2^m (x - 1)^m}{m}$

23. Suppose that the power series $\sum_{k=0}^{\infty} a_k x^k$ converges when $x = -3$ and diverges when $x = 7$. Indicate which of the following statements **must** be true, which **might** be true, and which **cannot** be true. Justify your answers.

(a) The power series converges when $x = -10$.

(b) The power series diverges when $x = 3$.

(c) The power series converges when $x = 6$.

(d) The power series diverges when $x = 2$.

(e) The power series diverges when $x = -7$.

(f) The power series converges when $x = -4$.

24. Suppose that the power series $\sum_{k=0}^{\infty} a_k (x + 2)^k$ converges when $x = -7$ and diverges when $x = 7$. Indicate which of the following statements **must** be true, which **might** be true, and which **cannot** be true. Justify your answers.

   (a) The power series converges when $x = -8$.

   (b) The power series converges when $x = 1$.

   (c) The power series converges when $x = 3$.

   (d) The power series diverges when $x = -11$.

   (e) The power series diverges when $x = -10$.

   (f) The power series diverges when $x = 5$.

   (g) The power series diverges when $x = -5$.

25. Consider a power series of the form $\sum_{k=0}^{\infty} a_k(x - 1)^k$. Indicate whether each of the following statements **must** be true, which **might** be true, and which **cannot** be true. Justify your answers.

   (a) The power series converges when $|x| > 2$.

   (b) The power series converges for all values of x.

   (c) If the radius of convergence of the power series is 3, the power series converges when $-2 < x < 4$.

   (d) The interval of convergence of the power series is $[-5, 5]$.

   (e) If the interval of convergence of the power series is $(-7, 9)$, the radius of convergence is 7.

26. The power series

$$\sum_{k=0}^{\infty} \frac{1}{k!} x^k = 1 + x + \frac{x^2}{2!} + \frac{x^3}{3!} + \cdots$$

converges, for any x, to e^x.

   (a) If $x = -1$ the series converges to $1/e$. For this case, find S_{10}. By how much does S_{10} differ from the *actual* value of $1/e$? (Use a calculator or a computer.)

   (b) If $x = -1$ the series is alternating. What does this say about the maximum possible error committed by S_{10} in estimating $1/e$? For which n must S_n estimate $1/e$ with error less than 10^{-10}?

27. Let $f(x) = \displaystyle\sum_{n=0}^{\infty} \dfrac{2x^n}{3^n + 5}$.

    (a) Show that $f(10)$ is undefined (i.e., the series that defines $f(x)$ diverges when $x = 10$).

    (b) Which of the numbers 0.5, 1.5, 3, and 6 are in the domain of f?

    (c) Estimate $f(1)$ to within 0.01 of its exact value.

28. Let $h(x) = \displaystyle\sum_{k=0}^{\infty} \dfrac{(x - 2)^k}{k! + k^3}$.

    (a) What is the domain of h?

    (b) Estimate $h(0)$ to within 0.005 of its exact value.

    (c) Estimate $h(3)$ to within 0.005 of its exact value.

29. Let $g(x) = \displaystyle\sum_{n=1}^{\infty} \dfrac{(x + 4)^n}{n^3 5^n}$.

    (a) What is the domain of g?

    (b) Estimate $g(0)$ to within 0.005 of its exact value.

    (c) Estimate $g(-5)$ to within 0.005 of its exact value.

30.   (a) Evaluate $\displaystyle\lim_{x \to 1^-} \sum_{k=0}^{\infty} (-1)^k x^k$.

    (b) Explain why the result in part (a) does *not* mean that $\displaystyle\sum_{k=0}^{\infty} (-1)^k$ converges.

11.6 Power series as functions

Any power series

$$S(x) = \sum_{k=0}^{\infty} a_k x^k = a_0 + a_1 x + a_2 x^2 + a_3 x^3 + \dots$$

can be thought of, naturally, as a *function* of x; its domain is the series' interval of convergence. As we saw in the last section, this domain is always an interval centered at 0. In this section we'll explore the remarkable—and useful—properties of functions defined by power series.

The convergence interval may or may not contain its endpoints; for most purposes, it doesn't matter much either way.

Calculus with power series

Given *any* power series $S(x)$, convergent or divergent, it's easy to differentiate or antidifferentiate "term by term" to produce *new* series $D(x)$ and $A(x)$:

$$S(x) \;=\; \sum_{k=0}^{\infty} a_k x^k = a_0 + a_1 x + a_2 x^2 + a_3 x^3 + \dots;$$

$$D(x) \;=\; \sum_{k=1}^{\infty} k a_k x^{k-1} = a_1 + 2a_2 x + 3a_3 x^2 + \dots;$$

$$A(x) \;=\; \sum_{k=0}^{\infty} a_k \frac{x^{k+1}}{k+1} = a_0 x + a_1 \frac{x^2}{2} + a_2 \frac{x^3}{3} + a_3 \frac{x^4}{4} + \dots.$$

Prudence dictates a modicum of caution; important questions remain to be answered:

- If the series S has radius of convergence r, is the same true of D and A?

- Even assuming that S, D, and A all converge on $(-r, r)$, is D really the derivative of S, in the ordinary calculus sense? Is A really an antiderivative of S?

The following convenient theorem answers all these questions in the affirmative—everything works just as we'd hope.

> **Theorem 12 (Derivatives and antiderivatives).** Let $S(x)$ be a power series with radius of convergence $r > 0$. Let $D(x)$ and $A(x)$ be defined as above. Then:
>
> - Both D and A have radius of convergence r;
> - For $|x| < r$, $D(x) = S'(x)$;
> - For $|x| < r$, $A(x) = \int_0^x S(t)\, dt$.

The theorem says, among other things, that a function S given by a power series is differentiable, and that its derivative is *another* power series, S', with the same radius of convergence. The same theorem applies to S', to S'', and so on, to show that S has *infinitely many* derivatives—all available by repeated term-by-term differentiation.

■ **Example 1.** For any x, $e^x = 1 + x + \dfrac{x^2}{2!} + \dfrac{x^3}{3!} + \dots$. Explain why.

Solution: Let $S(x)$ represent the series above; we saw in the previous section that $S(x)$ converges for *all* x.➤

By Theorem 12, S' can be found by differentiating S term by term. Notice, however:➤

Differentiating S term by term leaves S unchanged.

But *every* differentiable function S for which $S' = S$ has the form $S(x) = Ce^x$. Since $S(0) = 1$ it follows that $C = 1$; hence $S(x) = e^x$, as claimed.   □

This S has infinite radius of convergence.

Try it!

Writing known functions as power series

In previous examples, including the last one, we've written various functions— $1/(1-x)$, e^x, $\ln(1-x)$, etc.—in power series form. It's natural to ask whether *other* functions can be "represented" as power series, and if so, how. Theorem 12 suggests several techniques for doing so. First, an even more natural question:

Why bother? Examples with the sine function

Why write a function as a power series? What good, for instance, is the equation

$$\sin x = x - \frac{x^3}{3!} + \frac{x^5}{5!} - \frac{x^7}{7!} + \frac{x^9}{9!} - \cdots,$$

which holds for all real numbers x?➤

It is an equation; assume so for now. We'll give convincing reasons below.

Good family values.   Transcendental functions—trigonometric functions, exponential functions, etc.—have no *finite* algebraic formulas. For such functions, power series are the next best thing. Using them, we can find extremely accurate—though approximate—values of many transcendental functions.

■ **Example 2.**   Use the sine series above to approximate sin 1 accurately.

Solution: Substituting $x = 1$ into the sine series gives

$$\sin 1 = 1 - \frac{1}{3!} + \frac{1}{5!} - \frac{1}{7!} + \frac{1}{9!} - \frac{1}{11!} + \cdots,$$

an *alternating* series with terms decreasing➤ to zero. By the alternating series theorem, any partial sum of such a series differs from the limit by no more than the *next* term. In particular, the partial sum $1 - 1/3! + 1/5! - \cdots - 1/11! \approx 0.8414709846$ differs from sin 1 by no more than $1/13! \approx 2 \times 10^{-10}$. In other words, our estimate is good to at least *nine* decimal places. This checks out numerically: $\sin 1 \approx 0.8414709848$.➤ Not bad for so little work.   □

Rapidly!

The estimate above is good through 9 decimal places.

Hard integrals made easy.   As we've seen repeatedly, many integrals—even simple-looking ones—can't be calculated in "closed form," i.e., by elementary antidifferentiation. Numerical methods➤ offer one recourse. Infinite series, being *easy* to integrate, offer another way to transcend such difficulties.

E.g., the midpoint rule.

■ **Example 3.**   Find, in *series* form, an antiderivative for $\sin(x^2)$. Use it to estimate $I = \int_0^1 \sin(x^2)\, dx$. (For comparison, the midpoint rule applied to I gives $M_{50} \approx 0.31025$.)

Solution: Even though the function $\sin(x^2)$ has no *elementary* antiderivative, it's easy to find an antiderivative in *series* form. Here's how.

Replacing x with x^2 in the sine series gives the new series

$$\sin(x^2) = x^2 - \frac{x^6}{3!} + \frac{x^{10}}{5!} - \frac{x^{14}}{7!} - \cdots .$$

(Because the original series converges for all x, so does this one.) Antidifferentiating term by term gives a *new* power series:

$$\int_0^x \sin(t^2)\,dt = \frac{x^3}{3} - \frac{x^7}{7 \cdot 3!} + \frac{x^{11}}{11 \cdot 5!} - \frac{x^{15}}{15 \cdot 7!} - \cdots .$$

It converges for all x, by Theorem 12.

The new series,◄ is not an elementary function, but it's a perfectly honest antiderivative for $\sin(x^2)$. We can find our definite integral, therefore, in the obvious way:

$$\begin{aligned}
\int_0^1 \sin(x^2)\,dx &= \left. \frac{x^3}{3} - \frac{x^7}{7 \cdot 3!} + \frac{x^{11}}{11 \cdot 5!} - \frac{x^{15}}{15 \cdot 7!} + \cdots \right]_0^1 \\
&= \frac{1}{3} - \frac{1}{7 \cdot 3!} + \frac{1}{11 \cdot 5!} - \frac{1}{15 \cdot 7!} + \cdots .
\end{aligned}$$

The alternating series theorem applies to the last series, so the estimate

$$\int_0^1 \sin(x^2)\,dx \approx \frac{1}{3} - \frac{1}{7 \cdot 3!} + \frac{1}{11 \cdot 5!} - \frac{1}{15 \cdot 7!} \approx 0.3102681578$$

I.e., the size of the next term.

is in error by no more than $1/19 \cdot 9! \approx 1.5 \times 10^{-7}$.◄ The result agrees through four decimal places with the midpoint rule estimate. □

New series from old: help from algebra and calculus

For many familiar functions, power series can be found by simple algebra or calculus operations, starting from a few standard known series. Differentiating the sine series, for instance, gives◄

Thanks to the previous theorem.

$$\cos x = 1 - \frac{x^2}{2!} + \frac{x^4}{4!} - \frac{x^6}{6!} + \cdots .$$

The new series, like the old, converges for all x.

With a little algebraic ingenuity, the famous series

$$\frac{1}{1-x} = 1 + x + x^2 + x^3 + x^4 + x^5 + \cdots ,$$

which converges for $|x| < 1$, yields many other useful series, all converging on the same set. Replacing x with $-x$, for instance, gives the alternating series

$$\frac{1}{1+x} = 1 - x + x^2 - x^3 + x^4 - x^5 + \cdots .$$

Replacing x with x^2 gives

$$\frac{1}{1+x^2} = 1 - x^2 + x^4 - x^6 + x^8 - x^{10} + \cdots .$$

Integrating *this* series term by term gives an even more striking result:

$$\arctan x = x - \frac{x^3}{3} + \frac{x^5}{5} - \frac{x^7}{7} + \frac{x^9}{9} + \cdots .$$

Setting $x = 1$ in this last series yields a remarkable and beautiful result, one discovered and rediscovered by some of the world's greatest mathematicians:

$$\frac{\pi}{4} = 1 - \frac{1}{3} + \frac{1}{5} - \frac{1}{7} + \frac{1}{9} + \ldots.$$

(A little caution is needed, however: the simple arguments above showed only that the series for arctan x is valid if $-1 < x < 1$. Showing carefully that the series converges to $\pi/4$ when $x = 1$ requires further argument.)➤

But it's true!

Multiplying power series. Convergent power series can be multiplied together, something like polynomials, to form new convergent series. As always with series, convergence is a question. Here's the answer: The product of two power series converges wherever both factors converge. We illustrate with an example.

■ **Example 4.** We showed above that

$$\frac{1}{1-x} = 1 + x + x^2 + x^3 + \ldots \quad \text{and} \quad \frac{1}{1+x} = 1 - x + x^2 - x^3 + \ldots.$$

Multiply these series. Where does the new series converge? What familiar function does the result represent?

Solution: Symbolically, the problem looks like this:

$$(1 + x + x^2 + x^3 + \ldots) \cdot (1 - x + x^2 - x^3 + \ldots) = a_0 + a_1 x + a_2 x^2 + a_3 x^3 + \ldots;$$

we want numerical values for the constants on the right.

Both factors have infinitely many summands, so ordinary expansion gets quickly out of hand. To avoid this, we'll collect like powers, right from the start. It's clear, for instance, that $a_0 = 1 \cdot 1 = 1$—no other combination of factors yields a constant result. Similarly, tracking powers of x gives

$$a_1 = 1 \cdot (-1) + 1 \cdot 1 = 0; \quad a_2 = 1 \cdot 1 + 1 \cdot (-1) + 1 \cdot 1 = 1.$$

Continuing this process➤ quickly shows a simple pattern:

$$(1 + x + x^2 + x^3 + \ldots) \cdot (1 - x + x^2 - x^3 + \ldots) = 1 + x^2 + x^4 + x^6 + \ldots.$$

Try the next term or two for yourself—there's no substitute for experience.

The result is a **geometric series in powers of** x^2, and so converges for $|x^2| < 1$, i.e., if $-1 < x < 1$.

What function does the product series represent? Since the two factors represent the functions $1/(1-x)$ and $1/(1+x)$, it follows that the *product series* represents the *product function* $1/(1-x^2)$.   □

The next example concerns yet another useful descendant of the geometric series.

■ **Example 5.** Find a power series for $\ln(1 + x)$; use it to estimate $\ln 1.5$ with error less than 0.0001.

Solution: Integrating the geometric series gives

$$\int \frac{1}{1+x}\,dx = \int \left(1 - x + x^2 - x^3 + \dots\right) dx$$

$$= x - \frac{x^2}{2} + \frac{x^3}{3} - \frac{x^4}{4} + \dots = \ln(1+x).$$

(We used $C = 0$ as the constant of integration, because our "target" function $\ln(1+x)$ has the value 0 when $x = 0$.) The new series, like the old, converges for x in $(-1, 1)$.

To estimate $\ln 1.5$, we plug $x = 0.5$ into our series expression:

$$\ln 1.5 = 0.5 - \frac{0.5^2}{2} + \frac{0.5^3}{3} - \frac{0.5^4}{4} + \dots = \sum_{k=1}^{\infty} (-1)^{k+1} \frac{0.5^k}{k}.$$

Now the alternating series theorem applies: to achieve our target accuracy, any partial sum S_n for which

$$\frac{0.5^{n+1}}{n+1} < 0.0001$$

will do. It's easy to see that $n = 10$ works, with room to spare. In fact, $S_{10} \approx 0.405435$; this compares favorably with the "exact" value $\ln 1.5 \approx 0.405465$. □

A brief atlas of power series

For ease of reference, here is a short list of "standard" power series for basic calculus functions. Each series appears with a suitable interval of convergence. In the next section we'll develop additional tools for showing rigorously that the limits are as stated.

A Power Series Sampler		
Function	**Series**	**Convergence Interval**
$\sin x$	$x - \frac{x^3}{3!} + \frac{x^5}{5!} - \frac{x^7}{7!} + \frac{x^9}{9!} - \dots$	$(-\infty, \infty)$
$\cos x$	$1 - \frac{x^2}{2!} + \frac{x^4}{4!} - \frac{x^6}{6!} + \frac{x^8}{8!} - \dots$	$(-\infty, \infty)$
$\exp x$	$1 + x + \frac{x^2}{2!} + \frac{x^3}{3!} + \frac{x^4}{4!} + \frac{x^5}{5!} + \dots$	$(-\infty, \infty)$
$\frac{1}{1-x}$	$1 + x + x^2 + x^3 + x^4 + x^5 + \dots$	$(-1, 1)$
$\frac{1}{1+x}$	$1 - x + x^2 - x^3 + x^4 - x^5 + \dots$	$(-1, 1)$
$\frac{1}{1+x^2}$	$1 - x^2 + x^4 - x^6 + x^8 - x^{10} + \dots$	$(-1, 1)$
$\arctan x$	$x - \frac{x^3}{3} + \frac{x^5}{5} - \frac{x^7}{7} + \frac{x^9}{9} - \dots$	$(-1, 1]$

What's next? A power series for "any" function

As we've seen, knowing a power series expression for one function can lead, *via* various manipulations, to power series versions of *related* functions.

A good question remains:

> *How, given any function f, can we find a power series "from scratch," without knowing a related series to begin with?*

We'll answer this question in the next section.

Exercises ──────────────────────────────

1. Let $f(x) = \displaystyle\sum_{k=0}^{\infty} \left(\frac{x}{2}\right)^k$.

   (a) What is the radius of convergence of the power series for f?

   (b) According to Theorem 12, $f'(x) = \displaystyle\sum_{k=1}^{\infty} \frac{kx^{k-1}}{2^k}$. What is the radius of convergence of the series for f'?

   (c) According to Theorem 12, $F(x) = \displaystyle\sum_{k=0}^{\infty} \frac{x^{k+1}}{(k+1)2^k}$ is an antiderivative of f. What is the radius of convergence of the series for F?

Use the power series representation of $(1-x)^{-1}$ to produce a power series representation of each of the following functions.

2. $f(x) = \dfrac{x^2}{1+x}$

3. $f(x) = \dfrac{1}{1-x^2}$

4. $f(x) = \dfrac{1}{(1+x)^2}$

5. $f(x) = \dfrac{x}{1-x^4}$

For each of the following functions, find a power series representation of the function and the radius of convergence of this power series. Then, plot the function and the fifth-order polynomial that is a partial sum of the power series on the same axes. [HINT: Write out the first few terms of the series before trying to find the form of the general term.]

6. $f(x) = \arctan(2x)$

7. $f(x) = \cos(x^2)$

8. $f(x) = x^2 \sin x$

9. $(x) = \dfrac{1}{2+x}$

10. $f(x) = \sin\left(\sqrt{x}\right)$

11. $f(x) = \sin x + \cos x$

12. $f(x) = 2^x = e^{x \ln 2}$

13. $f(x) = \ln\left(1 + x^2\right)$

14. $f(x) = (x^2 - 1) \sin x$

15. $f(x) = \ln\left(\dfrac{1+x}{1-x}\right)$

16. $f(x) = \cos^2 x = \frac{1}{2}\left(1 + \cos(2x)\right)$

17. $f(x) = (x^2 - 1) \sin x$

Find the first four non-zero terms in the power series representation of each of the following functions.

18. $\dfrac{e^x}{1-x}$

19. $\dfrac{\cos x}{1+x^2}$

20. $e^{2x}\ln(1+x^3)$

21. $\arctan x \sin(4x)$

22. $e^{\sin x}$

23. $\ln(1+\sin x)$

Find the elementary function represented by each of the following power series by manipulating a more familiar power series (e.g., the series for $\cos x$, $\sin x$, $(1-x)^{-1}$).

24. $\displaystyle\sum_{k=1}^{\infty} k x^{k-1}$

25. $\displaystyle\sum_{k=0}^{\infty} \dfrac{x^k}{(k+1)!}$

26. $\displaystyle\sum_{k=1}^{\infty} (-1)^{k+1} x^k$

27. $\displaystyle\sum_{k=1}^{\infty} \dfrac{(2x)^k}{k}$

28. Show that $\dfrac{1}{x-1} = \displaystyle\sum_{k=1}^{\infty} \dfrac{1}{x^k}$ when $|x| > 1$.
    [HINT: $1/(x-1) = x/(x-1) - 1$.]

Use power series to evaluate each of the following limits. Compare your answers with those obtained via L'Hôpital's rules.

29. $\displaystyle\lim_{x\to 0} \dfrac{\sin x}{x}$

30. $\displaystyle\lim_{x\to 0} \dfrac{e^x - 1}{x}$

31. $\displaystyle\lim_{x\to 0} \dfrac{1 - \cos x}{x}$

32. $\displaystyle\lim_{x\to 0} \dfrac{1 - \cos x}{x^2}$

33. $\displaystyle\lim_{x\to 0} \dfrac{\arctan x}{x}$

34. $\displaystyle\lim_{x\to 0} \dfrac{e^x - e^{-x}}{x}$

35. $\displaystyle\lim_{x\to 0} \dfrac{\ln(1+x) - x}{x^2}$

36. $\displaystyle\lim_{x\to 0} \dfrac{x - \arctan x}{x^3}$

37. $\displaystyle\lim_{x\to 0} \dfrac{\ln x}{x - 1}$

38. $\displaystyle\lim_{x\to 0} \dfrac{1 - \cos^2 x}{x}$
    [HINT: $1 - \cos^2 x = \left(1 - \cos(2x)\right)/2$.]

39. Let $f(x) = \dfrac{1}{1+x^4}$.

    (a) Find a power series representation of f.

    (b) What is the interval of convergence of the series in part (a)?

    (c) Use the series found in part (a) to evaluate $\displaystyle\int_0^{0.5} f(x)\, dx$ with an error no greater than 0.001.

40. (a) Find the power series representation of an antiderivative of e^{-x^2}.

(b) Use the result from part (a) to estimate $\int_0^1 e^{-x^2}\, dx$ to within 0.005 of its exact value.

41. Use a power series to estimate $\int_0^1 \cos\left(x^2\right)\, dx$ with an error no greater than 0.005.

42. Use a power series to estimate $\int_0^1 \sqrt{x}\, \sin x$ with an error no greater than 0.001.

43. Use a power series to show that $x - \frac{1}{2}x^2 < \ln(1+x) < x$ for all $x \in (0, 1)$.

44. Use power series to show that $1 - \cos x \ln(1+x) < \sin x$ when $0 < x < 1$.

45. Let $f(x) = \ln(1+x)$. Show that the power series for f and f' have the same *radius* of convergence, but not the same *interval* of convergence.

46. Use power series to show that $y = e^x$ is a solution of the differential equation $y' = y$.

47. Use power series to show that $y = 2e^x$ is the solution of the initial value problem $y' = y$, $y(0) = 2$.

48. Use power series to show that $y = e^{3x}$ is the solution of the initial value problem $y' = 3y$, $y(0) = 1$.

49. Use power series to show that $y = \sin x$ is a solution of the differential equation $y'' = -y$.

50. Use power series to show that $y = (1-x)^{-1}$ is the solution of the initial value problem $y' = y^2$, $y(0) = 1$.

51. Determine the coefficients a_k such that $\dfrac{1}{1-x} = \displaystyle\sum_{k=0}^{\infty} a_k (x-2)^k$.

[HINT: $\dfrac{1}{1-x} = -\dfrac{1}{1+(x-2)}$.]

52. Use the fact that $\cos x = -\cos(x + \pi)$ to write $\cos x$ as a series in powers of $x - \pi$.

53. Show that $\displaystyle\int_0^{\infty} e^{-t} \sin(xt)\, dt = \dfrac{x}{1+x^2}$ when $|x| < 1$. (Use the fact that $\int_0^{\infty} t^n e^{-t}\, dt = n!$.)

54. (a) Show that $f(x) = \tan x$ is the solution of the initial value problem $f'(x) = 1 + \left(f(x)\right)^2$, $f(0) = 0$.

(b) Use part (a) to find the first four non-zero terms in the power series representation of $\tan x$.

55. Let r be a fixed number and define the function f by

$$f(x) = \sum_{n=0}^{\infty} \frac{r(r-1)(r-2)\cdots(r-n+1)}{n!} x^n.$$

(a) Show that the series defining f converges when $|x| < 1$.

(b) Show that $(1+x)f'(x) = rf(x)$.

(c) Let $g(x) = (1+x)^{-r} f(x)$. Show that $g'(x) = 0$.

(d) Show that part (c) implies that $f(x) = (1+x)^r$.
   [NOTE: The power series for f is known as the **binomial series**.]

Use the series in exercise 55 to find the first four nonzero terms of a power series representation of each of the following functions.

56. $f(x) = \sqrt{1+x}$ 58. $g(x) = (1+x^2)^{-3/2}$

57. $g(x) = \sqrt[3]{1-x^2}$ 59. $g(x) = \arcsin x.$

11.7 Maclaurin and Taylor series

In the last section we saw some of the practical advantages of writing a function as a power series. We also saw how to use a power series for *one* function to derive➤ power series for related functions.

Using algebra, calculus, and other devices.

In this section we show, given a suitable function, how to find its power series "from scratch," i.e., without starting from a related series. In the process we'll draw important connections among a function, its derivatives, and its series. First, however, a note of caution.

Does every function have a power series?

Mathematical life would be simpler if every function $f(x)$ could be written as a power series—ideally, a series that converges to $f(x)$ for all x. Alas, it isn't so. At least two things can go wrong:

Smaller domains. A power series may have a smaller domain than the function it represents. For instance, the series equation

$$\frac{1}{1 + x^2} = 1 - x^2 + x^4 - x^6 + x^8 - x^{10} + \dots$$

holds if—but only if—$|x| < 1$, even though $f(x) = 1/(1 + x^2)$ is defined➤ for *all* real numbers x.

And well behaved.

No series at all. A function may have no series at all. Theorem 12 of the previous section showed that every power series can be differentiated again and again on its interval of convergence.➤ Thus any function that *has* a power series must itself be repeatedly differentiable at $x = 0$. This fact rules out functions like $f(x) = |x|$, which is continuous everywhere but not differentiable at $x = 0$.

Each derivative is another power series.

Despite these cautions, many important functions *can* be written as power series.

Coefficients and derivatives at zero

A simple equation relates the *coefficients* and the *derivatives at zero* of a power series. If

$$S(x) = a_0 + a_1 x + a_2 x^2 + a_3 x^3 + a_4 x^4 + \dots,$$

then, simplest of all, $S(0) = a_0$. Differentiating S repeatedly➤ gives

*We **can** differentiate repeatedly, by Theorem 12 of the previous section.*

$$\begin{aligned} S(x) &= a_0 + a_1 x + a_2 x^2 + a_3 x^3 + a_4 x^4 + \dots; \\ S'(x) &= a_1 + 2a_2 x + 3a_3 x^2 + 4a_4 x^3 + \dots; \\ S''(x) &= 2a_2 + 6a_3 x + 12a_4 x^2 + \dots; \\ S'''(x) &= 6a_3 + 24a_4 x + \dots \\ &\vdots \end{aligned}$$

The equations above show that

$$S(0) = a_0; \quad S'(0) = a_1; \quad S''(0) = 2a_2; \quad S'''(0) = 6a_3 \dots.$$

In general:

> **Fact:** If $S(x) = \displaystyle\sum_{k=0}^{\infty} a_k x^k$, then $a_k = \dfrac{S^{(k)}(0)}{k!}$ for all $k \ge 0$.

This important fact connects coefficients and derivatives—knowing either, we can find the other. We illustrate one of its uses by an example.

For now.

■ **Example 1.** Assuming[◄] that $f(x) = \sin x$ has a power series, find that power series.

Solution: To find the desired series $f(x) = a_0 + a_1 x + a_2 x^2 + a_3 x^3 + \dots$ we need "only" find the coefficients a_0, a_1, a_2, a_3, etc.

In principle, finding infinitely many of *anything* sounds difficult. In practice, it's often easy; for the sine function, as for many functions of interest, the coefficients follow a simple pattern.

The first few derivatives of f show the pattern:

$$a_0 = \frac{f(0)}{0!} = \sin 0 = 0; \qquad a_1 = \frac{f'(0)}{1!} = \cos 0 = 1;$$

$$a_2 = \frac{f''(0)}{2!} = \frac{-\sin 0}{2} = 0; \qquad a_3 = \frac{f'''(0)}{3!} = \frac{-\cos 0}{6} = -\frac{1}{6};$$

$$a_4 = \frac{f^{(4)}(0)}{4!} = \frac{\sin 0}{4!} = 0; \qquad a_5 = \frac{f^{(5)}(0)}{5!} = \frac{\cos 0}{5!} = \frac{1}{5!}.$$

Therefore, written as a series,

$$\sin x = x - \frac{x^3}{3!} + \frac{x^5}{5!} - \frac{x^7}{7!} + \dots.$$

But beside the main point, here.

It's not hard[◄] to show that this series converges for *all* x, as we'd hope.

Better yet, the series converges to $\sin x$. We haven't *proved* this fact yet, but numerical evidence is readily available. If, say, $x = 1$, $S_{10} \approx 0.8414710096$, while $\sin 1 \approx 0.8414709848$. □

Maclaurin series

Using the fact above as a recipe for the coefficients, we can write a power series for *any* function f that has repeated derivatives at $x = 0$.[◄] Here's the formal definition, named for the 17th-century Scottish mathematician Colin Maclaurin:

*The **zeroth derivative** of f is f itself.*

> **Definition:** (**Maclaurin series.**) Let f be any function with infinitely many derivatives at $x = 0$. The **Maclaurin series for f** is the series $\sum_{k=0}^{\infty} a_k x^k$ with coefficients given by
> $$a_k = \frac{f^{(k)}(0)}{k!}, \quad k = 0, 1, 2 \dots.$$

Finding Maclaurin series: help from technology

Computing the necessary derivatives to find a Maclaurin series can be tedious and error-prone by hand. Fortunately, *Maple* and other software programs will find any *finite* partial sum of the Maclaurin series, quickly and easily. (Such sums are called either **Maclaurin polynomials** or—see below—**Taylor polynomials**.) Here's how the process works for the sine function:

```
> taylorpoly( sin(x), x=0, 7 );
```

$$x - 1/6\ x^3 + 1/120\ x^5 - 1/5040\ x^7$$

The output is the **7th-degree Taylor polynomial for** $\sin x$, based at $x = 0$.

Taylor series: expansion about $x = a$

A Maclaurin series

$$\sum_{k=0}^{\infty} \frac{f^{(k)}(0)}{k!} x^k$$

is said to be **expanded about** $x = 0$, because all derivatives of f are calculated there. This choice—though often convenient—isn't necessary. A similar series can be found by expanding about *any* point $x = a$. The **Taylor series for f, expanded about** $x = a$ has the form

$$\sum_{k=0}^{\infty} \frac{f^{(k)}(a)}{k!} (x - a)^k.$$

For some functions, expansion about a point other than zero is convenient.

■ **Example 2.** The function $f(x) = \ln x$ isn't defined at $x = 0$. Expand f about $x = 1$.➤

We'll let Maple help.

Solution: Derivatives of $f(x) = \ln x$ are easy—for a human or for a machine—to calculate.➤ Here are several:

See for yourself.

$$f'(x) = x^{-1} \quad f''(x) = -x^{-2} \quad f'''(x) = \frac{2}{x^3}$$
$$f^{(4)}(x) = -\frac{6}{x^4} \quad f^{(5)}(x) = \frac{24}{x^5} \quad f^{(6)}(x) = -\frac{120}{x^6}$$

At $x = 1$, therefore:

$$f'(1) = 1 \quad f''(1) = -1 \quad f'''(1) = 2$$
$$f^{(4)}(1) = -6 \quad f^{(5)}(1) = 24 \quad f^{(6)}(1) = -120$$

The general pattern is now clear. The Taylor series for the function $\ln x$, expanded about $x = 1$, has the form

$$\sum_{k=1}^{\infty} \frac{(-1)^{k+1}}{k} (x - 1)^k = (x - 1) - \frac{(x - 1)^2}{2} + \frac{(x - 1)^3}{3} - \frac{(x - 1)^4}{4} + \cdots.$$

For the record, *Maple* agrees:➤

To seventh degree, anyway.

```
> taylorpoly(ln(x),x=1,7);
                   2              3              4
    x - 1 - 1/2 (x - 1)  + 1/3 (x - 1)  - 1/4 (x - 1)
                  5              6              7
      + 1/5 (x - 1)  - 1/6 (x - 1)  + 1/7 (x - 1)        □
```

Converging to the right place: Taylor's theorem

Any function f that's infinitely differentiable at $x = 0$ has a Maclaurin series—ideally, one with a large radius of convergence. One possible problem remains: the series might conceivably converge at x, but perhaps to a limit *other than* $f(x)$.

Taylor's theorem guarantees that this unfortunate event almost never occurs. In the bargain, it lets us predict in advance how closely a Maclaurin polynomial approximates the "target" function f.

> **Theorem 13 (Taylor's theorem).** Suppose that f is repeatedly differentiable on an interval I containing 0. Let $\sum_{k=0}^{\infty} a_k x^k$ be the Maclaurin series of f, and let
>
> $$P_n(x) = \sum_{k=0}^{n} a_k x^k = a_0 + a_1 x + a_2 x^2 + \cdots + a_n x^n$$
>
> be the nth degree Maclaurin polynomial. Suppose further that for all x in I,
>
> $$\left| f^{(n+1)}(x) \right| \leq K_{n+1}.$$
>
> Then
>
> $$|f(x) - P_n(x)| \leq \frac{K_{n+1}}{(n+1)!} |x|^{n+1}.$$

Notice:

- Taylor's theorem estimates the *error* committed by $P_n(x)$ in estimating $f(x)$. Unless K_{n+1} grows very quickly with n, this error tends to 0 as $n \to \infty$, so the series converges, as it should, to $f(x)$.

- The general appearance of the theorem involves familiar elements. Like our numerical integral error estimates, this one depends on bounding a higher derivative of the function in question.

- A rigorous proof of Taylor's theorem is beyond the scope of this book. The underlying ideas, however, are based firmly on the mean value theorem. Indeed, Taylor's theorem is sometimes thought of as a general form of the mean value theorem. We pursued this idea briefly at the end of Section 4.3, in deriving error bounds for linear approximation.

■ **Example 3.** For $f(x) = \sin x$, show that the Maclaurin series converges to $\sin x$ for *any* value of x.

Solution: For $f(x) = \sin x$, *all* derivatives are sines, cosines, or their opposites. Thus, for *any* n, the inequality

$$\left| f^{(n+1)}(x) \right| \leq 1$$

holds for all x. By Taylor's theorem,

$$|P_n(x) - \sin x| \leq \frac{1 \cdot |x|^{n+1}}{(n+1)!}.$$

The last quantity tends to 0 as n tends to infinity, so the series converges, for any x, to $\sin x$. The graphs below show what it means, geometrically, for the Maclaurin polynomials to converge to $\sin x$:

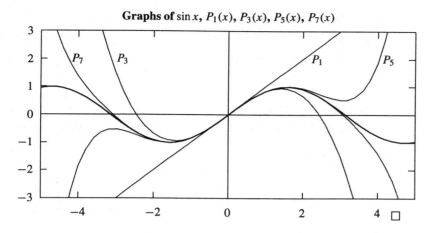

Graphs of $\sin x$, $P_1(x)$, $P_3(x)$, $P_5(x)$, $P_7(x)$

Exercises

1. Let $f(x) = x^4 - 12x^3 + 44x^2 + 2x + 1$.

   (a) Find the Maclaurin series representation of f.

   (b) Find the Taylor series representation of f expanded about $x = 3$.

2. Let $p(x) = (1 + x)^n$ where n is a positive integer. Explain why the Maclaurin series for p is the polynomial p itself.   [HINT: Use Theorem 13.]

3. Use Theorem 13 to show that the Maclaurin series for e^x converges to e^x for all x.

4. Use Theorem 13 to show that the Maclaurin series for $\dfrac{1}{1 + x}$ converges to $\dfrac{1}{1 + x}$ when $-1/2 < x < 1$.

5. (a) Find the Maclaurin series representation of $f(x) = \dfrac{1}{2 + x}$.
    [HINT: $\frac{1}{2+x} = \frac{1}{2} \cdot \frac{1}{1+(x/2)}$.]

    (b) Find $f^{(259)}(0)$ exactly.

6. Let $f(x) = \sqrt{1 + x}$.

   (a) Find the first three nonzero terms in the Maclaurin series for f.

   (b) Use Theorem 13 to bound the approximation error made in using the Maclaurin polynomial from part (a) to estimate $f(1)$.

7. Repeat the previous problem with $f(x) = \tan x$.

8. Let $f(x) = x^{-1} \sin x$ when $x \neq 0$ and define $f(0) = 0$.

   (a) Find the Maclaurin series representation of f.

   (b) What is the interval of convergence of the power series found in part (a)?

   (c) Use the series in part (a) to estimate $f'''(1)$ with an error no greater than 0.005.

9. Explain why the power series $1 - x + \dfrac{x^2}{2} - \dfrac{x^4}{8} + \dfrac{x^5}{15} - \dfrac{x^6}{240} + \cdots$ **cannot** be the Maclaurin series representation of the function f shown in the graph below.

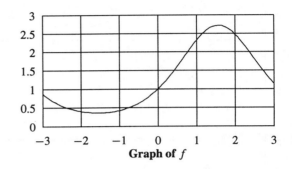

Graph of f

10. Suppose that f is a function such that $f^{(n)}$ exists for all $n \geq 1$.

    (a) Explain why the Maclaurin series for f converges to f if $\left| f^{(n)}(x) \right| \leq n$ for all $n \geq 1$.

    (b) Does Theorem 13 guarantee that the Maclaurin series for f converges to f if $\left| f^{(n)}(x) \right| \leq 2^n$ for all $n \geq 1$?

11. During an encounter with a friendly extraterrestial, it is revealed to you that the answer to life, the universe, and everything is $f(1)$ for some function f. It is also revealed that $f(0) = 26$, $f'(0) = 22$, $f''(0) = -16$, $f'''(0) = 12$, and $\left| f^{(4)}(x) \right| \leq 7x^4$ when $|x| \leq 2$.

    (a) Find an upper bound on the value of $f(1)$.

    (b) Find a lower bound on the value of $f(1)$.

11.8   Chapter summary

In this long and relatively technical chapter we study the sophisticated topic of infinite series. Infinite series are formed by adding—in a special sense, and with due concern for convergence—infinitely many numbers or functions. Making sense of such infinite summation requires special care: hence the subtlety of the subject.

Convergence.   A infinite series converges if its partial sums—formed from only finitely many terms—converge as a sequence. Alas, the sequence of partial sums, being so defined, is often hard to understand or handle directly.

Convergence tests.   Among the first questions to ask about any series is whether it converges or diverges. As with improper integrals, the answer may be far from obvious at a glance. To address this problem, several tests for convergence and divergence are discussed, including the **integral test**, the **comparison test**, and the **ratio test**. All these tests apply, in raw form, only to series of positive terms. Series that contain both positive and negative terms need special care; given that care, they too can be tested for convergence.

Estimating limits.   Many infinite series—even ones known to converge—are difficult to evaluate exactly. The problem arises naturally, therefore, of *estimating* their limits. As with integration, technology proves helpful, both for calculating partial sums and for estimating the error committed in using partial sums to approximate integrals themselves.

Power series: series as functions.   Power series, or "infinite polynomials," are among the most useful and convenient functions of calculus. Many calculus functions, moreover, can be written in power series form. Finding such a power series, and understanding its functional properties, is the subject of the last two sections of the chapter.

Taylor and Maclaurin series; Taylor's theorem.   Taylor's theorem—one of the great theorems of calculus—guarantees that under appropriate conditions a function *can* be written in series form, usually as a "Maclaurin series." The partial sums of a Maclaurin series, being polynomials, are easy to handle, and they offer useful approximations to less convenient functions.

Exercises

1. It is known that $\displaystyle\sum_{n=1}^{\infty} \frac{1}{n^2} = \frac{\pi^2}{6}$. Use this fact to evaluate $\displaystyle\sum_{n=4}^{\infty} \frac{1}{n^2}$.

2. It is known that $\displaystyle\sum_{n=1}^{\infty} \frac{(-1)^{n+1}}{n} = \ln 2$. Use this fact to evaluate $\displaystyle\sum_{n=3}^{\infty} \frac{(-1)^{n+1}}{n}$.

3. A certain series $\displaystyle\sum_{k=1}^{\infty} a_k$ has partial sums $S_n = \displaystyle\sum_{k=1}^{n} a_k = 2 - \frac{5}{n}$.

   (a) Find the real number L such that $\displaystyle\sum_{k=1}^{\infty} a_k = L$.

(b) Show that $\displaystyle\sum_{k=1}^{500} a_k$ underestimates L by 0.01.

(c) Find an integer N such that $\displaystyle\sum_{k=N+1}^{\infty} a_k \leq 0.0001$.

(d) Let N be the integer found in part (c). Explain why $m \geq N$ implies that $\displaystyle\sum_{k=1}^{m} a_k$ approximates $\displaystyle\sum_{k=1}^{\infty} a_k$ within 0.0001.

4. A certain series $\displaystyle\sum_{j=1}^{\infty} b_j$ has partial sums $S_n = \displaystyle\sum_{j=1}^{n} b_j = \dfrac{3n+2}{2n+1}$.

(a) Find the real number L such that $\displaystyle\sum_{j=1}^{\infty} b_j = L$.

(b) Does S_{100} underestimate or overestimate L? By how much? Justify your answers.

(c) Find an integer N such that S_N approximates L within 0.0001.

(d) Let N be the integer in part (c) and let m be an integer such that $m > N$. Explain why S_m approximates $\displaystyle\sum_{j=1}^{\infty} b_j$ within 0.0001.

5. Does the series $\displaystyle\sum_{k=0}^{\infty} \dfrac{1}{2+\sin k}$ converge? Justify your answer.

Find the Maclaurin series for each of the following functions.

6. $f(x) = \dfrac{\sin x}{x}$

7. $f(x) = x^2 e^{-x^2}$

8. $f(x) = \dfrac{\ln(1+x)}{x^2}$

9. $f(x) = \displaystyle\int_0^x \dfrac{e^t - 1}{t}\, dt$

Let $\{a_k\}$ denote a sequence of real numbers. Indicate whether each of the following statements is true or false. Justify your answers.

10. If $a_k > 0$ for all k and $\displaystyle\lim_{k \to \infty} a_k = L$, then $L > 0$.

11. If $a_k \geq 0$ for all k and $\displaystyle\lim_{k \to \infty} a_k = L$, then $L \geq 0$.

12. If $a_k \leq 0$ for all k and $\{a_k\}$ is increasing, then $\{a_k\}$ converges.

13. If $a_k \geq 1$ for all k and $\{a_k\}$ is decreasing, then $\{a_k\}$ converges.

14. If $a_k \geq 1$ for all k and $\{a_k\}$ is decreasing, then $\displaystyle\lim_{k \to \infty} a_k = 1$.

15. If $a_k \geq 0$ for all k and $\displaystyle\lim_{k \to \infty} (-1)^k a_k = L$, then $L = 0$.

16. If $\{a_k\}$ is neither increasing nor decreasing, then it diverges.

17. If $\{a_k\}$ is not bounded, then it diverges.

18. If $\{a_k\}$ is bounded, then it converges.

19. If $\{a_k\}$ diverges, then it is unbounded.

20. If $\{a_k\}$ converges, then it is bounded.

21. If $\lim\limits_{k \to \infty} (a_k)^2 = L$, then $\lim\limits_{k \to \infty} a_k = \pm\sqrt{L}$.

22. If $\lim\limits_{k \to \infty} |a_k| = 0$, then $\lim\limits_{k \to \infty} a_k = 0$.

23. If $\lim\limits_{k \to \infty} |a_k|$ converges, then $\lim\limits_{k \to \infty} a_k$ converges.

24. If $\lim\limits_{k \to \infty} a_k = 0$, then $\lim\limits_{k \to \infty} 1/a_k = \pm\infty$.

25. If $\lim\limits_{k \to \infty} a_k = 0$, then $\lim\limits_{k \to \infty} 1/a_k$ diverges.

26. If $\lim\limits_{k \to \infty} a_k = \lim\limits_{k \to \infty} b_k = L$, then $\lim\limits_{k \to \infty} a_k/b_k = 1$.

27. If $\lim\limits_{k \to \infty} a_k = L$, then $\lim\limits_{k \to \infty} (a_{k+1} - a_k) = 0$.

28. If $\lim\limits_{k \to \infty} (a_{k+1} - a_k) = 0$, then there is a real number L such that $\lim\limits_{k \to \infty} a_k = L$.

29. If $\sum a_k$ converges, then $\{a_k\}$ converges.

30. If $\{a_k\}$ converges, then $\sum a_k$ converges.

31. If $\lim\limits_{n \to \infty} a_n = 0$, then $\sum a_n$ converges.

32. If $\lim\limits_{n \to \infty} a_n = 2$, then $\sum a_n$ converges.

33. If $\sum a_k$ converges and $\sum b_k$ diverges, then $\sum (a_k + b_k)$ diverges.

34. If $\sum a_k$ diverges and $\sum b_k$ diverges, then $\sum (a_k + b_k)$ diverges.

35. If $\sum (a_k)^2$ converges, then $\sum a_k$ converges.

36. If $\sum a_k$ converges and $a_k \neq 0$, then $\sum 1/a_k$ diverges.

37. If $\sum a_k$ converges and $a_k \neq 0$, then $\{1/a_k\}$ diverges.

38. If $\sum a_k$ diverges and $a_k \neq 0$, then $\sum 1/a_k$ converges.

39. If $1 \leq \sum\limits_{k=1}^{n} a_k \leq 2^n$, then $\sum\limits_{k=1}^{\infty} a_k$ diverges.

40. If $S_n = \sum\limits_{k=1}^{n} a_k \leq 3$ and $S_n < S_{n+1}$, then $\sum\limits_{k=1}^{\infty} a_k = 3$.

41. If $\lim\limits_{n \to \infty} a_n = 2$, then $\sum\limits_{n=1}^{\infty} a_n = 2$.

Chapter 12

Differential Equations

12.1 Differential equations: the basics

Ideas, definitions, and examples: a quick review

(The basic ideas, definitions, and terminology of differential equations were introduced in Chapter 4. In this subsection we present only a quick summary. For more details, definitions, and examples, see Section 4.1.)

DE's and solutions to DE's

A **differential equation** (or **DE**) is any equation that involves at least one derivative. Each equation below—

$$y' = 0; \quad y' = y; \quad y' = 2xy; \quad \frac{dP}{dt} = \frac{K}{\sqrt{P}}; \quad y'' = -y$$

is a differential equation.

This bears repeating: solutions of differential equations are functions, not numbers.

A **solution** of a differential equation is any *function* that *satisfies* the differential equation. ◄ As the following example illustrates, a DE normally has *many* solution functions. To *solve* a DE means to find at least one—but preferably all—of these solution functions.

■ **Example 1.** Find solution(s) of the DE $y' = y$.

Solution: A solution of the DE $y' = y$ is a function $y(x)$ that is *its own derivative*. The **exponential function** $y(x) = e^x$ is famous for just this reason:

$$y = e^x \implies y' = e^x = y.$$

Thus $y(x) = e^x$ is a **solution** of the DE.

C may be positive, negative, or zero.

In fact, *every* function of the form $y(x) = Ce^x$, where C is a constant, ◄ is a solution of the same DE, because

$$y = Ce^x \implies y' = Ce^x = y.$$

This DE, in other words, has an *infinite family of solutions*. Several of these solution functions appear below; each is labeled by C:

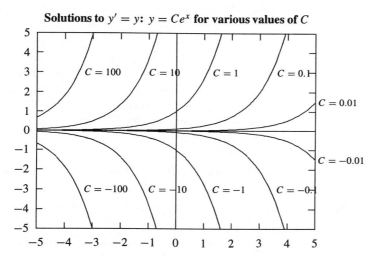

Solutions to $y' = y$: $y = Ce^x$ for various values of C

Each graph represents one solution of the DE $y' = y$. Not *all* solutions are shown, of course—by choosing C correctly we can find a solution curve passing through *any* point of the xy-plane. (Approximately what value of C would cause the solution curve to pass through the point $(2, 2)$, for example?)   □

IVP's vs. DE's

A DE typically has *infinitely many* solution functions. Adding an **initial condition** (or specified "starting value") to a DE usually lets us choose a *single* desired solution. The combination of an initial condition and a DE is called an **initial value problem**, or **IVP**. The next example illustrates the idea.

■ **Example 2.**   Find a function $y = y(x)$ that satisfies both the DE $y' = y$ and the initial condition $y(0) = -1$.

Solution: We saw above that *every* function of the form $y(x) = Ce^x$ satisfies the DE. By choosing C judiciously we'll select the one solution that satisfies both the DE *and* our initial condition. Here's how:

$$-1 = y(0) = Ce^0 = C \implies C = -1.$$

Thus $y = -1e^x$ is our desired solution—this time, the *only* solution function.

The curve $y = -1e^x$ is among those shown in Example 1. Notice, especially, that this curve *passes through the point* $(0, -1)$—as it should, because $y(0) = -1$.□

DE's and IVP's: vocabulary and notation

Every area of mathematics develops its own convenient technical terms and notations. The list below contains several such terms and shorthand forms related to DE's and IVP's.

Order. The **order** of a DE refers to its highest-order derivative. Thus $y' = y$ is a **first-order DE**; $y'' = 0$ is a **second-order DE**.

Omitting arguments. Consider this typical calculus sentence:

If $f(x) = e^x$, then f is its own derivative.

The function f is mentioned twice, first as $f(x)$—*with* the **argument** x—then simply as f—*without* an argument.

t, in this case.

DE terminology allows the same option. For example, the two DE's $y'(t) = y(t)$ and $y' = y$ mean the same thing: each describes a function that is its own derivative. Whether to include or to omit the argument[◄] depends on the situation. By convention, DE's are usually written *without* explicit function arguments.[◄] Omitting "understood" arguments simplifies a DE's appearance, but it can take some getting used to. The DE

For typographical convenience, perhaps.

$$y' = y + t,$$

for example, relates variables y and t; y is *understood* to denote some (unknown) function $y(t)$.

Variable names: t, x, etc. In writing DE's the symbol t, rather than x, often denotes the input variable. This choice is made because the letter t suggests *time*. A solution y (aka $y(t)$), then, is naturally thought of as a quantity that varies with *time*.

Instead of x, u, v, or any other letter.

We hasten to add that using t[◄] is *only* an aid to intuition. Mathematically speaking, one variable name is as good as another. DE's, moreover, can model physical situations that have nothing to do with time.

What's "initial" about an IC? Terminology notwithstanding, an "initial" condition need not refer specifically to time $t = 0$—though doing so is often most convenient. In DE-speak *any* condition of the form $y(a) = b$ is called an initial condition. In modeling the growth of a bank account, for instance, the DE $y' = 0.06y$ describes a 6% interest rate.[◄] The initial condition $y(1993.753) = 100$ could describe an initial deposit of \$100 on October 1, 1993.

With continuous compounding.

"Reading and writing" DE's and IVP's: the language of change

If $y = y(x)$ is *any* function of x, then the derivative function y' (aka dy/dx) represents the *instantaneous rate of change* of y with respect to x. A differential equation, therefore, can[◄] be understood as a highly compressed statement about the *rate* at which some unknown function "grows." To *solve* a differential equation, therefore, means to find a function[◄] that "grows" in the way the DE stipulates. The DE $y' = y$, for instance, is a succinct way of saying:

And should!

Or family of functions.

> *The quantity y grows at a rate equal to y itself. Thought of as a function, y is its own derivative.*

As remarked above, functions of the form $y = Ce^x$ (but no others!) have this important property.

Translating to and from DE language

An ability to "translate" back and forth[◄] between the symbolic language of DE's and ordinary English statements about changing quantities and derivatives is important, both for understanding what DE's say and for applying them to solve worthwhile problems. The table below illustrates several such translations. In each entry y represents an unknown function $y(t)$ of a variable t;[◄] k denotes a *constant*.

I.e., in both directions.

If the input variable t denotes *time*, then $y = y(t)$ represents a quantity that varies with time; $y' = y'(t)$ (aka dy/dt) tells *how fast*[◄] y grows. (Thinking of the input variable as time is intuitively convenient, but it's not necessary. Sometimes the input variable denotes distance, position, or some other physical quantity.)

We might have used x instead of t, but t naturally suggests time.

In the ordinary sense of the word "fast."

Translating between DE language and English	
DE version	**English version**
$y' = 0$	The growth rate of y is zero. (I.e., y remains constant.)
$y' = 0.06y$	The growth rate of y is *proportional*, with proportionality constant 0.06, to y itself. (I.e., y behaves like a bank account drawing 6% interest.)
$y' = kt$	The growth rate of y at time t is *proportional* to t.
$y'' = k$	y has constant *second derivative*. (I.e., y has constant "acceleration".)
$y' = k(y - 65)$	y varies at a rate proportional to the *difference* between y and 65 .

Interpreting DE's in real contexts

Each DE in the table above describes how some (unknown) function y varies with time. Does any familiar, real-life quantity $y(t)$ vary in the manner described by each DE?

A DE chosen purely at random need not "model" any familiar real-world phenomenon. However, many simple DE's—including all those in the table above—*do* describe important, naturally-occurring phenomena. The DE $y' = 0$, for instance, describes *any* quantity that remains unchanged over time: the position of an immovable object, the value of a financial "deposit" under one's mattress, etc.

In a similar vein, the DE $y' = 0.06y$ describes the value $y(t)$ at time t of a deposit that bears 6% interest, compounded continuously. As we discussed in Section 4.2, any function of the form $y(t) = Ce^{0.06t}$ is a solution.➤

See Section 4.2 for more information on interest rates and derivatives.

The DE's $y' = kt$ and $y'' = k$ are closely related; every solution to the first automatically satisfies the second.➤ In physical language, both DE's describe quantities $y(t)$ with *constant acceleration*. For instance, $y(t)$ might represent the varying *height* of a free-falling object, because earth's gravity exerts a constant downward force, and so causes a constant downward acceleration. (Newton's second law of motion says so.)

Do you see why?

The DE $y' = k(y - 65)$, finally, can be read as an instance of **Newton's law of cooling.**➤ If $y(t)$ is the *temperature* at time t of an object in 65-degree room, then the DE says (à la Newton) that the object warms or cools at a rate *proportional to the difference between its own temperature and room temperature*.

See Section 4.1 for more information.

(The interpretations given above aren't unique: other real-world quantities vary according to the same differential equations. See the exercises.)

Interpreting initial conditions

A DE, such as $y' = y$, tells how *fast* a varying quantity $y(t)$ grows, but not how *large* y is at any specific time. An initial condition, such as $y(0) = 1$, adds a point of reference. It tells, in effect, the value from which y *starts*.

What initial conditions mean in real-world settings depends on the situation. In the various scenarios described above an initial condition might specify, at some specific time, the value of a bank deposit, the vertical position of a falling object, or the temperature of a cooling object. The graphs below, for instance, describe the different temperature "histories" of several forgotten cups of coffee in a 70-degree room:

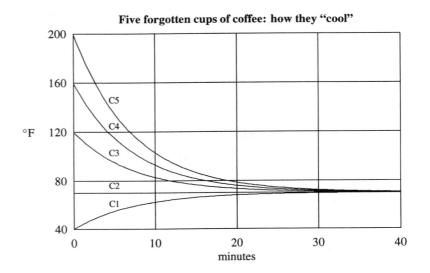

Five forgotten cups of coffee: how they "cool"

DE's, IVP's, and graphs

So far in this section we've interpreted the derivatives in DE's and IVP's mainly as *rates of growth*. From this point of view a solution to an IVP is a function with some prescribed growth behavior and also some specified value at a specified time.

Derivatives can also be thought of in terms of *slope*: for any input x, $y'(x)$ is the *slope of the y-graph at the point* (x, y). (Because time is no longer of the essence we'll revert, for the moment, to using x as the input variable.) From this geometric point of view, a first-order DE or IVP prescribes the *slopes* of solution curves at given points.

■ **Example 3.** Interpret the DE $y' = y$ in terms of slope. Do the same for the IVP $y' = y$; $y(0) = 1$.

Solution: Here's what it means, geometrically, for a function $y(x)$ to satisfy the DE $y' = y$:

At any point (x, y), the y-graph has slope y.

In Example 1 we plotted *several* solution functions to the DE in question. A close look shows that each of these curves does indeed have the claimed slope property, at each of its points.

Consider, for instance, the curve $y = e^x$ as it passes through the point $P = (\ln 2, 2) \approx (0.693, 2)$; our claim above is that the slope at P is 2. To estimate this slope graphically we can "zoom in" on P, as shown bulleted below:

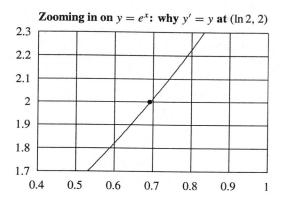

The graph shows what it should: at $P = (\ln 2, 2)$, $y = y' = 2$.

The initial condition $y(0) = 1$ selects, from among all the solution curves to $y' = y$, the one passing through the point $(0, 1)$. Understood geometrically, therefore, an IVP specifies both a single point through which a solution curve must pass *and* the slope of such a curve at each point along its length. □

Checking guesses

This section is mainly about the idea and meaning of differential equations, not the mechanics of solving them. We'll consider the problem of finding solutions soon. It's worth remarking now, however, that while *finding* solutions to DE's can be difficult, *checking* candidates for solutions is usually quite easy. By now, this pattern should be no surprise: we've already seen the closely related fact that checking antiderivatives is much easier, as a rule, than finding them. The next example illustrates this principle.

Antiderivatives and DE's. Finding antiderivatives and solving DE's are closely related problems. Indeed, antiderivatives are nothing more than special types of DE's. The antiderivative statement

$$\int \cos x \, dx = \sin x + C,$$

for example, means, in DE language, that every function of the form $f(x) = \sin x + C$ is a solution of the DE

$$y' = \cos x.$$

More generally, *every* antiderivative expression $\int f(x)\,dx$ is equivalent to the DE $y' = f(x)$.

Some DE's can be readily translated into antiderivative problems. DE's with this property are called **separable**; we'll study some in a later section. Not all DE's, on the other hand, can be solved by antidifferentiation. The DE $y' = \sin y + t^2$, for instance, has no convenient antiderivative version. But all is not lost. As with antiderivatives we couldn't solve symbolically, we'll see graphical and numerical methods for solving almost any first-order DE approximately.

■ **Example 4.** Is $y = \sin x$ a solution to the second-order DE $y'' = -y$? Is $y = 2 \sin x - 3 \cos x$ another solution? Why or why not?

Solution: The answer is "yes" in both cases, as direct computations show. Here's the computation for the latter solution:

$$y = 2\sin x - 3\cos x \quad \Longrightarrow \quad y' = 2\cos x + 3\sin x$$
$$\Longrightarrow \quad y'' = -2\sin x + 3\cos x = -y.$$

(A similar calculation shows that $y = a\sin x + b\cos x$ solves the DE for *any* constants a and b. Setting $a = 1$ and $b = 0$ shows that $y = \sin x$ is another solution.)   □

■ **Example 5.**   Verify that $y = \sqrt{x}$ solves the IVP

$$y' = -\frac{1}{2y}; \qquad y(1) = 1.$$

Solution: Direct calculation is enough. If $y = \sqrt{x}$, then

$$y' = -\frac{1}{2\sqrt{x}} = -\frac{1}{2y} \quad \text{and} \quad y(1) = \sqrt{1} = 1.$$ □

Exercises

1. For any function of the form $y = Ce^x$, $y(0) = Ce^0 = C$. What does this mean about the *graphs* of such functions?

   Graphs of several such functions appear in Example 1 of this section. How do the graphs there exhibit the property you identified above?

2. A small-town charity fund drive aims to raise \$65,000; updated current totals are posted in the town square. According to Alfred E. Neuman's Law of Cooling of Enthusiasm, the rate at which people contribute to such a drive is proportional to the difference between the current total and the announced target amount.

   Let $y(t)$ represent the current total, in thousands of dollars, t weeks after the start of the drive.

   (a) Does Neuman's Law of Cooling sound reasonable? Why or why not?

   (b) Express Neuman's Law of Cooling as a DE.

   (c) Why is the *name* Neuman's Law of Cooling appropriate?

3. It's a fact that for any constants k, T, and A, the function $y(t) = T + Ae^{kt}$ is a solution to the DE $y' = k(y - T)$.

   (a) Show by differentiation that $y(t) = T + Ae^{kt}$ *really* solves the DE $y' = k(y - T)$.

   (b) Consider the fund drive in the previous problem, which satisfies Neuman's Law of Cooling of Enthusiasm. Assume that the drive starts with an \$10,000 gift, and that after 5 weeks the total is \$45,000. How long does it take to raise \$60,000? \$64,900?

4. Look again at the cooling coffee graphs on page 291. The curves C1–C5 represent five different solutions to a DE of the form

$$y' = k(y - 70).$$

(a) Use any convenient graph to estimate the value of k. (Hint: Choose any convenient point on a graph; at that point, estimate both the slope (y') and the height (y). Use your results and the DE to solve for k.)

(b) For each of the curves C1–C5, write an IVP of which the graph is the solution. (Hint: Look at each graph to choose an appropriate initial condition.)

12.2   Slope fields: solving DE's graphically

The most basic principle in calculus, perhaps.

Derivatives can always be understood in terms of *slope*. This fundamental principle[*] holds for *all* derivatives. The derivatives in DE's and IVP's are no exception. As everywhere else in calculus, the geometric viewpoint is vital to understanding what DE's and IVP's are really "about." Interpreting DE's graphically also leads to very simple but powerful techniques for finding *approximate* solutions of almost any first-order DE or IVP. We alluded briefly in the last section to the graphical view of DE's; now we take up the subject in more detail.

DE's and slopes of solution curves

Any first-order DE can be understood as a statement about *slopes* of solution curves. Doing so is simplest if the DE is given in the form

$$y' = \text{an expression involving } y, t, \text{ or both,}$$

as is each DE below:

$$y' = y; \quad y' = 2t - 5; \quad y' = t + y; \quad y' = \sin(t^2 - 3y).$$

(Many DE's come ready-made in this form; of those that don't, most can be rewritten to do so.)

■ **Example 1.**   Discuss the DE $y' = y$ in terms of slope.

Solution: In "slope language," the DE $y'(t) = y(t)$) means the following:

> *At any point* (t, y) *on a solution curve* $y = y(t)$, *the* slope $y'(t)$ *is equal to* y.

Pun intended.

In Example 3, p. 292, we looked closely[*] at the specific point $(\ln 2, 2)$. We found what the DE requires: at that point, the solution curve has slope 2.

There's no t on the right-hand side.

The simple form of the particular DE $y' = y$ says something else about its solution curves. That y' depends *only* on y[*] (rather than on t or on both y and t) means that all solution curves have the *same* slope for a given value y. A close look at several solution curves bears this out:

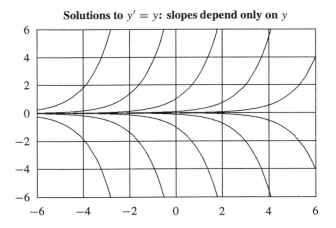

Solutions to $y' = y$: slopes depend only on y

■ **Example 2.**   Interpret the DE $y' = 2t - 5$ graphically.

Solution: The DE says, in slope-talk, that all solution curves must have slope $2t - 5$ at the point (t, y). For this simple DE it's easy to check directly—by *finding* solutions—that this slope property holds as claimed. Whether by antidifferentiation or simple guessing, one sees that

$$y'(t) = 2t - 5 \iff y(t) = t^2 - 5t + C.$$

All solution curves, therefore, have the form $y = t^2 - 5t + C$; C may be any constant. Several such curves appear below, labeled by C:

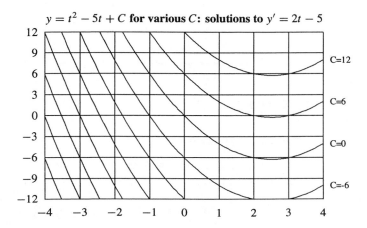

Compare these curves with those in the previous example. For the DE $y' = y$, all solution curves have the same slope at a given y-value. For the DE $y' = 2t - 5$, all solution curves have the same slope at a given t-value. This difference stems from the fact that for the first DE, y' depends only on y; for the second, y' depends only on t. □

In many DE's, such as $y' = y + t$, y' depends on *both* y and t. The next example shows what this means about solution curves.

■ **Example 3.** Interpret the DE $y' = y + t$ graphically.

Solution: The DE means, this time, that the solution curve through any point (t, y) has, at that point, slope $t + y$. In particular, solution curves should become steeper as *either* t or y increases. In a moment we'll plot some solution curves and *see* that their slopes behave this way.

But what *are* some solution functions for the current DE? In the last two examples the DE's were simpler; it wasn't hard to *guess* solution functions. With the current DE, guessing a solution is considerably harder.➡ We'll simply state, therefore, that solutions are of the form

But don't let that stop you. Try guessing anyway.

$$y = Ce^t - t - 1,$$

where C is any constant. As usual, *checking* that such functions *are* solutions is a routine exercise.➡ Here are graphs of several solution functions, labeled with C:

So we leave it as an Exercise.

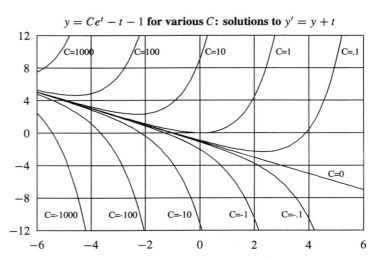

$y = Ce^t - t - 1$ **for various C: solutions to $y' = y + t$**

This time, we notice, slopes of solution curves vary both with y and with t.   □

First-order IVP's, graphically

A first-order IVP is a DE together with an initial condition. In symbolic garb, the package usually has the form

$$y' = f(t, y); \qquad y(a) = b,$$

where $f(t, y)$ is an expression involving y, t, or both. In graphical terms, the initial condition $y(a) = b$ requires that a solution curve pass through (a, b). In all cases of interest to us, only *one* such curve exists—the solution curve for the IVP. An initial condition, in other words, lets us choose, from among all the solution curves for a DE, the single curve that solves the IVP.

■ **Example 4.** Among the solution curves shown in Example 3, which one solves the IVP

$$y' = y + t; \qquad y(0) = -2.$$

For this curve, find $y(2)$, $y(10)$, and $y(-10)$.

Solution: A look at the graphs shows that the curve labeled $C = -1$ passes through $(0, -2)$. Therefore $y = -1e^t - t - 1$ solves the IVP, and

$$y(2) \approx -10.4; \quad y(10) \approx -22037.5; \qquad y(-10) \approx -8.99995.$$

(Only the first result appears on the graph. The second is far, far outside the window shown. The third, as the graph suggests, lies very near the line $y = -t - 1$.)   □

Slope fields

In the examples above we illustrated what DE's mean graphically by plotting families of solution curves. Doing so requires, of course, that we *know* (or be told) formulas for the solution functions we plot. Another graphical approach to a first-order DE, called a **slope field**◄ uses only the DE itself—no solution formulas are needed. Better yet, slope fields offer a simple and natural method of *approximating* solution curves.

*Some authors use the term **direction field**.*

Slope fields and solution curves

A first-order DE $y' = f(t, y)$▸▸ prescribes the *slope* of a solution curve through the point (t, y) in the ty-plane. A **slope field** captures this information by drawing, centered at each of *many* "grid points" (t, y), a short line segment with the appropriate slope. These segments, in effect, "point the way" along solution curves. We illustrate the idea of slope fields using three by-now familiar DE's.

$f(t, y)$ can be any expression in one or more of the variables t and y, such as y, $2t - 5$, or $y + t$.

■ **Example 5.**   Draw slope fields for the DE's $y' = y$, $y' = 2t - 5$, and $y' = y + t$. How are the results related to the pictures in Examples 1–3?

Solution: The first slope field is for the DE $y' = y$:

Slope field for the DE $y' = y$

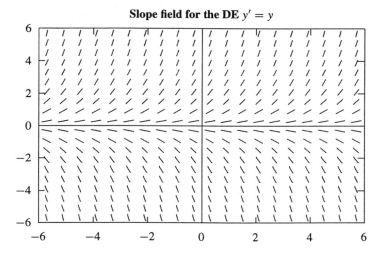

Plotting solution curves and the slope field *together* shows clearly how the two are related:

Slope field and solution curves for the DE $y' = y$

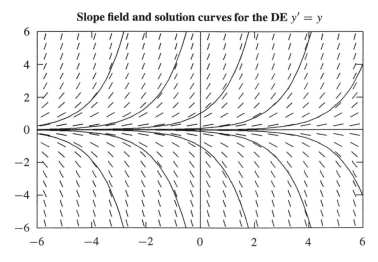

Solution curves (quite literally!) "go with the flow" that a slope field describes: at any point on a solution curve, the slope field is *tangent* to the curve.▸▸

For the DE's $y' = 2t - 5$ and $y' = y + t$ the respective slope fields and solution curves "agree" in the same sense:

I.e., the field "points along" the curve.

Slope field and solution curves for $y' = 2t - 5$

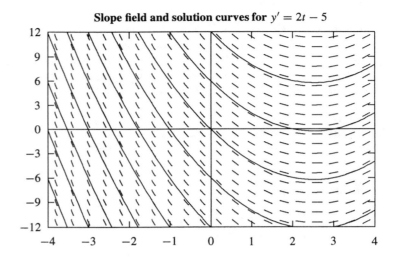

Slope field and solution curves for $y' = y + t$

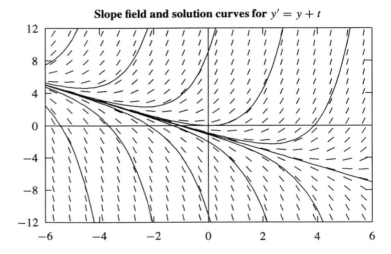

Notice, in particular, how the first slope field varies only with y, the second only with t, and the third with *both* y and t, as we expect.   □

Drawing slope fields

How, given a first-order DE $y' = f(t, y)$ and a rectangle in the ty plane, would a person (or a computer) *draw* a slope field? The *idea* is simple:

Step 1. Choose a "grid" of conveniently spaced points within the rectangle. (For each picture above we used a 20×20 grid—400 points in all.)

Step 2. At each grid point $P(t, y)$, use the DE to calculate y'. Use the result to draw a short⬲ line segment, centered at P, with slope y'.

There's a problem, of course: no sane person wants to repeat Step 2 hundreds of times. Computers, luckily, thrive on such simple, repetitive tasks. We'll use computer-drawn slope fields, therefore, except in exercises.⬲

Short enough not to interfere with each other.

And then only for simple DE's and "coarse" grids.

Go with the flow: how to solve DE's and IVP's graphically

Solving DE's.   The slope field for any first-order DE $y' = f(y, t)$ suggests a natural method of *drawing* solution curves: start *anywhere* in the direction field and "go with the flow," always following where the field ticks lead. (Solution curves are usually drawn in *both* directions from the starting point, corresponding both to increasing *t*—the "future"—and to decreasing *t*—the "past.")➤ This step produces *one* solution curve; to produce others, start somewhere else and repeat the process, as often as desired. The result is a collection of solution curves for the given DE—one curve for each starting point. Sometimes (but not always) it's even possible to use such curves to guess *formulas* for the solution curves. Even without formulas, solution curves can be extremely useful in practice.➤

"Future" and "past" should be understood metaphorically—DE's need not always involve time.

See the next example.

Solving IVP's.   Given the appropriate slope field, IVP's are even easier to solve➤ than DE's, because just *one* solution curve➤ needs to be drawn. If the IVP has the form

$$y' = f(t, y); \qquad y(a) = b,$$

for instance, then the **initial condition**➤ $y(a) = b$ tells *where* in the slope field to start drawing—at the point (a, b). The resulting curve represents *the* solution to the IVP.

Approximately, of course.

Rather than a whole family.

A good name, from this point of view.

■ **Example 6.**   Hot coffee in a 70-degree room cools at a rate proportional to the difference between the coffee temperature and room temperature.➤ In symbols (letting $y(t)$ denote coffee temperature at time t): $y'(t) = k(y - 70)$. At a certain time, direct measurement➤ showed the coffee at 190 degrees, with its temperature dropping at the rate of 12 degrees per minute. How much later did the temperature reach 130 degrees? How hot would coffee have had to be initially to be 130 degrees after 10 minutes?

Newton's Law of Cooling says so.

With a thermometer.

Solution: Let $t = 0$ denote the "certain time" mentioned above. The facts➤ that $y(0) = 190$, $y'(0) = -12$, and the DE $y' = k(y - 70)$ let us solve for k:

$$-12 = k(190 - 70) = 120 \implies k = -0.1.$$

Lightly disguised above.

Putting everything together, we have the IVP

$$y' = -0.1(y - 70); \qquad y(0) = 190.$$

Starting with a slope field on a convenient rectangle,➤ we'll draw a plausible solution curve emanating from the initial point (0, 190):➤

We chose $0 \le t \le 20$, $60 \le y \le 210$; other choices are possible, of course.

I.e., initial condition.

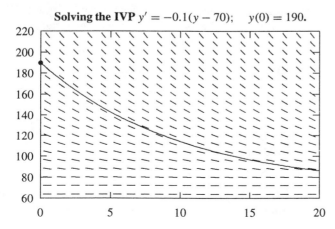

Solving the IVP $y' = -0.1(y - 70)$; $y(0) = 190$.

A close look at the solution curve suggests that the temperature reaches 130 degrees around $t = 7$ minutes.

Coffee at 130 degrees at $t = 10$ suggests *another* IVP, this time with initial value $y(10) = 130$. The solution curve through $(10, 130)$ looks like this:

Solving the IVP $y' = -0.1(y - 70); \quad y(10) = 130.$

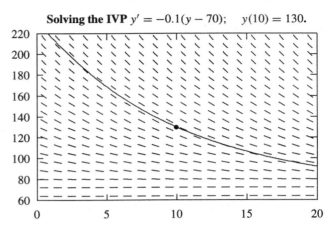

Following the curve *backward* in time shows the difficulty: at time $t = 0$, the coffee would have to be above 220 degrees—a physical impossibility.   □

Does every DE have solutions? How many?

For every slope field we've seen in this section, one solution curve—and only one—passes through *any* given point in the plane. In other words, every first-order DE we've seen has many solution functions.◄ Specifying an initial condition narrows the choice to just *one* solution function.

Infinitely many, in fact.

Geometric intuition suggests,◄ and it can be shown, that the same principle holds for *any* DE of the form $y' = f(t, y)$, as long as $f(t, y)$ is a well-behaved function of t and y. Given such a DE and any starting point (a, b), the slope field near (a, b) prescribes exactly one curve through (a, b).

Think about it. Do you agree?

The fine points of this theory (some are very fine indeed) comprise the **existence and uniqueness theory of DE's**, subjects of more advanced study in mathematics.

Exercises

1. Look again at Example 1 and its accompanying graphs. It's observed there: *All solution curves for the DE $y' = y$ have the same slope for a given value y.*

   (a) How do the shapes of the various solution curves reflect this property? (Answer in a sentence or two.)

   (b) Five of the curves shown in Example 1 pass through the level $y = 3$. Do all four curves really appear to have the same slope at that level? Use a ruler and any curve you like to *estimate* this common slope as closely as you can. (The grid should help.) Could your answer have been predicted in advance?

   (c) Repeat the previous part, but at the level $y = -4$.

   (d) Repeat the previous part, but at the level $y = 0$. (Hint: Despite appearances, only one solution curve touches the line $y = 0$.)

2. In Example 3 we claimed (but didn't show) that all functions of the form
   $y = Ce^t - t - 1$, where C is any constant, are solutions of the DE $y' = y + t$.

   (a) Verify the claim above by differentiation.

   (b) Use the work above to find *one* function (there is *only* one!) that solves
       the IVP

       $$y' = y + t; \qquad y(0) = 1.$$

       What's $y(2)$?

   (c) Is the graph of the solution function you found in the previous part one
       of those shown in Example 3? If so, which one? If not, *draw* a graph of
       your solution function.

3. Below are graphs of several *more* functions of the form $y = Ce^t - t - 1$,
   i.e., more solutions of the DE $y' = y + t$. (Compare them to the graphs in
   Example 3.)

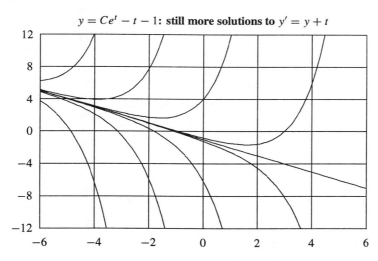

$y = Ce^t - t - 1$: **still more solutions to** $y' = y + t$

   (a) One of the "curves" is a straight line. Which straight line? Why? Label
       this curve with the appropriate value of C.

   (b) One of the curves passes through the point $(0, 4)$. What value of C
       corresponds to this curve? How do you know? Label this curve with the
       appropriate value of C.

   (c) Estimate (use the graph) the slope at $(0, 4)$ of the curve mentioned in the
       previous part. Does your answer agree with what the DE predicts?

   (d) On the axes above, draw—very carefully, with a ruler and sharp pencil—
       the line $y + t = 0$ (aka $y = -t$). This line crosses four of the solution
       curves at points of special interest. At *what* points of special interest?
       Why does this occur? (Hint: What does the DE $y' = y + t$ say about
       points on the line $y + t = 0$?)

   (e) Draw—very carefully, with a ruler—the line $y + t = -3$ (aka $y = -t - 3$)
       on the axes above. This line crosses four of the solution curves. What
       do these crossing points have in common? Explain your answer.

   (f) Draw—very carefully, with a ruler—several lines with slope -1 on the
       axes above. Each such line crosses several solution curves. What can be
       said about the points at which these crossings occur? Why?

(g) The curves shown correspond to the C-values $C = \pm 500$, $C = \pm 50$, $C = \pm 5$, $C = \pm 0.2$. Label each curve with its appropriate C-value.

4. All solution curves shown in Example 3 and in the previous problem have one property in common: they seem to "converge," in the *leftward* direction, toward the line $y = -t - 1$. To explain this phenomenon, show that if $y(t)$ is any solution function,

$$\lim_{t \to -\infty} (y(t) - (-t - 1)) = 0.$$

5. In Example 5 we remarked that "the first slope field varies only with y, the second only with t, and the third with *both* y and t." What visual property of the first slope field corresponds to "varying only with y?" What visual property of the second slope field corresponds to "varying only with t?" Explain briefly; the word "parallel" should come in handy.

12.3   Euler's method: solving DE's numerically

In the previous section we showed➤ how to solve a first-order IVP

$$y' = f(t, y); \qquad y(a) = b$$

See Example 6—especially the pictures.

graphically, using its slope field:

> *Starting at (a, b), move "along" the slope field. The result is a solution curve➤ for the IVP.*

I.e., the graph of a solution function.

Euler's method: the idea by example

The *numerical* version of the same idea is known as **Euler's method**. It describes, in concrete numerical terms, exactly how to move, *step by step*, through a slope field. We illustrate with another look at the situation of Example 6.

■ **Example 1.**   Let $y(t)$ denote the temperature (in degrees Fahrenheit) of a cup of coffee at time t (in minutes).➤ Room temperature is 70 degrees; the coffee starts at 190 degrees. The coffee's temperature is described by the IVP

Time $t = 0$ can denote any convenient reference time.

$$y' = -0.1(y - 70); \qquad y(0) = 190.$$

How hot is the coffee after 10 minutes?

Solution: Here, again, are the slope field and a plausible solution curve:➤

The bulleted starting point represents the initial condition.

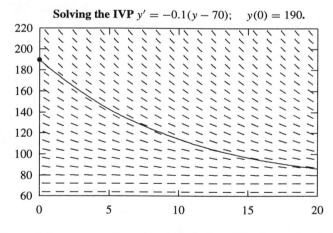

Solving the IVP $y' = -0.1(y - 70); \quad y(0) = 190.$

Judging from the solution curve, the temperature after 10 minutes is about 115 degrees.➤   □

Because $y(10) \approx 115$.

The idea of drawing a solution curve by "going with the flow" is simple, natural, and easy to carry out. Any two humans, given the slope field and initial condition shown above, would draw very similar➤ solution curves, and thus arrive at similar estimates for $y(10)$.

But not quite identical: no two humans draw exactly alike.

This graphical approach, for all its virtues, has two main flaws: (1) it gives inexact answers; (2) "drawing a curve" isn't precisely defined as a mathematical process. The second flaw is much more serious than the first. Approximate answers are perfectly respectable; sometimes they're the best available. To have any idea how *accurate* estimates may be, however, we must first describe precisely, in mathematical terms, how we *produce* estimates. A precise description is also essential if we want➤ to use technology to produce (and improve) our estimates.

We do want!

Integration revisited. The old problem of estimating definite integrals, such as

$$I = \int_0^1 \sin(x^2)\, dx,$$

illustrates the pros and cons of graphical and numerical methods. To say that I is the shaded area below—

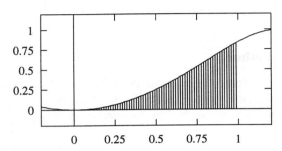

is helpful intuitively, but it doesn't generate practical estimates for I. A precisely-described numerical method, such as the left rule—

$$L_{100} = 0.30607$$

is easier to use and more amenable to accuracy-checks.

Let's describe (in terms even a computer could understand) one sensible way to move through the slope field, starting from $(0, 190)$ and ending when we reach $t = 10$. We'll take *ten straight-line steps, each of horizontal length 1*.◄ The *slope* of each linear step is determined◄ by the slope field at its *left* end. To make a "curve" we join successive linear steps. Below is the graphical result:◄

Each step corresponds to a time interval of 1 second.

Fittingly enough . . .

Look closely: the "curve" is really ten straight segments.

Ten Euler steps through a slope field

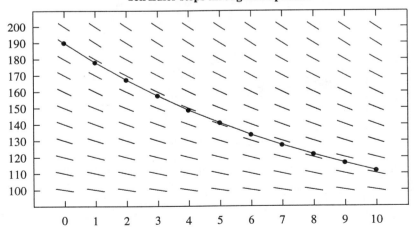

Here's how we found the successive Euler points (shown bulleted):

Step 1. The first linear step starts at $(0, 190)$ and ends one unit to the right, at $(1, 178)$. Its *slope* is given by the DE: if $t = 0$ and $y = 190$, then

$$y'(0) = -0.10(y(0) - 70) = -0.10(190 - 70) = -12.$$

Our estimate, therefore, is

$$y(1) \approx y(0) + y'(0) \cdot 1 = 190 - 12 \cdot 1 = 178.$$

Step 2. The second step begins at $(1, 178)$ and moves one unit to the right. Its slope is determined, as before, by the DE—but with the updated y-value. Since $y = 178$,

$$y'(1) = -0.10(178 - 70) = -10.8.$$

Therefore

$$y(2) \approx y(1) + y'(1) \cdot 1 \approx 178 - 10.8 \cdot 1 = 167.2.$$

Each step described above is called an **Euler step with step size 1**. To reach $t = 10$ took ten Euler steps. The following table (entries rounded to 2 decimals) summarizes our numerical work:

	From $t = 0$ to $t = 10$ in 10 Euler steps										
step	0	1	2	3	4	5	6	7	8	9	10
t	0	1	2	3	4	5	6	7	8	9	10
y'	-19.00	-10.80	-9.72	-8.75	-7.87	-7.09	-6.34	-5.74	-5.17	-4.65	-4.18
y	190.00	178.00	167.20	157.48	148.73	140.86	133.77	127.40	121.66	116.49	111.84

We estimate, therefore, that the coffee is about 112 degrees after 10 minutes.   □

How good is Euler's method?

Both Euler's method and the left rule for definite integrals^{➤➤} are systematic and precisely defined, but both find only *approximate* answers. The source of error in each case, moreover, is exactly the same. While the left rule "pretends" that an *integrand* remains constant over short intervals, Euler's method "pretends" that the *slope of a solution curve* remains constant over short intervals. Neither is usually true.

The two are closely related. See the exercises.

In the coffee example we can tell exactly how "good" our Euler's method results are.^{➤➤} An easy calculation^{➤➤} shows that the function $y(t) = 70 + 120e^{-0.1t}$ is an *exact* solution of the IVP $y' = -0.1(y - 70)$; $y(0) = 190$. In particular,

$$y(10) = 70 + 120e^{-1} \approx 114.15;$$

We couldn't do this for a harder IVP.

Do it!

significantly different from what Euler's method predicted. Plotting both the Euler "curve" and the exact solution curve shows how Euler errors *accumulate* over the time interval:

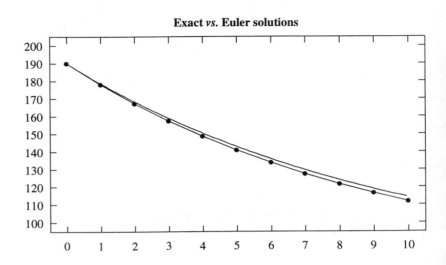

Exact vs. Euler solutions

Shorter steps: improving Euler's method

As a rule—exceptions exist.

We saw in earlier work that the left rule's accuracy generally improves ← as the number of subdivisions *increases*, i.e., as the left rule's "step size" Δx *decreases*. Euler's method behaves similarly: smaller steps usually produce more accurate results. At a price, of course: smaller steps mean more work, whether for person or machine. ←

For very small steps, roundoff errors may also accumulate.

What happens in the coffee situation with steps of size 0.1, not 1? Carrying out 100 steps by hand would be tedious, but computers don't complain. Below are some sample results; *exact* values of y appear in the last column, for comparison:

From $t = 0$ to $t = 10$ in **100 Euler steps**								
step	0	1	2	3	10	20	30	90	100
t	0	0.1	0.2	0.3	1.0	2.0	3.0	9.0	10.0
y'	−12.00	−11.89	−11.76	−11.64	−10.85	−9.82	−8.88	−4.86	−4.39
y_{euler}	190.00	188.80	187.61	186.43	178.53	168.15	158.76	118.57	113.92
y_{exact}	190.00	188.81	187.62	186.45	178.58	168.25	158.90	118.79	114.15

The Euler estimates are better this time.

We might, in theory, use even smaller step sizes, with thousands or millions of Euler steps. As with numerical integration, however, this is computationally foolish. For one thing, roundoff error eventually becomes significant. For another, better and more efficient numerical strategies are available, for DE's as for integrals. We won't pursue such "smart" strategies, except to mention that one of the most popular, called the **Runge-Kutta method**, is related to Simpson's rule as Euler's method is to the left rule.

Euler's method in general

Suppose that we're given a first-order DE $y' = f(t, y)$ and an initial condition $y(a) = y_0$, and that we use Euler's method with n steps on the interval $a \leq t \leq b$.

What Euler's method produces: graphs and numbers

What, exactly, does Euler's method give? The answer depends on our point of view:

Graphically speaking: Euler's method says how to move step by step through the DE's slope field, starting at (a, y_0) and ending when $t = b$. Every Euler step has *horizontal* length $\Delta t = (b - a)/n$, called the **step size.** Each step goes in the *direction* determined by the slope field at the *beginning point* of the step. Joining successive **Euler points**➤ with line segments produces a **piecewise-linear**➤ curve. This curve is the graph of an **approximate solution function** $Y(t)$ for the DE. (We'll call the approximate solution $Y(t)$,➤ reserving $y(t)$➤ for the *exact* solution.)➤

Connecting the dots, in other words.

Math-speak for a "curve" built from linear pieces.

Big Y.

Little y.

With any luck, $Y(t) \approx y(t)$!

Numerically speaking: Euler's method produces a list

$$Y(t_0), \ Y(t_1), \ Y(t_2), \ Y(t_3), \ \ldots \ Y(t_n)$$

of numerical values of an approximate solution function $Y(t)$, for equally-spaced inputs

$$a = t_0 < t_1 < t_2 < t_3 < \cdots < t_n = b;$$

each t_i is Δt units to the right of its predecessor. (Plotting these values and joining successive points with line segments produces the graph mentioned above.)

What Euler's method doesn't produce: formulas

Euler's method uses—and produces—only numerical information. We can use Euler's method, in other words, to produce a graph or table of values, but not a "formula" for a solution function.

The same principle holds if the left rule is applied to $\int_a^b f$. Because the left rule "sees" only isolated values of f,➤ it can produce only numbers, not formulas.

It doesn't "see" the formula for f.

This property of Euler's method➤ is a mixed bag. To its credit, Euler's method can be used with almost any DE, no matter how complicated. To its debit, Euler's method produces only approximate results, even for simple DE's that might otherwise be solved exactly.

And of the left rule.

Euler's method step by step

How, in general, does Euler's method produce the successive values

$$Y(t_0), \ Y(t_1), \ Y(t_2), \ Y(t_3), \ \ldots \ Y(t_n)?$$

Step 1: The starting point $Y(t_0) = y_0$ comes from the initial condition. By the DE, a solution curve through (t_0, y_0) has slope $m_0 = f(t_0, y_0)$, so the first Euler step has *slope m_0*. Its rise (or fall), therefore, is $m_0 \cdot \Delta t$ vertical units. In other words,

$$Y(t_1) = Y(t_0) + m_0 \Delta t.$$

Step 2: The second Euler step starts at $(t_1, Y(t_1))$.➤ At $(t_1, Y(t_1))$, says the DE, a solution curve has slope $m_1 = f(t_1, Y(t_1))$. Thus the second Euler step rises (or falls) $m_1 \cdot \Delta t$ vertical units:

Where the previous step ended.

$$Y(t_2) = Y(t_1) + m_1 \Delta t.$$

Continue: The pattern should now be clear. At each step we (1) use the DE to "update" the slope m_i; (2) move with slope m_i from $(t_i, Y(t_i))$ to $(t_{i+1}, Y(t_{i+1}))$. After n steps we arrive at $t_n = b$.

Reduced entirely to symbols, Euler's recipe for assembling an approximate solution $Y(t)$ looks like this:

$$\begin{aligned}
Y(t_0) &= y_0; \\
Y(t_1) &= Y(t_0) + f(t_0, Y(t_0)) \cdot \Delta t; \\
Y(t_2) &= Y(t_1) + f(t_1, Y(t_1)) \cdot \Delta t; \\
&\vdots \\
Y(t_{i+1}) &= Y(t_i) + f(t_i, Y(t_i)) \cdot \Delta t; \\
&\vdots \\
Y(t_n) &= Y(t_{n-1}) + f(t_{n-1}, Y(t_{n-1})) \cdot \Delta t.
\end{aligned}$$

Simple but powerful. It may seem surprising that Euler's method—the idea of which could hardly be simpler—should be named after one of the greatest mathematicians of all time, Leonhard Euler (1707-1783). One lesson may be the surprising power of simple ideas. Another lesson may be that powerful ideas, properly understood, *become* simple.

Modeling logistic population growth: Euler's method and the wise flies

Look back if necessary, but what follows is self-contained.

Euler's method looks best in action. To show it off we'll model, again, that prudent fruit fly population from Section 4.2.◄ That fly population, we assume, grows **logistically**. In other words:

The population's growth rate is proportional both to the population itself and to the difference between the carrying capacity and the population.

In symbols:

$$P' = kP(C - P),$$

P and P' vary with time; k and C are constants.

where P represents the population, P' the population's growth rate, C the carrying capacity of the environment, and k is the constant of proportionality.◄

■ **Example 2.** Through empirical measurements, researchers determine that their captive fly population, initially 1000 members strong, satisfies the logistic IVP

$$P' = 0.00000556P(P - 10000); \qquad P(0) = 1000.$$

(Here t measures *time* (in days) since the original measurement; $P = P(t)$ is the population at time t; $P' = P'(t)$ is the growth rate (in flies per day) at time t.)

Biologists like to say "hence."

What will the population be 10 days hence?◄ How will the population evolve?

Solution: To estimate $P(10)$ we'll take 10 (one-day) Euler steps. The numbers are clumsy, so we'll get technological help.◄

But follow along with a calculator.

Step 1: When $t = 0$, $P = 1000$, so

$$P'(0) = 0.00000556 \cdot 1000 \cdot 9000 \approx 50.04 \text{ flies per day.}$$

After one day, therefore, we estimate $P(1) \approx 1000 + 50 = 1050.04$ flies.

Step 2: When $t = 1$, $P \approx 1050.04$, so

$$P'(1) = 0.00000556 \cdot 1050.04 \cdot (10000 - 105.04) \approx 52.25 \text{ flies per day.}$$

Therefore we estimate

$$P(2) \approx P(1) + P'(1) \cdot 1 \approx 1050.04 + 52.25 = 1102.29 \text{ flies.}$$

The idea should now be clear, so we'll let the computer do the rest. Here are the results:➤➤

Numbers rounded to 2 decimal places.

Flies: from $t = 0$ to $t = 10$ in 10 Euler steps			
step	t	P'	P
0	0	50.04	1000.00
1	1	52.25	1050.04
2	2	54.53	1102.29
3	3	56.88	1156.82
4	4	59.29	1213.70
5	5	61.77	1272.99
6	6	64.31	1334.76
7	7	66.91	1399.07
8	8	69.56	1465.97
9	9	72.27	1535.53
10	10	75.02	1607.80

Thus we'd guess a population of about 1608 flies on day 10. ☐

A computer program (optional). Below is the computer program that produced the table of values above. Language details➤➤ don't matter, but try to understand how the program works—it shows clearly how Euler's method works.

It happens to be written in ISETL.

A technical remark might help. The statement starting with `for` indicates a **loop**. The next three commands (up to `end for;`) are *repeated* ten times, once for each `t` from 1 to 10.

```
program flies;
    population := 1000.0;
    rate := 50.0;
    t := 0;
    writeln t, rate, population;
    for t in [1..10] do
        rate := 0.00000556*population*(10000-population);
        population := population + rate*1;
        writeln t, rate, population;
    end for;
end program;
```

Caveats. The estimate $P(10) \approx 1608$ must be viewed cautiously:

- Those decimals look impressive, but flies don't come in fractions. Models only *approximate* reality.

- The figure 1608 *underestimates* the true population at day 10. The reason has to do with *step size*: taking one-day Euler steps amounts, in effect, to pretending that the growth rate *stays constant* over one-day periods. In fact, the growth rate *increases* along with the population.

Each is 0.1 day long.

- A smaller step size would update the growth rate more often, and so better estimate the population on day 10. With, say, 100 Euler steps[◄] the computer gives

$$P(0.1) \approx 1005.00, \quad P(0.2) \approx 1010.03, \quad \ldots \quad P(10) \approx 1621.43.$$

With other techniques it can be shown that $P(10) = 1638.09$. See the next section.

The last estimate is *still* low, but better than before.[◄]

Flies in the long run. With machine help it's easy to use Euler's method[◄] to project the fly population far into the future—remembering, of course, that errors accumulate over time.

A slight variant of the program would work.

We did so for a 150-day period, using step size 0.1 day. The graph below shows our results.[◄] Each dot represents one day; *between* every two dots, the machine took 10 Euler steps.[◄] Reassuringly, the graph's shape looks right:

Tabulating results would take a lot of space!

The machine worked hard.

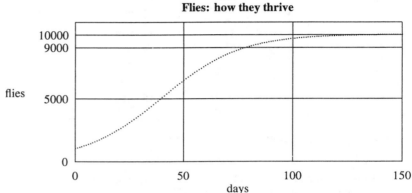

Flies: how they thrive

Notice:

- On day 100, the population was over 9500; room is fast running out. Day 200 isn't shown, but the graph's shape suggests that at $t = 200$, the population will *still* be just below 10,000.

- In about 40 days, the population reaches 5000, half the upper limit. In about 80 days, the population reaches 9000—90% capacity. Around day 130, 99% capacity is reached.

Do you agree? Look carefully.

- The fly population seems to increase *fastest* around day 40, when the population is 5000.[◄]

Intuition says that under logistic conditions a population increases slowly at first (when there are few breeding members), more quickly as the population rises, and then more slowly as the upper limit nears. Thus we might expect what we see on the graph: population grows fastest when it is *halfway* to its carrying capacity.

Exercises ————————————————————————————

1. This problem explores the connection between Euler's method and the left integral rule.

   Consider the IVP $y' = \sin(t)$ and $y(0) = 0$. Suppose we want to find (or estimate) $y(1)$.

   (a) Carry out Euler's method with 4 steps to *estimate* $y(1)$. Do all steps by hand.

   (b) Use the left rule with 4 subdivisions to estimate $I = \int_0^1 \sin(t)\,dt$.

   (c) Solve the IVP above *exactly*—i.e., find an explicit function $y(t)$. What's the *exact* value of $y(1)$?

   (d) Find $\int_0^1 \sin(t)\,dt$ exactly. How much error did L_4 commit?

   (e) Let $y(t)$ be the *exact* solution function found above, let $Y(t)$ be the *approximate* solution function constructed by Euler's method. Complete the following table:

t	0.00	0.25	0.50	0.75	1.00
$Y(t)$					
$y(t)$					

   (f) Plot both $y(t)$ and $Y(t)$ on the same axes. What do you see?

2. Repeat the previous problem using the IVP $y' = e^t$; $y(0) = 0$.

3. Here's a very simple IVP: $y' = 3$; $y(0) = 0$.

   (a) Just by guessing, find an *exact* solution $y(t)$ to the IVP above. (Hint: What functions $y(t)$ have *constant* derivatives?) Then list $y(1)$, $y(2)$, $y(3)$, $y(4)$, $y(5)$.

   (b) Use Euler's method, with 5 steps of size 1, to *estimate* $y(1)$, $y(2)$, $y(3)$, $y(4)$, $y(5)$. How close are your estimates?

   (c) Explain what you found in part (b). When would you expect Euler's method to behave this way?

4. The number e is sometimes defined as $y(1)$, where $y(t)$ is a solution to the IVP $y' = y$, $y(0) = 1$. In this problem we'll find some numerical estimates for e.

   (a) Use Euler's method with 1 subdivision to estimate e. Is your answer an under- or over-estimate? Can you tell even without knowing the value of e?

   (b) Repeat the previous part with 4 subdivisions.

   (c) (With computing.) Repeat the previous part with 100 subdivisions.

   (d) (With a calculator.) Imagine that Euler's method is to be carried out on the IVP above, with 10000 subdivisions, to estimate $y(1)$. Write out the results of the first 3 steps. Convince yourself that the result after 10000 steps is $(1.0001)^{10000}$. Use a calculator to estimate this number. Does it look familiar?

5. On page 307 we said:

   > An easy calculation shows that the function $y(t) = 70 + 120e^{-0.1t}$
   > is an *exact* solution of the IVP $y' = -0.1(y - 70)$; $y(0) = 190$.

   Do this "easy calculation." What's $y(5)$?

6. Another population of flies breeds according to the same (logistic) DE $P' = 0.00000556P(10000 - P)$, but starts with only 100 original members, so that $P(0) = 100$.

   (a) Carry out by hand the first three Euler steps; use step size 1 day.

   (b) (With computing.) The original fly population was about 9600 on day 100. (See the graph on p. 312.) Use step size 1 to estimate *this* fly population on day 100.

7. Consider again the flies' logistic DE $P' = 0.00000556P(10000 - P)$, but this time with the new initial condition $P(0) = 0$.

   (a) Carry out by hand three Euler steps; use step size 1 day. Estimate the population on day 200.

   (b) Explain in biological terms what's going on.

12.4 Separating variables: solving DE's symbolically

In earlier sections we approached DE's and IVP's *graphically*, with slope fields, and *numerically*, with Euler's method. In this section we'll solve DE's *symbolically*, by **separating variables**. Before describing the method in general terms, we'll illustrate how it works—when it works—with another look at the situation of Example 1, p. 305.

The idea—an example and remarks

■ **Example 1.** Let $y(t)$ denote➤ the temperature (in degrees Fahrenheit) of a cup of coffee at time t (in minutes).➤ Room temperature is 70 degrees; the coffee starts at 190 degrees. The coffee's temperature is described by the IVP

$$y' = -0.1(y - 70); \qquad y(0) = 190.$$

How hot is the coffee after 10 minutes? How hot is the coffee at *any* time t?

Again!

Time $t = 0$ can denote any convenient reference time.

Solution: First we rewrite the DE slightly:➤

$$\frac{dy}{dt} = -0.1(y - 70).$$

Next comes the key step:➤

Separate the y's and dy's from the t's and dt's.

Here's the result:

$$\frac{dy}{y - 70} = -0.1 dt.$$

Antidifferentiating both sides gives a new equation:

$$\int \frac{dy}{y - 70} = \int -0.1 \, dt.$$

Both antiderivatives are straightforward;➤ a simple calculation gives

$$\ln|y - 70| + C_1 = -0.1t + C_2.$$

Because *both* C_1 and C_2 are arbitrary constants, it's convenient➤ to combine them into one:

$$\ln|y - 70| = -0.1t + C.$$

To solve for y we exponentiate both sides:➤

$$\ln|y - 70| = -0.1t + C \implies |y - 70| = e^{-0.1t+C} \implies y = \pm e^C e^{-0.1t} + 70.$$

One last simplification helps. Because C is an arbitrary constant, so is $K = \pm e^C$, so y has the form

$$y = Ke^{-0.1t} + 70.$$

The last step is to use the initial condition to assign K a numerical value:

$$y(0) = 190 \implies 190 = Ke^0 + 70 \implies K = 120.$$

Our problem is now completely solved: the function

$$y(t) = 120e^{-0.1t} + 70$$

solves the IVP *exactly*, for *any* time t.➤ At $t = 10$, for example,

$$y(10) = 120e^{-1} + 70 \approx 114.146 \text{ degrees.}$$

Using the Leibniz notation for y'.

Hence the technique's name.

Convince yourself that the antiderivatives are correct.

And legal!

Check each step.

See $y(t)$ plotted on p. 305.

□

A legal separation?

By differentiation; try it!

It's easy to check[←] that $y(t) = 120e^{-0.1t} + 70$ *does* solve the IVP above. But does the method really make sense? Can dy really be "separated" from dt as we did above?

A good question: dy/dt usually denotes *one* quantity—the derivative function—not the quotient of two separate quantities. Care with symbols *is* wise, but things work out all right here. The precise reasons are mathematically subtle. Roughly speaking, the antiderivative notations $\int f(t)\,dt$ and $\int g(y)\,dy$ are *designed* to ensure that dt and dy can be handled separately.

As we saw in Chapter 6.

Our best defense of the separation method, however, is that it works—*and can be verified* to work by a straightforward, direct check. The same important principle applies to solving DE's and to finding antiderivatives:[←] *finding* answers may be difficult, but *checking* answers is easy.

Symbolic solutions: from one, many

Solving the DE above by symbolic antidifferentiation was easy. Solving more complicated DE's symbolically can be *much* harder, if not impossible. Symbolic methods—when they work—have two important advantages over numerical and graphical approaches:

1. They produce *exact* solutions, not approximations.

I.e., symbols that can stand for any *constant.*

2. They handle DE's that include **parameters**[←] as well as numerical constants.

The second property means that solving *one* DE symbolically can amount, in effect, to solving whole families of DE's. The next example illustrates this disadvantage.

■ **Example 2.** Let $y(t)$ denote the temperature (in degrees Fahrenheit) of a cup of coffee at time t (in minutes).[←] Room temperature is T_r degrees; at time $t = 0$ the coffee is at T_0 degrees. The coffee's temperature is described by the IVP

Time $t = 0$ can denote any convenient reference time.

$$y' = -0.1(y - T_r); \qquad y(0) = T_0.$$

How hot is the coffee after 10 minutes? How hot is it at *any* time t?

Solution: (Notice first that neither slope fields nor Euler's method would be any use here; both require *numerical* data.)

This example and the last are virtual clones. So are their solutions; we need only replace 70 with T_r, 190 with T_0, and 0 with t_0 wherever they appear in the *previous* solution. Here's the result:[←]

Check that this is the result.

$$y(t) = (T_0 - T_r)e^{-0.1t} + T_r.$$

The beauty of such a solution is that it applies for *any* values of the parameters T_0 and T_r. Several solution curves, all with $T_r = 70$ but with various values of T_0, appear on page 291. Another set of solution curves—each with $T_0 = 190$ but with various values of T_r—appears below:[←]

See the exercises for more on these curves.

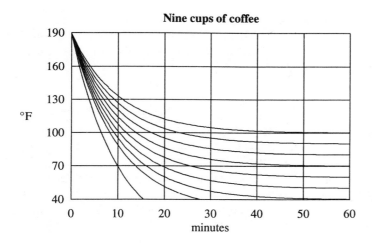

Nine cups of coffee

The method in general

When it works, when it doesn't

Separable and inseparable DE's.   Separating variables can "work" only with **separable** DE's—those in which the t and y variables *can* be separated. Below are several separable DE's, each with its "separated" form:

$$\frac{dy}{dt} = ty \quad \Longrightarrow \quad \frac{dy}{y} = t\, dt;$$

$$\frac{dy}{dt} = \frac{\sin t}{y} \quad \Longrightarrow \quad y\, dy = \sin t\, dt;$$

$$\frac{dy}{dt} = \frac{\sin(t^2)}{y} \quad \Longrightarrow \quad y\, dy = \sin(t^2)\, dt.$$

Separable DE's, typically, have➤ one of the following forms: *Or can be written in.*

$$\frac{dy}{dt} = f(t)g(y) \quad \text{or} \quad \frac{dy}{dt} = \frac{f(t)}{g(y)},$$

so that either

$$\frac{dy}{g(y)} = f(t)\, dt \quad \text{or} \quad g(y)\, dy = f(t)\, dt.$$

In some DE's (called **inseparable**) the t and y variables *can't* be separated. Here are three examples:

$$\frac{dy}{dt} = t + y; \quad \frac{dy}{dt} = \sin(t + y); \quad \frac{dy}{dt} = \frac{\sec(y^2) + \sin(t + y)}{\sqrt{t^2 + y + \cos y}}.$$

Antiderivative problems.   Even if t and y *can* be separated, the antiderivative problems

$$\int \frac{dy}{g(y)} = \int f(t)\, dt \quad \text{or} \quad \int g(y)\, dy = \int f(t)\, dt$$

still remain. Again, trouble may loom: some elementary functions can't be antidifferentiated in elementary symbolic form. One of the separable DE's above poses just this problem: the right-hand antiderivative below—

$$\int y\, dy = \int \sin(t^2)\, dt$$

has no elementary solution.

The big picture and two morals. Separation of variables is only one symbolic method; it applies only to separable DE's. DE's can take many other symbolic forms, and many other symbolic methods exist to solve them. For example, many practically useful DE's have the form

$$y' = a(t)y + b(t);$$

We won't study them systematically; that's done in DE courses.

they're called **first-order linear DE's**. General methods[*] exist for solving first-order (and higher-order) linear DE's. If $a(t)$ and $b(t)$ are suitably simple functions, solutions can be found explicitly, in terms of elementary functions.

What's $a(t)$? What's $b(t)$?

For example, the DE $y' = y + t$ is linear, but *not* separable.[*] With linear DE methods one can produce the solution $y = Ce^t - t - 1$; C can be any constant.

There are two main morals. First, if separating variables fails, other methods may

Like finding antiderivatives symbolically.

succeed. Second, solving DE's symbolically[*] is art as well as science; no fail-safe route to a solution exists.

Rumors, separable DE's and logistic growth

In earlier sections we used the **logistic DE**

$$\frac{dP}{dt} = kP(C - P) = kP(t)\,(C - P(t))$$

Flies in a lab; see the previous section.

to model populations that grow logistically.[*] Both constants k and C can be interpreted in biological terms: C represents the environment's long-run carrying capacity; k measures the population's reproduction rate.

Biological populations come first to mind, but the logistic DE works elsewhere, too. Translating the logistic DE into words gives the key property of *any* quantity that grows logistically toward a long-run value C:

> *The P grows at a rate that is proportional* both *to P itself* and *to C − P, the "room available" for further growth.*

Solving the logistic DE

For later reference let's solve the logistic DE in general form. Separating variables sets up the antiderivative problem:

$$\frac{dP}{dt} = kP(C - P) \implies \int \frac{dP}{P(C - P)} = \int k\,dt.$$

The right side is easy. The left side is stickier, but it can be found in an integral table. Do-it-yourselfers might prefer *this* approach, based on splitting the integrand into **partial fractions**:

$$\int \frac{dP}{P(C - P)} = \frac{1}{C} \int \left(\frac{1}{P} - \frac{1}{C - P} \right) dP = \frac{1}{C} \ln \left| \frac{P}{C - P} \right|.$$

Check for yourself.

However it's obtained, the antiderivative above *is* correct, as differentiation shows.[*] (We can drop the absolute value signs, moreover, if we assume—as we will—that $0 < P < C$.)[*]

Both inequalities hold naturally in all cases below.

Putting everything together and solving for P solves the logistic DE:[*]

D is a constant of integration; watch for it below.

$$\frac{dP}{dt} = kP(C-P) \implies \int \frac{dP}{P(C-P)} = \int k\,dt$$

$$\implies \ln\frac{P}{C-P} = Ckt + CD$$

$$\implies \frac{P}{C-P} = e^{Ckt}e^{CD}$$

$$\implies \frac{C-P}{P} = e^{-Ckt}e^{-CD}$$

$$\implies \frac{C}{P} = e^{-Ckt}e^{-CD} + 1$$

$$\implies P = \frac{C}{de^{-Ckt}+1}.$$

(Since D is arbitrary, so is $d = e^{-CD}$.) The result is prettier than the calculation:

Fact: For positive constants C, d, and k, the function

$$P(t) = \frac{C}{de^{-Ckt}+1}$$

solves the logistic DE $P' = kP(C-P)$.

To apply this result we need only choose appropriate values for the constants.

Rumors and logistic growth

A rumor spreads as "tellers" pass it on to "hearers." (Once told, a hearer becomes a teller.) The rumor spreads slowly at first, when tellers are few. It spreads faster when both tellers and hearers are plentiful, but slows down again as hearers become scarce, and it stops when everyone knows the rumor. The teller population, in other words, grows as shown:

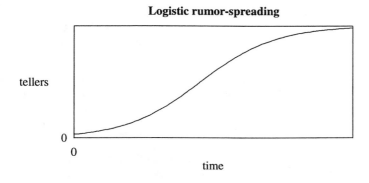

Logistic rumor-spreading

tellers

0

0

time

As the shape above suggests, logistic growth is one➤ plausible model for rumor-spreading.   *Not the only one; see below.*

■ **Example 3.** Riverdale High has 1000 students. On day 0 Archie, Jughead, Betty, Veronica, and 16 of their friends start a rumor, which spreads logistically. A day later, 50 students know it. What happens over the next 10 days? When is the rumor spreading fastest?

I.e., those who know the rumor.

Solution: Let $P(t)$ denote the number of "tellers"[◄] after t days. Then $P(t)$ satisfies the logistic DE and two additional conditions:

$$\frac{dP}{dt} = kP(1000 - P); \qquad P(0) = 20; \qquad P(1) = 50.$$

Set $C = 1000$.

By the fact above,[◄]

$$P(t) = \frac{1000}{de^{-1000kt} + 1},$$

$P(0) = 20$ is an initial condition.

for appropriate constants d and k. The other conditions[◄] let us evaluate d and k:

$$P(0) = \frac{1000}{d+1} = 20 \quad \Longrightarrow \quad d = 49;$$

$$P(1) = \frac{1000}{49e^{-1000k} + 1} = 50 \quad \Longrightarrow \quad k \approx 0.000947.$$

We've found our solution function:

$$P(t) = \frac{1000}{49e^{-0.947t} + 1}.$$

Here's the graph:

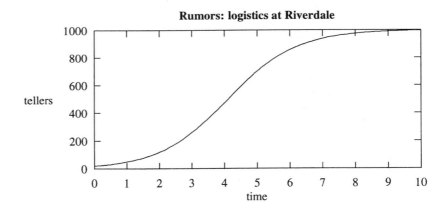

Rumors: logistics at Riverdale

The graph is steepest there.

The answer is $t \approx 4.11$—see the exercises.

By 10 days almost *everyone* knows the rumor. The rumor seems to spread fastest, moreover, when $P = 500$, just after day 4.[◄] (With elementary calculus we can find this time *exactly*.)[◄]   □

◼ **Example 4.** Chagrined that bad news travels so fast, Archie, Jughead, Betty, and Veronica start the next rumor all by themselves. It travels through Riverdale with the same "transmission coefficient" $k = 0.000947$.[◄] What happens this time?

See the exercises for more on this terminology.

Solution: Except for the new initial condition, things are exactly as in the previous example. Thus

$$P(0) = \frac{1000}{d+1} = 4 \quad \Longrightarrow \quad d = 249 \quad \Longrightarrow \quad P(t) = \frac{1000}{249e^{-0.947t} + 1}.$$

Here is the new rumor graph; the old one appears for comparison:

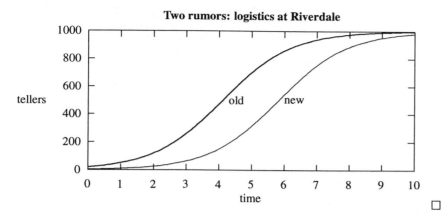

Two rumors: logistics at Riverdale

Yesterday's news: non-logistic rumors

Not all rumors spread logistically. One possible flaw in the logistic model $P' = kP(C - P)$ concerns the "transmission coefficient" k. In the logistic model, k remains *constant* over time. In practice, however, k probably *shrinks* over time; as a rumor ages, people care less and less.

A more realistic rumor model might somehow reflect this "staling" process. One approach is to replace the constant k in the logistic DE with a *decreasing function of* t. Here's one possibility: if

$$k(t) = \frac{K}{1+t},$$

where K is any positive constant, then (as we'd hope) $k(t) \to 0$ as $t \to \infty$. Then the new differential equation would have the form

$$\frac{dP}{dt} = k(t)P(t)(C - P(t)) = \frac{K}{1+t}P(C - P).$$

■ **Example 5.** Another Riverdale rumor spreads according to the DE above. When the principal hears the rumor, at time 0, 100 students know the rumor and it's spreading at the rate of 100 students per day. What happens as time goes on?

Solution: As before, let $P(t)$ denote the number of students who know the rumor at time t. The new DE, like the old, is separable:

$$\frac{dP}{dt} = \frac{K}{1+t}P(1000 - P) \implies \int \frac{dP}{P(1000 - P)} = \int \frac{K}{1+t}\,dt.$$

Both sides can be antidifferentiated, and the resulting equation solved for P.➤ When the dust settles, the result is

It takes some effort.

$$P(t) = \frac{1000}{d(1+t)^{-1000K} + 1}.$$

The additional information lets us find the constants K and d. From $P(0) = 100$ it follows➤ that $d = 9$. To find K we use the DE itself:

Check it!

$$P'(0) = 100 = \frac{K}{1+0}100(1000 - 100) \implies K = \frac{1}{900}.$$

All the numbers are now in place. Here's the formula:

$$P(t) = \frac{1000}{9(1+t)^{-10/9} + 1};$$

here's the graph—over a 100-day interval:

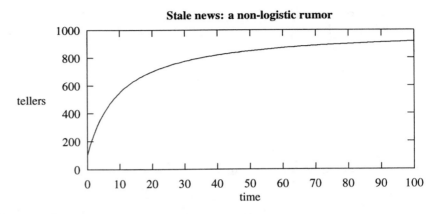

Clearly, stale news travels slowly. □

Exercises

1. Consider the nine coffee curves shown in Example 2, p. 316. Label them C_1–C_9, from top to bottom. Each curve describes the temperature "evolution" of a different cup of coffee.

   (a) All curves "start" at $(0, 190)$. What does this mean about the cups of coffee?

   (b) Each curve corresponds to a different value of T_r. What does this mean in terms of the coffee situation?

   (c) What values of T_r correspond to C_1, C_4, and C_9?

   (d) Which cup of coffee is cooling fastest at $t = 10$? Why?

   (e) Curve C_4 solves the IVP $y' = -0.1(y - 70)$; $y(0) = 190$. What IVP does curve C_1 solve? What IVP does curve C_9 solve?

   (f) *Use the curve C_1* to estimate the rate at which coffee cup C_1 is cooling at time $t = 10$.

   (g) *Use the DE for C_1* to find the rate at which coffee cup C_1 is cooling at time $t = 10$.

   (h) *Use the curve C_9* to estimate the rate at which coffee cup C_9 is cooling at time $t = 10$.

   (i) *Use the DE for C_9* to estimate the rate at which coffee cup C_9 is cooling at time $t = 10$.

   (j) Still another cup of coffee, initially at 190 degrees, is at 100 degrees after 20 minutes. *Use the graphs* to estimate the room temperature.

2. In Example 2, p. 316 we solved the IVP $y' = -0.1(y - T_r)$; $y(0) = T_0$. We found the general solution $y(t) = (T_0 - T_r)e^{-0.1t} + T_r$.

   (a) Check (by differentiation) that $y(t) = (T_0 - T_r)e^{-0.1t} + T_r$ really does solve the IVP as claimed.

   (b) Suppose that $T_0 = 200$ and $y(10) = 100$. Find T_r. Then find $y(20)$. Interpret everything in coffee terms.

   (c) Suppose that $T_r = 80$ and $y(10) = 120$. Find T_0. Then find $y(20)$. Interpret everything in coffee terms.

   (d) Suppose that $y(10) = 100$ and $y(20) = 80$. How hot was the coffee at time $t = 0$? What's the room temperature? How hot will the coffee be at $t = 40$?

   (e) It's true that $\lim\limits_{t \to \infty} e^{-0.1t} = 0$. Use this fact to find $\lim\limits_{t \to \infty} y(t)$. Interpret the result in coffee terms.

3. A population P that satisfies a DE of the form $P'(t) = kP(t)(C - P(t))$ (k and C are constants) is said to grow **logistically**.

   (a) If $P < C$, then (says the DE) $P' > 0$. What does this mean in population terms?

   (b) If $P > C$, then (says the DE) $P' < 0$. What does this mean in population terms?

   (c) If $P = C$, then (says the DE) $P' = 0$. What does this mean in population terms?

4. We showed in this section that non-zero solutions to the DE $P'(t) = kP(t)(C - P(t))$ (k and C are constants) have the form

$$P(t) = \frac{C}{Ke^{-Ckt} + 1}$$

   for some constant K. Assuming that C and k are *positive* constants, show that $P(t) \to C$ as $t \to \infty$. What does this mean in biological terms?

5. In Example 3, p. 319, we said that

$$\frac{1000}{49e^{-1000k} + 1} = 50 \implies k \approx 0.00947.$$

   Verify this claim. (First solve for k *exactly*; then find a decimal approximation.)

6. In Example 3, p. 319, we said that (i) the rumor spreads fastest when the population is 500; (ii) this occurs just after day 4.

   (a) The DE $P' = kP(C - P)$ gives the *slope* of the P-graph as a function of P. By differentiating the right side with respect to P, show that P' is maximum when $P = C/2$. (In the present case, this means $P = 500$.)

   (b) For what value of t is $P(t) = 500$? (Hint: Solve the equation for t.)

7. Below are plotted several variants on the logistic rumor graph of Example 4, p. 320. In each case, 50 people know the rumor on day 0.

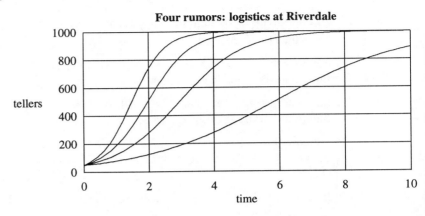

Four rumors: logistics at Riverdale

(a) Call the curves P_1 through P_4, reading from top to bottom on the picture. Each curve's equation is of the form $P(t) = \dfrac{1000}{19e^{-kt} + 1}$ for some value of k. Use the graph to estimate a value of k for each curve P_1 through P_5.

(b) In this context, the parameter k can be thought of as the rumor's "transmission coefficient." It measures how likely a given teller is to pass the rumor on to a given hearer. Alternatively, k can be thought of as a measure of how "interesting" the rumor may be. Which rumor curve corresponds to the hottest rumor? Which corresponds to the dullest rumor? Why?

8. Throughout this problem assume that rumors spread through a 1000-student school in the manner described in Example 5. In each part below, use the given information to calculate an appropriate population function $P(t)$. (Hint: Find appropriate values for K and d.)

   (a) Assume that $P(0) = 100$ and $P'(0) = 50$.

   (b) Assume that $P(0) = 50$ and $P'(0) = 100$.

   (c) Assume that $P(0) = 800$ and $P'(0) = 50$.

   (d) Assume that $P(0) = 800$ and $P'(0) = 100$.

   (e) On the same axes, plot your results from the previous parts. (Use a machine!) Label your graphs.

   (f) Discuss the differences among the rumors above, using such terms as "hot," "cool," "long-running," and "short-running."

9. Solve the flies' IVP

$$P' = 0.00000556P(P - 10000); \qquad P(0) = 1000$$

exactly, by separating variables. Find $P(10)$. How does the answer compare with what we found in the previous section, using Euler's method?

12.5   Chapter summary

Differential equations—equations that contain derivatives—permit the most important and useful applications of calculus. This chapter illustrates the meaning, uses, and techniques of differential equations.

Differential equations: the idea   We first broached the idea of a DE in Chapter 4, where we used DE's to model various growth phenomena. Key to understanding DE's is the notion of a **solution**. A function—not a number—is a solution to a DE if it "satisfies" the DE's condition. Every function of the form $y = Ce^x$, for instance, solves the DE $y' = y$. Adding an **initial condition** to a DE produces an **initial value problem**, or IVP. Here's one:

$$y' = y; \qquad y(0) = 1.$$

The same functions $y = Ce^x$ still solve the DE, but only one function—$y = 1e^x$—solves the IVP.

Solving differential equations graphically: slope fields   A **slope field** is associated with every first-order DE $y' = f(t, y)$; at each point of a rectangle in the ty-plane, a small segment points in the direction of a solution curve. Given a slope field, a solution curve (i.e., the graph of a solution function) can be drawn through any initial point by "going with the flow."

Solving differential equations numerically: Euler's method   Euler's method is the numerical version of the graphical technique described just above. To solve a DE $y' = f(t, y)$ with initial condition (t_0, y_0), Euler's method starts at the initial point (t_0, y_0) and proceeds, step by step, toward an estimate for $y(t)$ at later values of y.

Euler's method, like the left rule (which it closely resembles), solves DE's only approximately. The error Euler's method commits (like that of the left rule) can usually be reduced by using a smaller Euler step size.

Solving differential equations symbolically: separation of variables.   Some DE's can be solved symbolically, by antidifferentiation. Among the simplest of these are **separable DE's**, those that can be written in the form

$$\frac{dy}{dt} = \frac{f(t)}{g(y)}.$$

"Separating" the t's and dt's from the y's and dy's produces a new equation, one that involves antiderivatives:

$$\int g(y)\, dy = \int f(t)\, dt.$$

Solving the new equation, by antidifferentiation, solves the original DE.

Chapter 13

Polar Coordinates

13.1 Polar coordinates and polar curves

Any point P in the xy-plane has a familiar and natural "address": its **rectangular** (or **Cartesian**) coordinates. The point P below left, for instance, has rectangular coordinates $(4, 3)$:

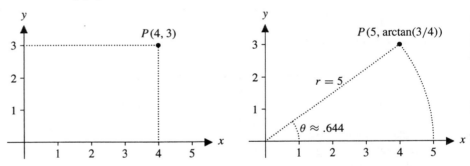

I.e., either positive or negative, depending on direction.

The x- and y-coordinates of P measure, respectively, directed[*] distances from P to the two perpendicular coordinate axes. To reach $P(4, 3)$ from the origin, one moves 4 units *right* and 3 units *up*.

Polar coordinates offer another way of "locating" a point in the plane. In the **polar coordinate system**, a point P has coordinates r and θ; they tell, respectively, the *distance from the origin O to P* and the *angle[*] from the positive x-axis to the ray from O to P*. The picture above right shows that the point P with rectangular coordinates $(4, 3)$ has **polar coordinates** $(5, \arctan(3/4)) \approx (5, 0.644)$.[*]

In radians!

Convince yourself that $\theta = \arctan(3/4)$.

Polar coordinate systems

A **rectangular coordinate system** in the Euclidean plane starts with an **origin** O and **two perpendicular coordinate axes**. Usually[*] the x-axis is horizontal and the y-axis vertical; x-coordinates increase to the *right* and y-coordinates increase *upward*.

But not invariably.

A **polar coordinate system** starts with different ingredients: an **origin** O, called the **pole**, and a ray,[*] beginning at the origin, called the **polar axis**. The polar axis normally points to the right, along the positive x-axis.

I.e., a half-line.

With these ingredients, and a unit for measuring distance, we can assign polar coordinates (r, θ) to any point P:

> r is the distance from O to P; θ is any angle from the polar axis to the segment $\overline{OP}$.

The picture below shows several points, with their polar coordinates, plotted on a **polar grid:**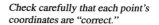

Check carefully that each point's coordinates are "correct."

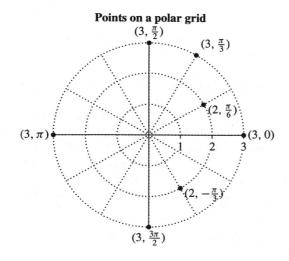

Points on a polar grid

Polar vs. rectangular grids. A rectangular coordinate system leads naturally to a **rectangular grid**, with vertical lines $x = a$ and horizontal lines $y = b$. In a polar system, holding the coordinates r and θ constant produces, respectively, concentric circles and radial lines. The result is a web-like **polar grid**. Here are grids of both types:

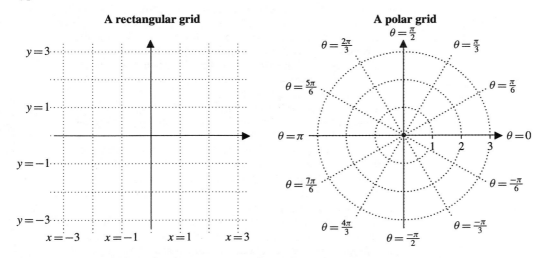

A rectangular grid   **A polar grid**

On the scales. In a rectangular coordinate system, the two axes may (and often do) carry different scales (i.e., units of measurement). On a graph, therefore, a vertical inch and a horizontal inch may represent different "distances." "Circles," in particular, may look far from round.

A polar coordinate system, by contrast, has just one axis—the polar axis. As a result, "distance" does *not* depend on direction. Circles look round.

Polar coordinates: not unique. A point in the plane—$P(4, 3)$, for instance—has just *one* possible pair of rectangular coordinates. Different rectangular coordinate pairs (x_1, y_1) and (x_2, y_2) correspond to different points in the plane.

And many others.

Polar coordinates, by contrast, are *not* unique. Every point in the plane has *many* possible pairs of polar coordinates. For example, all of the following polar coordinate pairs—

$$\left(2, \frac{\pi}{4}\right), \left(2, \frac{9\pi}{4}\right), \left(2, -\frac{7\pi}{4}\right), \left(2, -\frac{15\pi}{4}\right), \left(-2, \frac{5\pi}{4}\right), \left(-2, -\frac{3\pi}{4}\right)$$

represent the *same* point: the one with rectangular coordinates $(\sqrt{2}, \sqrt{2})$. Notice especially the last two pairs: a *negative* r-coordinate means that to locate the point P, one moves r units *opposite* to the θ-direction. The point $(-2, 5\pi/4)$, for instance,

That's the ray opposite to $\theta = \pi/4$.

lies two units from the origin on the ray $\theta = \pi/4$). The origin O allows even more freedom: it's represented by *any* pair of the form $(0, \theta)$—regardless of θ.

E.g., $\theta = \pi/4, \theta = 9\pi/4$, $\theta = -7\pi/4$, etc.

This "ambiguity" of polar coordinates arises for a simple reason. All angles that differ by integer multiples of 2π determine the same direction. In practice this ambiguity can present some annoyance, but seldom a serious problem. Two simple rules will help:

See the exercises for more on these rules.

Multiples of 2π: For any r and θ, the pairs (r, θ) and $(r, \theta + 2\pi)$ describe the same point.

Negative r: For any r and θ, the pairs (r, θ) and $(-r, \theta + \pi)$ describe the same point.

Polar coordinates on Earth. The polar grid somewhat resembles an overhead view of Earth, looking "down" at the North Pole. In cartographers' language, lines of **longitude** (aka **meridians**) converge at the pole; the concentric circles are lines of **latitude** (aka **parallels**). The **prime meridian** (or polar axis, in calculus language), for which $\theta = 0$, has been taken for hundreds of years to be the line of longitude that passes through the Greenwich Observatory, just east of London, England. For the same reason, Greenwich Mean Time (the time of day along the prime meridian) is used worldwide was a reference point.

What makes Greenwich "prime," rather than, say, India or Arabia, where navigation and time-keeping flourished even in antiquity? Nothing intrinsic: Greenwich just happened to be a center of attention when the terms were defined—an early (and quite literal!) instance of Eurocentrism.

Polar coordinates in the *plane*, it should also be said, aren't perfectly suited to measuring the (almost) spherical earth. In practice geographers use a related system, similar in some respects, called *spherical* coordinates.

Polar graphs

The ordinary graph of an equation in x and y is the set of points (x, y) whose coordinates satisfy the equation. The graph of $x^2 + y^2 = 1$, for instance, is the circle of radius 1 about the origin. The point $(2, 3)$ does *not* lie on this graph because $2^2 + 3^2 \neq 1$.

Because $3 = 2 + \cos 0$.

The idea of a **polar graph** is similar—but not quite identical. The graph of an equation in r and θ is the set of points whose *polar coordinates* r and θ satisfy the equation. For instance, the polar point $(3, 0)$ lies on the graph of $r = 2 + \cos\theta$ but the polar point $(2, \pi)$ does not.

A warning.   The fact that a point in the plane has more than one pair of polar coordinates means that polar plotting requires extra care. At first glance, for instance, the point P with polar coordinates $(-3, \pi)$ seems *not* to satisfy the polar equation $r = 2 + \cos\theta$. A closer look, shows, however, that P can *also* be written with polar coordinates $(3, 0)$—which *do* satisfy the given equation.➤ Here's the moral:

As we saw above.

> *A point P lies on the graph of a polar equation if P has* any *pair of polar coordinates that satisfy the equation.*

Drawing polar graphs

The simplest polar graphs come from *functions*, usually of the form $r = f(\theta)$. Given such a function and a specific θ-domain, it's a routine matter to tabulate points, and then plot them. We illustrate by example.

■ **Example 1.**   Plot the equation $r = 2 + \cos\theta$, for $0 \le \theta \le 2\pi$.

Solution: Let $f(\theta) = 2 + \cos\theta$; we want the r-θ graph of f. First we'll tabulate some values:

\multicolumn{13}{c}{**Values of** $r = f(\theta) = 2 + \cos\theta$}
θ	0	$\frac{\pi}{6}$	$\frac{\pi}{3}$	$\frac{\pi}{2}$	$\frac{2\pi}{3}$	$\frac{5\pi}{6}$	π	$\frac{7\pi}{6}$	$\frac{4\pi}{3}$	$\frac{3\pi}{2}$	$\frac{5\pi}{3}$	$\frac{11\pi}{6}$	2π
r	3	2.87	2.5	2	1.5	1.14	1	1.14	1.5	2	2.5	2.87	3

Next we plot the data (polar "graph paper" makes the job easier) and fill in the gaps smoothly. Here's the result:

A polar graph: $r = 2 + \cos\theta$

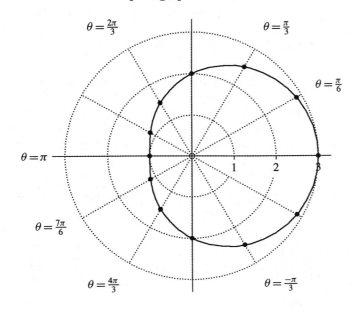

Polar graphs: a sampler

Several polar graphs are shown below. The simplest polar graphs of all—those of the equations $r = a$ and $\theta = b$—we've seen already.

I.e., "heart-like".

Cardioids and limaçons. Graphs of the form $r = a \pm b \cos\theta$ and $r = a \pm b \sin\theta$, where a and b are positive numbers, are called **limaçons**; if $a = b$, the term **cardioid** ◄ is used. (The graph in the previous example is a limaçon.) The graphs below illustrate the variety of limaçons, and show the effects of the constants a and b.

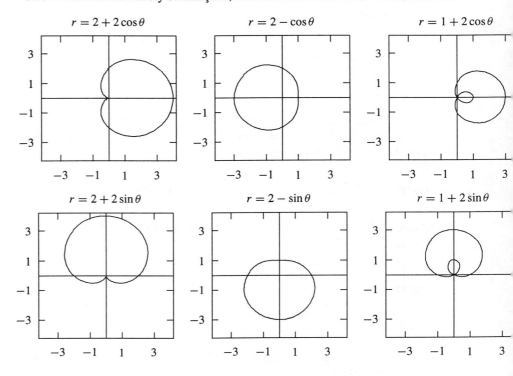

$r = 2 + 2\cos\theta$ $\qquad$ $r = 2 - \cos\theta$ $\qquad$ $r = 1 + 2\cos\theta$

$r = 2 + 2\sin\theta$ $\qquad$ $r = 2 - \sin\theta$ $\qquad$ $r = 1 + 2\sin\theta$

Notice:

We might, instead have used the interval $-\pi \le t \le \pi$.

What θ-range? The graphs above were drawn by letting θ vary through the interval $[0, 2\pi]$. Since all functions involved are 2π-periodic, any other interval of length 2π ◄ would produce the same result.

Symmetry. Three of the limaçons above—those that involve the cosine function—are symmetric about the x-axis, i.e., the line $\theta = 0$. (The other three are symmetric about the y-axis.) This symmetry occurs because the cosine function is *even*: for any θ, $\cos\theta = \cos(-\theta)$. The other graphs are symmetric about the y-axis because the sine function is odd. (For more on symmetry, see the exercises.)

Inner loops. Each of the limaçons $r = 1 + 2\cos\theta$ and $r = 1 + 2\sin\theta$ has an **inner loop**. A close look at the graphs and the formulas shows that these loops correspond to *negative* values of r. For $r = 1 + 2\cos\theta$, for instance, we have $r = 0$ when $\theta = 2\pi/3$ or $\theta = 4\pi/3$, and $r < 0$ for $2\pi/3 < \theta < 4\pi/3$. For these θ-values, therefore, the curve is drawn on the *opposite* side of the origin.

Roses. Equations of the form $r = a\cos(k\theta)$ and $r = a\sin(k\theta)$, where a is a constant and k is a positive integer, produce graphs called **roses**. The following pictures explain the name:

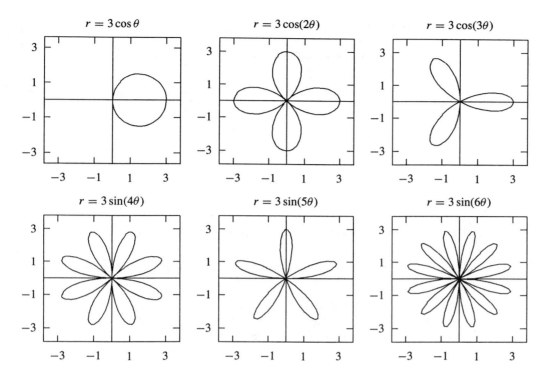

Notice:

Symmetry. Like limaçons (and for the same reason), all roses are symmetric about an axis—"cosine roses" about the x-axis and "sine roses" about the y-axis.

The rose's radius. The coefficient a in $r = a\cos(k\theta)$ and $r = a\sin(k\theta)$ determines the rose's "radius."

How many petals? The coefficient k in $r = a\cos(k\theta)$ and $r = a\sin(k\theta)$ determines the number of "petals": k if k is *odd*, $2k$ if k is even. But here's a subtlety, best revealed by plotting some roses by hand:

If k is odd, then each petal is traversed *twice* for $0 \leq \theta \leq 2\pi$.

In other words: for odd k, the rose $r = a\cos(k\theta)$ (or $r = a\sin(k\theta)$) has k *double petals*.

Trading polar and rectangular coordinates

How are the polar coordinates (r, θ) of a point P related to the rectangular coordinates (x, y) of the *same* point? How can either type of coordinates be found from the other?

Here's a useful picture:

Relating polar and rectangular coordinates

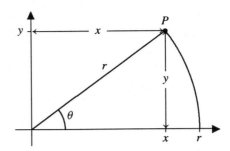

Convince yourself of each.

The picture illustrates many relations⁺ among x, y, r, and θ. Among the simplest:

$$x = r\cos\theta; \quad y = r\sin\theta; \quad r^2 = x^2 + y^2; \quad \tan\theta = \frac{y}{x}.$$

(The last equation holds only if $x \neq 0$, of course.)

These relations let us convert from one type of coordinates to the other. Equations in x and y, for instance, are easy to rewrite in terms of r and θ. The next two examples show the process in action.

■ **Example 2.** Find a polar equation for the straight line $y = mx + b$.

Solution: Substituting $x = r\cos\theta$ and $y = r\sin\theta$ into the equation for the line gives

$$y = mx + b \iff r\sin\theta = m(r\cos\theta) + b \iff r = \frac{b}{\sin\theta - m\cos\theta}.$$

(One moral: rectangular coordinates are better-suited to straight lines than polar coordinates!) □

Look again. Do you agree?

■ **Example 3.** The graph of $r = 3\cos\theta$ *looks* like a circle.⁺ Is it a circle? Which circle?

Solution: It *is* a circle. Changing to rectangular coordinates shows why:

$$r = 3\cos\theta \implies r^2 = 3r\cos\theta \implies x^2 + y^2 = 3x.$$

Check details.

The last equation does, as expected, define a circle. To decide *which* circle, complete the square:⁺

$$x^2 + y^2 = 3x \implies x^2 - 3x + y^2 = 0 \implies \left(x - \frac{3}{2}\right)^2 + y^2 = \frac{9}{4}.$$

Do you agree? Look again at the picture.

The circle, therefore, has radius 3 and center at $(0, 3/2)$—just as the picture suggests.⁺ □

Exercises

1. Several points are given below in *rectangular* coordinates. In each part, plot the given point; then give three different pairs of *polar* coordinates for the same point. For at least one pair, r should be *negative*. **Note**: All parts of this problem involve "familiar" angles θ—so give *exact* answers, not decimal approximations.

   (a) $(1, 1)$
   (b) $(-1, 1)$
   (c) $(1, \sqrt{3})$

   (d) $(-\sqrt{3}, 1)$
   (e) $(\pi, 0)$
   (f) $(0, \pi)$

2. Several points are given below in *rectangular* coordinates. In each part, plot the given point; then give three different pairs of *polar* coordinates for the same point. For at least one pair, r should be *negative*. **Note**: Angles in this problem are not necessarily "familiar," so use a calculator (in radian mode!). Round answers to 3 decimal places.

   (a) $(1, 2)$
   (b) $(-1, 2)$
   (c) $(1, 4)$

   (d) $(1, 100)$
   (e) $(0.356, 0.478)$
   (f) $(-0.356, -0.478)$

3. Several points are given below in *polar* coordinates. In each part, plot the given point; then give *rectangular* coordinates for the same point. **Note**: All parts of this problem involve "familiar" angles θ—so give *exact* answers, not decimal approximations.

   (a) $(2, \pi/4)$
   (b) $(-2, 5\pi/4)$
   (c) $(1, 13\pi/6)$

   (d) $(42, 0)$
   (e) $(a, 0)$ (a any positive number)
   (f) $(-a, 0)$ (a any positive number)

4. Several points are given below in *polar* coordinates. In each part, plot the given point; then give *rectangular* coordinates for the same point. **Note**: Angles in this problem are not necessarily "familiar," so use a calculator (in radian mode!). Round answers to 3 decimal places.

   (a) $(1, 1)$
   (b) $(-1, 1)$
   (c) $(2, 2)$

   (d) $(2, -2)$
   (e) $(1, \arctan 1)$
   (f) $(1, \arctan 2)$

5. The point $P(1, 0)$ has *identical* rectangular and polar coordinates. What other points have this property? Why?

6. Imitate Example 1, page 329, to plot the limaçon $r = 2 + \sin\theta$. Proceed as follows:

   (a) Make a table of values like that in Example 1—let θ range from 0 to 2π in steps of $\pi/6$. Round r-values to 2 decimals.

(b) Copy the polar grid shown in Example 1. On it, plot the points calculated in the previous part. Join them with a smooth curve.

(c) Discuss the symmetry of the resulting limaçon. What is its axis of symmetry?

(d) Look carefully at the table of values in Example 1; notice how the r-values are "symmetric" about $\theta = \pi$. (The graph in Example 1 is also symmetric about $\theta = \pi$.) What similar type of symmetry does your table from (a) above show? Does your graph agree?

7. Imitate Example 1, page 329, to plot the cardioid $r = 1 + \cos\theta$. Proceed as follows:

(a) Make a table of values like that in Example 1—let θ range from 0 to 2π in steps of $\pi/6$. Round r-values to 2 decimals.

(b) Copy the polar grid shown in Example 1. On it, plot the points calculated in the previous part. Join them with a smooth curve. Why does the name "cardioid" fit?

(c) Discuss the symmetry of the resulting cardioid. What is its axis of symmetry? How does the table of values in (a) reflect the cardioid's symmetry?

8. Plot and discuss the limaçon $r = 1 - 2\cos\theta$, as follows:

(a) Make a table of values like that in Example 1—let θ range from 0 to 2π in steps of $\pi/6$. Round r-values to 2 decimals. Plot the points; join them with a smooth curve.

(b) For what values of θ is $r = 0$? How do these values appear on the graph?

(c) On what θ-interval is $r < 0$? How do this interval show up on the graph?

9. In each part below, draw a quick plot of the given polar equation. To avoid tedious point-plotting, use the models of cardioids and limaçons beginning on page 330. (A calculator or computer is OK, but shouldn't be necessary. If you use one, be sure to label graphs with appropriate units.)

(a) $r = 3 + 3\cos\theta$

(b) $r = 3 - \cos\theta$

(c) $r = 1 + \sqrt{3}\cos\theta$

(d) $r = 4 + 4\sin\theta$

(e) $r = 4 - 2\sin\theta$

(f) $r = 2 - 4\sin\theta$

10. In each part below, quickly sketch the given polar rose. To avoid tedious point-plotting, use the models of roses beginning on page 330. (A calculator or computer is OK, but shouldn't be necessary. If you use one, be sure to label graphs with appropriate units.)

(a) $r = 2\sin\theta$

(b) $r = 2\sin(2\theta)$

(c) $r = 2\sin(3\theta)$

(d) $r = 2\cos(4\theta)$

(e) $r = 2\cos(5\theta)$

(f) $r = 2\cos(1001\theta)$ (rough is OK!)

11. We claimed in this section that for any numbers r and θ, the pairs (r, θ), $(r, \theta + 2\pi)$, and $(-r, \theta + \pi)$ all describe the same point in the plane.

   (a) What does the claim above say if $r = 1$ and $\theta = 0$? Is it true? Why? Explain in your own words.

   (b) What does the claim above say if $r = -1$ and $\theta = \pi/4$? Is it true? Why? Explain in your own words.

   (c) The point with rectangular coordinates $(1, 0)$ can be written in polar coordinates as $(1, 2k\pi)$, where k is any integer, or as $(-1, (2k - 1)\pi)$, where k is any integer. In the same sense, describe all the possible polar coordinates of the point with rectangular coordinates $(1, 1)$.

12. In each part below, change the given equation in r and θ to an equivalent equation in x and y. Then plot the result—use whichever form seems simpler.

   (a) $r = 2 \sec \theta$ (c) $\tan \theta = 1$

                                                 (d) $r = 2 \sin \theta$
   (b) $r = 4$

13. In each part below, change the given equation in x and y to an equivalent equation in r and θ. Then plot the result—use whichever form seems simpler.

   (a) $x^2 + y^2 = 9$ (c) $y = 2x$

                                                 (d) $(x - 1)^2 + y^2 = 1$
   (b) $y = 4$

14. (Use a graphing calculator or other polar plotting tool for this problem.) This problem concerns limaçons of the form $r = 1 + a \cos \theta$, where a is any real constant. The questions below are open-ended—answer them by experimenting with plots for various values of a: positive, negative, large, small, etc.

   (a) For what *positive* values of a does the graph have an inner loop?

   (b) What happens at $a=1$?

   (c) What happens as $a \to 0$?

   (d) How are the graphs for a and $-a$ (e.g., $r = 1 + 0.5 \cos \theta$ vs. $r = 1 - 0.5 \cos \theta$) related to each other?

   (e) What happens as $a \to \infty$?

15. In each case below, the graph is a some sort of *spiral*. Plot each either by hand or by calculator (if the latter, be sure to adjust the θ-range appropriately).

   (a) $r = \theta$, for $0 \le \theta \le 2\pi$ (an ordinary spiral)

   (b) $r = \theta$, for $-2\pi \le \theta \le 2\pi$ (another ordinary spiral)

   (c) $r = \ln(\theta)$, for $1 \le \theta \le 4\pi$ (a logarithmic spiral)

   (d) $r = \exp(\theta)$, for $-\pi \le \theta \le \pi$ (an exponential spiral)

13.2   Calculus in polar coordinates

In the last section we explored the polar coordinate system, polar equations, and the geometry of polar curves. This section is about *calculus* on polar curves. As for rectangular curves, two main problems stand out for polar curves: finding the *slope* at a point and finding the area enclosed.

To illustrate these questions, consider these pictures:

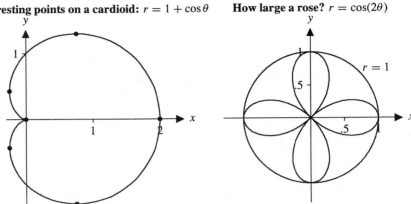

Interesting points on a cardioid: $r = 1 + \cos\theta$ **How large a rose?** $r = \cos(2\theta)$

Among the questions we'll address:

Horizontal and vertical tangents. Where—exactly—does the cardioid have horizontal and vertical tangent lines? (The bulleted points on the cardioid seem to be of interest, but what are their coordinates?)◄

Slope at any point. What's the slope of either curve at *any* point?

Areas. How much area does one petal of the rose enclose? What fraction of the *circular* area does the rose enclose?◄

Slopes on polar curves

We'll consider curves of the form $r = f(\theta)$, where f is a differentiable function.◄ For given θ_0, we want the slope at the polar point (r_0, θ_0).

An obvious—but wrong—guess. Let's acknowledge, just for the record, that the desired slope is *not* $f'(\theta_0)$. A look at either graph above quickly confirms this. For $f(\theta) = 1 + \cos\theta$, for instance, $f'(\theta) = -\sin\theta$, which varies only between -1 and 1. Yet it's clear at a glance that slopes on the cardioid take *all* real values; at three points, moreover, the cardioid has *vertical* tangent lines. Thus $f'(\theta)$ is *not* the slope we seek, though it will play a role.

Polar curves and parametric equations

We showed in earlier work how to find the slope at a point on a curve defined by parametric equations. Here, for reference, is the result:◄

> **Fact: (Slope of a parametric curve.)** Let a smooth curve C be given by parametric equations $x = f(t)$, $y = g(t)$, with $a \le t \le b$. If $f'(t) \ne 0$, then the slope dy/dx of C at (x, y) is given by
>
> $$\frac{dy}{dx} = \frac{g'(t)}{f'(t)} = \frac{dy/dt}{dx/dt}.$$

Make a guess. Record it here in the margin.

Make a guess. Record it here in the margin.

Both curves shown above are of this type.

See Section 4.9 for more details and fine print.

To *use* the result, we'll need first to write a polar curve $r = f(\theta)$ in parametric form. This is surprisingly easy to do. The key facts*➤* are the "conversion" formulas from polar to rectangular coordinates:

We explained them in the last section.

$$x = r \cos \theta; \qquad y = r \sin \theta.$$

For points on our polar curve, $r = f(\theta)$; for such points, therefore,

$$x = r \cos \theta = f(\theta) \cos \theta; \qquad y = r \sin \theta = f(\theta) \sin \theta.$$

These are our desired parametric equations. We've rewritten our curve, originally given in r-θ form, as a pair of parametric equations, with θ as parameter. All that's left is to apply the slope formula above. We'll need $dx/d\theta$ and $dy/d\theta$ below; let's compute them now. By the product rule,

$$x = f(\theta) \cos \theta \quad \Longrightarrow \quad \frac{dx}{d\theta} = f'(\theta) \cos \theta - f(\theta) \sin \theta;$$

$$y = f(\theta) \sin \theta \quad \Longrightarrow \quad \frac{dy}{d\theta} = f'(\theta) \sin \theta + f(\theta) \cos \theta.$$

Everything is now in place. Here's the result, stated with appropriate technical hypotheses:*➤*

See the exercises for more on these hypotheses.

Fact: (Slope of a polar curve.) Let a curve C be given in polar coordinates by a function $r = f(\theta)$, $\alpha \leq \theta \leq \beta$, where f and f' are continuous on (α, β), and not simultaneously zero. Then for θ in (α, β), the slope of C at $(r, \theta) = (f(\theta), \theta)$ is given by

$$\frac{dy}{dx} = \frac{dy/d\theta}{dx/d\theta} = \frac{f'(\theta) \sin \theta + f(\theta) \cos \theta}{f'(\theta) \cos \theta - f(\theta) \sin \theta}$$

wherever the denominator is not zero.

Using this Fact we can answer the questions on tangents raised at the beginning of this section. The next example shows how; it illustrates, too, some of the caution needed in working with polar coordinates.

■ **Example 1.**   Consider the cardioid $r = 1 + \cos \theta$, shown on page 336.*➤* Where is the curve horizontal? Where is it vertical? What happens at the origin?

Take a close look.

Solution: Here, $f(\theta) = 1 + \cos \theta$ and $f'(\theta) = -\sin \theta$, so

$$\frac{dy/d\theta}{dx/d\theta} = \frac{f'(\theta) \sin \theta + f(\theta) \cos \theta}{f'(\theta) \cos \theta - f(\theta) \sin \theta} = \frac{(1 - 2 \cos \theta)(1 + \cos \theta)}{(2 \cos \theta + 1) \sin \theta}.$$

(We used the Fact above and a little algebra.)*➤*

Check our work.

The cardioid can have a **horizontal tangent line** only if the *numerator* $dy/d\theta$ is zero, i.e., if

$$(1 - 2 \cos \theta)(1 + \cos \theta) = 0 \iff \cos \theta = \frac{1}{2} \quad \text{or} \quad \cos \theta = -1.$$

For θ in $[0, \pi]$, one or the other condition holds only if $\theta = \pi/3$ or $\theta = \pi$. (By symmetry, it's enough to look only on the *upper half* of the cardioid.)

The picture shows that, in fact, the upper half of the cardioid is horizontal *only* at $\theta = \pi/3$, i.e., at the point with polar coordinates $(3/2, \pi/3)$, and rectangular coordinates $(3/4, 3\sqrt{3}/4) \approx (0.75, 1.30)$. Symmetry dictates that the cardioid is also horizontal at $\theta = 5\pi/6$.

What happens at $\theta = \pi$? Isn't the cardioid horizontal there, too? Not necessarily—at $\theta = \pi$, *both* $dx/d\theta$ and $dy/d\theta$ are zero** and so the slope expression is undefined. (In fact, the cardioid has a "cusp" at the origin, and hence no well-defined slope there.)

See for yourself!

A **vertical tangent line** can occur only where the *denominator* $dx/d\theta$ is zero, i.e., if

$$-\sin\theta\,(2\cos\theta + 1) = \iff \sin\theta = 0 \quad \text{or} \quad \cos\theta = -\frac{1}{2}.$$

For θ in $[0, \pi]$, one or the other holds if $\theta = 0$ or $\theta = 2\pi/3$. As the picture shows, both $\theta = 0$ and $\theta = 2\pi/3$ correspond to points of vertical tangency on the cardioid. These points have polar coordinates $(2, 0)$ and $(1/2, 2\pi/3)$, respectively.**

What are their rectangular coordinates?

Superimposing the cardioid on a polar grid supports all the calculations above:

Horizontal and vertical points on a cardioid: $r = 1 + \cos\theta$

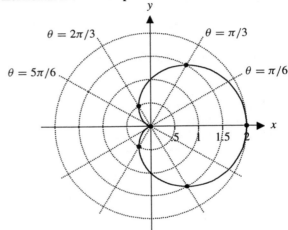

■ **Example 2.** Let $a > 0$ be any positive constant. What does the slope formula say about the circle $r = a$?

Solution: For the circle $r = a$, $f(\theta) = a$ and $f'(\theta) = 0$. Now the Fact says

$$\begin{aligned}
\frac{dy}{dx} &= \frac{f'(\theta)\sin\theta + f(\theta)\cos\theta}{f'(\theta)\cos\theta - f(\theta)\sin\theta} \\
&= \frac{a\cos\theta}{-a\sin\theta} \\
&= -\cot\theta = -\frac{1}{\tan\theta}.
\end{aligned}$$

Why perpendicular? Because the slopes are negative reciprocals.

Note what the last expression means: the tangent line to the circle at any point $P(a, \theta)$ is *perpendicular* to the ray from the origin to P.**   □

■ **Example 3.** Let $r = f(\theta)$ describe a polar curve C; suppose that $f(\theta_0) = 0$. What does the formula

$$\frac{dy}{dx} = \frac{f'(\theta)\sin\theta + f(\theta)\cos\theta}{f'(\theta)\cos\theta - f(\theta)\sin\theta}$$

say about the slope of C at the point $(0, \theta_0)$?

Solution: The slope formula becomes much simpler if $f(\theta_0) = 0$. In that case, ➤➤ *Remember: f and f' aren't simultaneously zero.*

$$\frac{dy}{dx} = \frac{f'(\theta_0) \sin \theta_0}{f'(\theta_0) \cos \theta_0} = \tan \theta_0.$$

Note what this means: if a smooth polar curve passes through the origin at $\theta = \theta_0$, its tangent line is simply the ray $\theta = \theta_0$. For the rose $r = \cos(2\theta)$, for example, $r = 0$ when $\theta = \pi/4$, $\theta = 3\pi/4$, $\theta = 5\pi/4$, and $\theta = 7\pi/4$. As the following picture shows, the curve passes through the origin in just these directions:

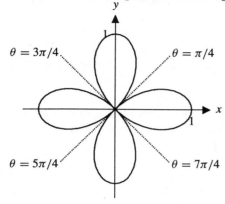

The rose $r = \cos(2\theta)$: **tangent lines at the origin**

$\square$

Finding area in polar coordinates

The standard area problem in *rectangular* coordinates concerns the area defined by an ordinary, $y = f(x)$-style graph, for $a \le x \le b$. In *polar* coordinates the standard problem is a little different: to find the area bounded by a *polar* curve $r = f(\theta)$, for $\alpha \le \theta \le \beta$. These pictures illustrate the "generic" situations; the areas in question are shown shaded:

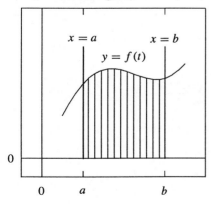

Area in rectangular coordinates

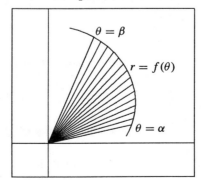

Area in polar coordinates

Areas by integration: rectangular vs. polar

The different styles of shading in the two figures above—one vertical, the other radial—were chosen intentionally. They reflect two slightly different approaches to finding areas. In order to emphasize both similarities and differences between the rectangular and polar situations, the next two paragraphs appear in "parallel."

Area, Cartesian style.   In rectangular coordinates, area is approximated by subdividing the region into thin *vertical strips*, each one based on the x-axis. Each strip corresponds to a small subinterval, say $[x_i, x_i + \Delta x]$, obtained by partitioning the domain interval $a \le x \le b$ into n equal pieces.

We've done this often before, so we'll omit some details.

If Δx is small, then◄ the strip is approximately a *rectangle*, with base Δx and height $f(x_i)$. Therefore:

$$\text{area of one strip} \approx f(x_i)\Delta x.$$

Adding all n strips gives

$$\text{total area} \approx \sum_{i=1}^{n} f(x_i)\Delta x.$$

Therefore, taking the limit as $\Delta x \to 0$,

$$\text{total area} = \lim_{n\to\infty} \sum_{i=1}^{n} f(x_i)\Delta x = \int_a^b f(x)\,dx.$$

Look carefully at the polar picture: it shows 20 radial wedges.

Area, polar style.   In polar coordinates, by contrast, area is approximated by subdividing the region into thin *pie-shaped wedges*, each with its vertex at the origin.◄ Each wedge corresponds to a small subinterval, say $[\theta_i, \theta_i + \Delta\theta]$, obtained by partitioning the domain interval $\alpha \le \theta \le \beta$ into n equal pieces.

Remember: $r = f(\theta)$ gives the radius for given θ.

If Δx is small, then each wedge is approximately a sector of a *circle*, with radius $f(\theta_i)$◄ and making the angle $\Delta\theta$ at the origin.◄

Now, a brief detour. Just ahead, we'll need to know the area of a **circular sector** as described above. The answer, though perhaps less familiar than the area of a rectangle, is easy to find. A *full* circle of radius R (for which θ runs from 0 to 2π) encloses total area πR^2. A wedge making angle $\Delta\theta$ at the origin represents the fraction $(\Delta\theta)/(2\pi)$ of the total circle, so

$$\text{area of wedge} = \pi R^2 \times \frac{\Delta\theta}{2\pi} = \frac{R^2}{2}\Delta\theta.$$

If, say, $R = f(\theta_i)$, then a circular wedge has area $f(\theta_i)^2\Delta\theta/2$.

Back to the main road. Our detour showed that in the "generic" case pictured above,

$$\text{area of one wedge} \approx \frac{f(\theta_i)^2}{2}\Delta\theta.$$

Adding all n wedges gives

$$\text{total area} \approx \sum_{i=1}^{n} \frac{f(\theta_i)^2}{2}\Delta\theta.$$

Taking the limit as $\Delta\theta \to 0$ gives our area formula:

$$\text{total area} = \lim_{n\to\infty} \sum_{i=1}^{n} \frac{f(\theta_i)^2}{2}\Delta\theta = \int_\alpha^\beta \frac{f(\theta)^2}{2}\,d\theta.$$

The formula is worth remembering:

Fact: (Area in polar coordinates.) Let f be a continuous function, and let R be the region in the xy-plane bounded by the polar curve $r = f(\theta)$ and the rays $\theta = \alpha$ and $\theta = \beta$. Then

$$\text{area of } R = \int_\alpha^\beta \frac{f(\theta)^2}{2}\,d\theta.$$

■ **Example 4.** What's the area of one leaf of the polar rose $r = f(\theta) = \cos(2\theta)$? What fraction of the circle $r = 1$ does the entire rose cover?

Solution: As the picture on page 339 shows, the eastward-pointing leaf lies between $\theta = -\pi/4$ and $\theta = \pi/4$. To ease calculations we'll integrate from $\theta = 0$ to $\theta = \pi/4$ and double the result. Here is the integral calculation—a table of integrals helps along the way:▶▶

But check our work.

$$\int_0^{\pi/4} \frac{\cos^2(2\theta)}{2}\, d\theta = \frac{\theta}{4} + \frac{\sin(4\theta)}{8} \Big]_0^{\pi/4} = \frac{\pi}{16}.$$

Thus *one* leaf has area $\pi/8$; all four leaves have area $\pi/2$—exactly *half* the area of the circle $r = 1$. □

A rose with any other *n* ... A more general (and more surprising) result than that of the last example is true:▶▶

See the exercises for more on this.

> For any even *n*, the rose $r = \cos(n\theta)$ encloses total area $\pi/2$—half the area of the enclosing circle.

■ **Example 5.** How much area does the cardioid $r = 1 + \cos\theta$ enclose?

Solution: With the integral formula, the answer is easy:

$$\begin{aligned}
\text{area} &= \frac{1}{2} \int_0^{2\pi} f(\theta)^2\, d\theta \\
&= \frac{1}{2} \int_0^{2\pi} \left(1 + 2\cos\theta + \cos^2\theta \right) d\theta \\
&= \frac{1}{2} \left(\theta + 2\sin\theta + \frac{\theta}{2} + \frac{\sin(2\theta)}{4} \right) \Big]_0^{2\pi} \\
&= \frac{3\pi}{2} \approx 4.71.
\end{aligned}$$

□

Exercises

1. We claimed in this section that

$$x = f(\theta)\cos\theta \implies \frac{dx}{d\theta} = f'(\theta)\cos\theta - f(\theta)\sin\theta;$$

$$y = f(\theta)\sin\theta \implies \frac{dy}{d\theta} = f'(\theta)\sin\theta + f(\theta)\cos\theta.$$

Use the product rule to show that these claims are valid.

2. We claimed in Example 1 that on the cardioid $r = 1 + \cos\theta$, the slope dy/dx at (r, θ) is

$$\frac{(1 - 2\cos\theta)(1 + \cos\theta)}{(2\cos\theta + 1)\sin\theta}.$$

Use the Fact on page 337 to verify this.

3. Consider the spiral $r = \theta$, for $0 \leq \theta \leq 4\pi$.

   (a) Draw the spiral.

   (b) At what points does the spiral have horizontal tangent lines? Vertical tangent lines?

   (c) Write an equation (in rectangular) form for the line tangent to the spiral at the polar point $(1, 1)$. Draw this tangent line on your graph.

4. Consider the limaçon $r = 1 + 2\sin\theta$.

   (a) Draw the limaçon.

   (b) Write an equation (in rectangular) form for the line tangent to the limaçon at the polar point $(1, 0)$. Draw this tangent line on your graph.

   (c) Find the slope of the line tangent to the limaçon at the polar point $(0, 7\pi/6)$. Draw this tangent line on your graph.

   (d) Find the slope of the line tangent to the limaçon at the polar point $(0, 11\pi/6)$. Draw this tangent line on your graph.

   (e) At what points does the limaçon have horizontal tangent lines?

5. This open-ended problem is about the family of limaçons of the form $r = 1 + a\cos\theta$.

   (a) After plotting example limaçons for various values of a, decide which ones have a "dimple" (i.e., a pushed-in section on either the right or the left) and which don't. Notice that any limaçons with a dimple has *three* vertical tangent lines.

   (b) The answer to (a) is that there's no dimple if $|a| \leq 1/2$. Show this using derivatives. One way is to decide which values of a lead to *three* vertical tangent lines.

6. In each part below, draw the region bounded by the given polar curves, and find its area.

   (a) $r = 1, \theta = 0, \theta = \pi$

   (b) $r = 3, \theta = 0, \theta = \pi$

   (c) $r = a, \theta = 0, \theta = \pi$

   (d) $r = 1, \theta = 0, \theta = \beta$

7. In each part below, draw the given region and find its area.

   (a) The outer loop of the limaçon $r = 1 + 2\sin\theta$.

   (b) The inner loop of the limaçon $r = 1 + 2\sin\theta$.

   (c) The region bounded by $r = \sec\theta, \theta = 0, \theta = \pi/4$.

   (d) The region bounded by $r = \sec\theta, \theta = 0, \theta = \arctan m$.

8. Consider the $2n$-leafed rose $r = \cos(n\theta)$, where n is even.

   (a) Find the area of one leaf.

   (b) Find the area of all $2n$ leaves. What fraction of the circle $r = 1$ does the entire rose fill up?

9. Consider the n-leafed rose $r = \cos(n\theta)$, where n is odd.

   (a) Find the area of one leaf.

   (b) Find the area of n leaves. What fraction of the circle $r = 1$ does the entire rose fill up?

10. Draw and then calculate the area bounded by one "turn" of the spiral $r = \theta$, i.e., from $\theta = 0$ to $\theta = 2\pi$.

11. Draw and then calculate the area bounded by one "turn" of the exponential spiral $r = e^{\theta}$, i.e., from $\theta = 0$ to $\theta = 2\pi$.

12. Draw and then calculate (or estimate numerically) the area bounded by one "turn" of the logarithmic spiral $r = \ln(\theta)$, i.e., from $\theta = 2\pi$ to $\theta = 4\pi$.

13. Use polar coordinates to find the area of the region inside the circle $r = 1$ and to the right of $x = 1/2$.

14. Let $0 < a < 1$. Use polar coordinates to find the area of the region inside the circle $r = 1$ and to the right of $x = a$.

15. In this section we used the idea that a *polar curve* can also be thought of as a *parametric* curve. Specifically, the curve defined by the polar function $r = f(\theta)$, for $\alpha \le \theta \le \beta$, can *also* be defined by the parametric equations

$$x = f(\theta)\cos\theta; \qquad y = f(\theta)\sin\theta,$$

   $\alpha \le \theta \le \beta$. Consider, for example, the unit circle $r = 1$, $0 \le \theta \le 2\pi$. Then $f(\theta) = 1$, so the parametric form is simply

$$x = f(\theta)\cos\theta = \cos\theta; \quad y = f(\theta)\sin\theta = \sin\theta; \quad 0 \le \theta \le 2\pi.$$

   In each part below, a curve is given in polar form. First rewrite the curve in parametric form and then plot it. (Use a graphing calculator or computer.) Do you see the "expected" results?

   (a) $r = 2; 0 \le \theta \le 2\pi$

   (b) $r = 2; 0 \le \theta \le \pi$

   (c) $r = \sec\theta; -\pi/4 \le \theta \le \pi/4$

   (d) $r = \csc\theta; \pi/4 \le \theta \le 3\pi/4$

   (e) $r = \theta; 0 \le \theta \le 2\pi$

   (f) $r = \cos\theta; 0 \le \theta \le \pi$

   (g) $r = \cos(2\theta); 0 \le \theta \le 2\pi$

   (h) $r = 1 + \cos\theta; 0 \le \theta \le 2\pi$

16. The formula for the slope of a polar curve requires that f and f' not be simultaneously zero. Show that if this condition holds, then $dy/d\theta$ and $dx/d\theta$ are not simultaneously zero. [HINT: Recall that $dy/d\theta = f'(\theta)\sin\theta + f(\theta)\cos\theta$ and $dx/d\theta = f'(\theta)\cos\theta - f(\theta)\sin\theta$. Look at $(dy/d\theta)^2 + (dx/d\theta)^2$.]

Chapter 3

More on Functions and their Derivatives

3.6   Inverse trigonometric functions and their derivatives

The natural logarithm and exponential functions are *inverses*; each "undoes" the other. In symbols:◄

$$\ln\left(e^x\right) = x \qquad \text{and} \qquad e^{\ln x} = x.$$

The first equation holds for all x, the second only for $x > 0$.

In this section we find and discuss inverses for several *trigonometric* functions.

Nomenclature: "arc" or "inverse"?

Two common forms are used to denote inverse trigonometric functions:

arcsin x, arccos x, *and* arctan x *are synonyms for* $\sin^{-1} x$, $\cos^{-1} x$, *and* $\tan^{-1} x$.

Each form has its advantages. The first makes explicit the connection with arcs on the unit circle; the second reminds us that function inversion is involved. We'll use both.◄

Don't confuse $\sin^{-1} x$ with $1/\sin x$. They're entirely different.

A technical problem and its solution: restricting domains

To be invertible, a function must be one-to-one; graphically speaking, it must pass the "horizontal line test." Alas, *no* trigonometric function has this property. Far from it—trigonometric functions, being 2π-periodic, repeat themselves on intervals of length 2π, "hitting" each output value *infinitely often*. The next example illustrates the problem.

We'll give a formal definition below.

■ **Example 1.** By any sensible definition,◄ arcsin(0.5) should be a number whose sine is 0.5. Find such a number. How many are there?

Solution: The picture below shows *four* inputs x for which $\sin x = 0.5$.

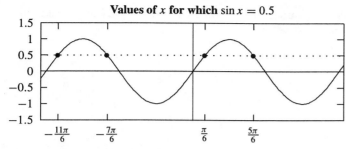

Values of x for which $\sin x = 0.5$

The bulleted points show that

$$0.5 = \sin\left(-\tfrac{11\pi}{6}\right) = \sin\left(-\tfrac{7\pi}{6}\right) = \sin\left(\tfrac{\pi}{6}\right) = \sin\left(\tfrac{5\pi}{6}\right).$$

Moreover, this list is far from exhaustive—infinitely many more solutions of the equation $\sin x = 0.5$ exist outside our viewing window.

Which of these numbers deserves to be called $\arcsin(0.5)$?[➤] None especially, but if *forced* to choose, we might reasonably select $\pi/6$, the solution nearest to zero. Any scientific calculator agrees. According to an HP-28S, for instance,

$$\sin^{-1}(0.5) = 0.523598775598 \approx \frac{\pi}{6}. \qquad \square$$

aka $\sin^{-1}(0.5)$

Surgery on domains

The problem just seen—choosing from among reasonable candidates—arises because the trigonometric functions aren't one-to-one. Our solution is crude but effective: amputate the offending part of each function's domain, and thus *make* it one-to-one. Graphs will show us where to cut.

Inverting the sine function

The sine function is *increasing*, and therefore one-to-one, on the interval $[-\pi/2, \pi/2]$. (There are, of course, many other intervals on which the sine function is one-to-one. Among all such intervals, this one, centered at 0, seems most natural.) The picture below shows both the ordinary sine curve (drawn dotted), and the special piece (drawn solid) in question here:

Restricting the sine function

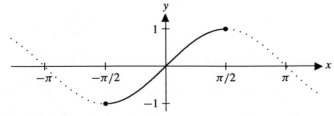

Restricted to this domain, the sine function *is* invertible; its inverse is defined formally as follows:

Definition: For x in $[-1, 1]$, $\sin^{-1} x$ (or $\arcsin x$) is defined by the conditions

$$x = \sin y \quad \text{and} \quad -\frac{\pi}{2} \le y \le \frac{\pi}{2}.$$

In words: $\arcsin x$ is the (unique) angle between $-\pi/2$ and $\pi/2$ whose sine is x.

Note:

Graphical symmetry. The graphs of $\sin x$ (restricted!) and $\sin^{-1} x$ show the expected reflective symmetry:

Sine and arcsine

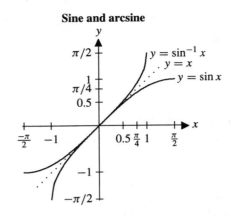

Domain and range. The sine function (as we've restricted it) has domain $[-\pi/2, \pi/2]$ and range $[-1, 1]$. The *inverse* sine function has it the other way around—its domain is $[-1, 1]$ and its range $[-\pi/2, \pi/2]$.

Collapsing equations. By the definition of inverse function, these equations hold:

$$\sin\left(\sin^{-1} x\right) = x \qquad \text{if } -1 \leq x \leq 1;$$

$$\sin^{-1}(\sin x) = x \qquad \text{if } -\pi/2 \leq x \leq \pi/2$$

Inverting the cosine function

The situation for $\cos x$ differs only in details from that for $\sin x$.

Restricting $\cos x$ to $[-\pi/2, \pi/2]$ doesn't work. Do you see why not?

The cosine function[*] becomes one-to-one, and therefore invertible, when restricted to the interval[*] $[0, \pi]$. This understood, the graphs and formal definition follow naturally:

Restricting the cosine function

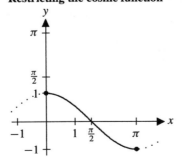

Graphs of cosine and arccosine

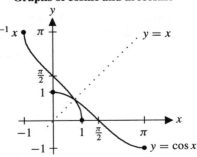

Definition: For x in $[-1, 1]$, $\cos^{-1} x$ (or $\arccos x$) is defined by the conditions

$$x = \cos y \quad \text{and} \quad 0 \leq y \leq \pi.$$

In words: $\arccos x$ is the (unique) angle between 0 and π whose cosine is x.

Note:

Graphical symmetry. The graphs of cos and $\cos^{-1}$ show the usual symmetry for inverse functions—they're symmetric with respect to the line $y = x$.

Domain and range. We restricted the cosine function to have domain $[0, \pi]$ and range $[-1, 1]$. The *inverse* cosine function, therefore, has domain $[-1, 1]$ and range $[0, \pi]$.

Collapsing equations. As with the arcsine, two equations express the inverse relationship between cos and arccos:

$$\cos\left(\cos^{-1} x\right) = x \qquad \text{if } -1 \le x \le 1;$$

$$\cos^{-1}\left(\cos x\right) = x \qquad \text{if } 0 \le x \le \pi.$$

Inverting the tangent function

The tangent function is one-to-one if restricted to any of its "branches." We'll use the "middle" branch, through the origin:

The tangent function: one-to-one on the middle branch

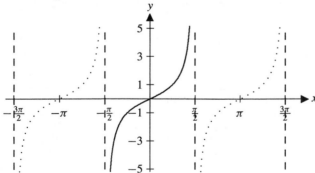

Unlike the sine and cosine functions, the tangent function "hits" *all* real numbers as outputs.[➤➤] The arctangent function, therefore, accepts *all* real inputs.

The graph shows this. Do you agree?

> **Definition:** Let x be any real number. Then $y = \tan^{-1} x$ (or $y = \arctan x$) means that
>
> $$x = \tan y \qquad \text{and} \qquad -\pi/2 < y < \pi/2.$$

Note:

Symmetry. The domain of $\arctan x$ (i.e., the range of $\tan x$) is infinite, so any picture of its graph is incomplete. The figure below gives the idea; notice the usual symmetry between a function and its inverse:

The tangent function and its inverse

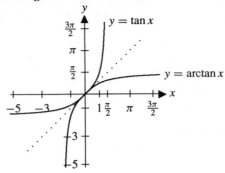

Infinite domain. The tangent function (as restricted) has *domain* $(-\pi/2, \pi, 2)$ and *range* $(-\infty, \infty)$. The arctangent function turns things around: it has *range* $(-\pi/2, \pi, 2)$ and *domain* $(-\infty, \infty)$.

Other inverse trigonometric functions

As your calculator knows.

All six trigonometric functions have inverses.[**] As for arcsin x, arccos x, and arctan x, defining arcsec x, arccsc x, and arccot x requires care with domains and ranges. Among these "other" functions, the arcsecant is most useful; it appears occasionally in antiderivatives.

There's trouble if $\cos t = 0$.

For any legal[**] input t, $\sec t = 1/\cos t$. The simplest definition of arcsec x—based on arccos x—reflects this fact:

> **Definition:** For any x with $|x| \geq 1$,
> $$\operatorname{arcsec}(x) = \sec^{-1}(x) = \cos^{-1}(1/x).$$

In particular,

$$y = \operatorname{arcsec}(x) = \arccos\left(\tfrac{1}{x}\right) \implies \cos y = \tfrac{1}{x} \implies \sec y = x,$$

Plot your own copy of the arcsecant function; experiment with various domains. See the exercises for more on the arcsecant function.

as we'd expect.[**]

Inverse trigonometric functions and the unit circle

We defined the sine, cosine, and tangent functions in terms of the unit circle. Their *inverses*, too, can be interpreted in the same way. Examples show how.

Starting from the point $(1, 0)$.

■ **Example 2.** A unit circle appears below; arclengths[**] appear in radians. Use the picture to estimate arcsin(0.5) and arctan 2.

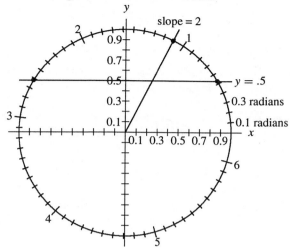

Arctangent, arcsine, and the unit circle

Solution: By definition, arctan 2 is an angle➤ whose tangent is 2. Recall that the tangent of *any* angle is the *slope* of the line that angle determines. Therefore, we want an angle—the one between $-\pi/2$ and $\pi/2$—that corresponds to a slope of 2. To find this angle, we draw a line from the origin, with slope 2, and observe the radian measure of the angle it determines. The picture shows that the line determines an arc of radian measure about 1.1; this is our estimate for arctan 2.➤

Or "arc."

Reading $\sin^{-1}(0.5)$ from the unit circle is similar. By definition, $\sin t$ is the y-coordinate of the point determined by an arc of length t on the unit circle. Therefore arcsin(0.5) is the (one and only) angle between $-\pi/2$ and $\pi/2$ (i.e., to the right of the y-axis) whose corresponding y-coordinate is 0.5. To find this angle, we draw the line $y = 0.5$ (as shown in the picture above). It crosses the circle at the bulleted point, the end of an arc with radian measure about 0.52; therefore, arcsin(0.5) ≈ 0.52.➤ □

According to an HP-28S calculator, arctan 2 ≈ 1.10714871779.

This estimate is surprisingly good: The "right" answer is about 0.523598775598.

Combining ordinary and inverse trigonometric functions

In practice, inverse trigonometric functions are often combined with ordinary trigonometric functions. Diagrams like the one below can be helpful in seeing how the process works.

Relating inverse trigonometric functions: pictorial aids

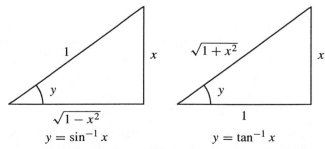

The first figure shows (among other things):

$$\cos(\sin^{-1} x) = \frac{\text{adjacent}}{\text{hypotenuse}} = \sqrt{1 - x^2}.$$

From the second figure, similarly:

$$\sin(\tan^{-1} x) = \frac{\text{opposite}}{\text{hypotenuse}} = \frac{x}{\sqrt{1 - x^2}}.$$

Derivatives of inverse trigonometric functions

We know derivatives of the ordinary trigonometric functions; with the chain rule, we can find derivatives of their inverses. After tabulating the derivatives, we'll derive them.

Derivatives of ordinary trigonometric functions are other trigonometric functions. Surprisingly, *inverse* trigonometric functions have *algebraic* derivatives:

Derivatives of inverse trigonometric functions				
Function	$\arcsin x$	$\arccos x$	$\arctan x$	$\text{arcsec } x$
Derivative	$\dfrac{1}{\sqrt{1 - x^2}}$	$\dfrac{-1}{\sqrt{1 - x^2}}$	$\dfrac{1}{1 + x^2}$	$\dfrac{1}{\lvert x \rvert \sqrt{x^2 - 1}}$

Notice:

See the exercises for more detail.

Restricted domains. The functions $\arcsin x$, $\arccos x$, and $\text{arcsec } x$ have restricted domains. So, therefore, must their derivatives.◄◄

Not much difference. The derivatives of $\arcsin x$ and $\arccos x$ differ only in *sign*. Therefore $(\arcsin x + \arccos x)' = 0$, for all legal inputs x. This is no accident, of course. For all x in $[-1, 1]$,

$$\arcsin x + \arccos x = \frac{\pi}{2}.$$

Graph both on the same axes to see why.

(The graphs of $\arcsin x$ and $\arccos x$ say the same thing.)◄◄

Graphical evidence. We'll see below, by calculation, *why* these formulas hold. Graphs show convincingly *that* they hold. Here, for instance, are graphs of the arctangent function and its claimed derivative:

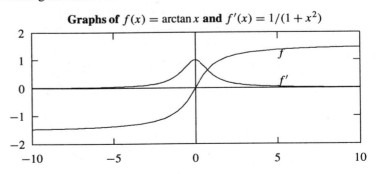

Graphs of $f(x) = \arctan x$ **and** $f'(x) = 1/(1 + x^2)$

■ **Example 3.** Explain why $(\arctan x)' = \dfrac{1}{1 + x^2}$.

Solution: If $f(x) = \tan x$ and $g(x) = \arctan x$, then f and g are inverses, and $f'(x) = \sec^2 x$. Unraveling the derivative formula above gives

$$(\arctan x)' = \frac{1}{f'(g(x))} = \frac{1}{\sec^2 (\arctan x)}.$$

Despite first appearances, the result is correct. We can simplify the last quantity using the trigonometric identity $\sec^2 t = 1 + \tan^2 t$:

$$\frac{1}{\sec^2(\arctan x)} = \frac{1}{1 + \tan^2(\arctan x)} = \frac{1}{1 + x^2}. \qquad \square$$

■ **Example 4.** Explain why $(\arcsin x)' = \dfrac{1}{\sqrt{1 - x^2}}$.

Solution: We play the same game as before, with slightly different twists. If $f(x) = \sin x$ and $g(x) = \arcsin x$, then $f'(x) = \cos x$, so ⟫

Watch for the combination of the cosine and arcsine functions below.

$$\begin{aligned}
(\arcsin x)' &= \frac{1}{f'(g(x))} \\
&= \frac{1}{\cos(\arcsin x)} \\
&= \frac{1}{\sqrt{1 - \sin^2(\arcsin x)}} \\
&= \frac{1}{\sqrt{1 - x^2}}. \qquad \square
\end{aligned}$$

The remaining derivative formulas can be found similarly.

Exercises

1. Find each of the following numbers exactly (in radians).

   (a) $\sin^{-1}(1)$

   (b) $\sin^{-1}\left(\sqrt{3}/2\right)$

   (c) $\cos^{-1}(-1)$

   (d) $\cos^{-1}\left(-\sqrt{2}/2\right)$

   (e) $\tan^{-1}(1)$

   (f) $\tan^{-1}\left(\sqrt{3}\right)$

2. Use the unit circle on page 349 to estimate each of the following numbers.

   (a) $\sin^{-1}(0.8)$

   (b) $\cos^{-1}(0.6)$

   (c) $\tan^{-1}(0.4)$

   (d) $\tan^{-1}(3)$

3. The table gives samples of what one calculator knows (and doesn't know) about inverse trigonometric functions:

Inverse Trigonometric Function Values											
x	-2	-1	-0.5	-0.25	0	0.25	0.5	1	2	5	100
$\sin^{-1} x$		-1.57	-0.52	-0.25	0.00	0.25	0.52	1.57			
$\cos^{-1} x$		3.14	2.09	1.82	1.57	1.31	1.04	0.00			
$\tan^{-1} x$	-1.11	-0.79	-0.46	-0.24	0.00	0.24	0.46	0.79	1.11	1.37	1.56

(a) Can the gaps in the table above be filled in? Why or why not? Answer using the word "domain."

(b) In each column, the entries for $\sin^{-1} x$ and $\cos^{-1} x$ add to 1.57. Surely this can't be coincidence! What's going on?

4. Use a calculator to find 4 values of x for which $\cos x = 0.5$

5. Use a calculator to find 4 values of x (2 positive, 2 negative) for which $\tan x = 2$.

6. The graph of the tangent function has *vertical* asymptotes. Where are they? Does the graph of the *arctangent* function have asymptotes? If so, where? Why?   [HINT: Look carefully at the graphs of the tangent function and its inverse. Explain what you see in terms of reflection across the line $y = x$.]

7. The equation $\sin^{-1} (\sin x) = x$ suggests that the graph of $y = \sin^{-1} (\sin x)$, over any interval, might be the straight line $y = x$. It isn't (see below).

Graphs of $y = \arcsin (\sin x)$ **and** $y = x$

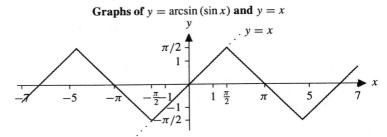

(a) Explain why $\arcsin(\sin 5) \neq 5$.

(b) For which values of x *does* the equation $\arcsin(\sin x) = x$ hold?

(c) Let $f(x) = \arcsin(\sin x)$. Show that $f'(x) = \dfrac{\cos x}{|\cos x|}$.

(d) How is the result in part (b) related to the graph shown above?

8. (a) The function $\operatorname{arcsec} x$ accepts as inputs only values of x for which $|x| \geq 1$. Why is this restriction necessary?   [HINT: What's the domain of $\arccos x$?]

(b) What's the *range* of $\operatorname{arcsec} x$?   [HINT: What's the range of $\arccos x$?]

9. Write each of the following as an algebraic expression. For example, $\cos(\arcsin x) = \sqrt{1 - x^2}$.

(a) $\sin(\arccos x)$ (c) $\sin(\arctan x)$

(b) $\tan(\arcsin x)$ (d) $\cos(\arctan x)$

10. (a) What is the domain of the derivative of the arcsine function? Give a geometric explanation for why it is not the same as the domain of the arcsine function.

(b) What is the domain of the derivative of the arctangent function? Is it the same as the domain of the arctangent function?

(c) What is the domain of the derivative of the arcsecant function? Is it the same as the domain of the arcsecant function?

11. Let $f(x) = \arctan x$

(a) Plot graphs of f and f' on the same axes.

(b) Evaluate $\lim\limits_{x \to \infty} f(x)$, $\lim\limits_{x \to -\infty} f(x)$, $\lim\limits_{x \to \infty} f'(x)$, and $\lim\limits_{x \to -\infty} f'(x)$. What do these results imply about the graph of the arctangent function?

(c) Is f even, odd or neither?

(d) Show that f is increasing on $(-\infty, \infty)$. How does the graph of f' reflect this fact?

(e) Where is f concave up? Concave down? Does f have any inflection points? If so, where?

12. Let $f(x) = \arcsin x$.

(a) Plot graphs of f and f' on the same axes.

(b) Evaluate $\lim\limits_{x \to 1^-} f(x)$, $\lim\limits_{x \to -1^+} f(x)$, $\lim\limits_{x \to 1^-} f'(x)$, and $\lim\limits_{x \to -1^+} f'(x)$. What do these results imply about the graph of the arcsine function?

(c) Is f even, odd or neither?

(d) Show that f is increasing on $(-1, 1)$. How does the graph of f' reflect this fact?

(e) Where is f concave up? Concave down? Does f have any inflection points? If so, where?

(f) How does the graph of f' reflect the information found in part (e)?

13. Let $f(x) = \arccos x$.

(a) Plot graphs of f and f' on the same axes.

(b) Evaluate $\lim\limits_{x \to 1^-} f(x)$, $\lim\limits_{x \to -1^+} f(x)$, $\lim\limits_{x \to 1^-} f'(x)$, and $\lim\limits_{x \to -1^+} f'(x)$. What do these results imply about the arccosine function?

(c) Is f even, odd or neither?

(d) Show that f is decreasing on $(-1, 1)$. How does the graph of f' reflect this fact?

(e) Where is f concave up? Concave down? Does f have any inflection points? If so, where are they located.

(f) How does the graph of f' reflect the information found in part (e)?

14. Let $f(x) = \operatorname{arcsec} x$.

(a) Plot graphs of f and f' using the plotting window $[-5, 5] \times [0, 5]$. [NOTE: If your calculator or computer doesn't recognize the arcsecant function, trick it by plotting $\arccos(1/x)$ instead.]

(b) Evaluate $\lim\limits_{x \to 1^+} f(x)$, $\lim\limits_{x \to -1^-} f(x)$, $\lim\limits_{x \to 1^+} f'(x)$, and $\lim\limits_{x \to -1^-} f'(x)$. What do these results imply about the graph of the arcsecant function?

(c) Evaluate $\lim\limits_{x \to \infty} f(x)$, $\lim\limits_{x \to -\infty} f(x)$, $\lim\limits_{x \to \infty} f'(x)$, and $\lim\limits_{x \to -\infty} f'(x)$. What do these results imply about the graph of the arcsecant function?

(d) Is f even, odd or neither?

(e) Show that f is increasing everywhere in the interior of its domain. How does the graph of f' reflect this fact?

(f) Where is f concave up? Concave down? Does f have any inflection points? If so, where are they located.

(g) How does the graph of f' reflect the information found in part (f)?

15. Find the derivative of each function below.

(a) $f(x) = \arctan(2x)$

(b) $f(x) = \arctan\left(x^2\right)$

(c) $f(x) = \sqrt{\arcsin x}$

(d) $f(x) = \arcsin(\sqrt{x})$

(e) $f(x) = \arcsin\left(e^x\right)$

(f) $f(x) = e^{\arctan x}$

(g) $f(x) = \arctan(\ln x)$

(h) $f(x) = \arccos(2x + 3)$

(i) $f(x) = \operatorname{arcsec}(x/2)$

(j) $f(x) = \arcsin x / \arccos x$

(k) $f(x) = x^2 \arctan\left(\sqrt{x}\right)$

(l) $f(x) = \ln(2 + \arcsin x)$

16. Find an antiderivative of each of the following function.

(a) $f(x) = 1/\left(1 + x^2\right)$

(b) $f(x) = 1/\sqrt{1 - x^2}$

(c) $f(x) = 2/\left(1 + 4x^2\right)$

(d) $f(x) = 3/\left(9 + x^2\right)$

(e) $f(x) = 1/\sqrt{9 - x^2}$

(f) $f(x) = 1/\sqrt{1 - 4x^2}$

(g) $f(x) = e^x/\left(1 + e^{2x}\right)$

(h) $f(x) = x/\sqrt{1 - x^4}$

(i) $f(x) = (2\arctan x)/(1 + x^2)$

(j) $f(x) = (\cos x)/\left(1 + \sin^2 x\right)$

(k) $f(x) = e^{\arcsin x}/\sqrt{1 - x^2}$

(l) $f(x) = 1/\left(x\left(1 + (\ln x)^2\right)\right)$

(m) $f(x) = 1/\left(\left(1 + x^2\right)\arctan x\right)$

17. (a) Show that $\arctan x + \arctan(1/x) = \pi/2$ for all $x > 0$.   [HINT: Start by differentiating both sides of the equation.]

(b) Find an equation similar to the one in part (a) that is valid when $x < 0$.

18. Use calculus to prove each of the following identities.   [HINT: See part (a) of the previous exercise.]

(a) $\arcsin(-x) = -\arcsin(x)$

(b) $\arccos(-x) = \pi - \arccos(x)$

(c) $\arccos x = \frac{\pi}{2} - \arcsin x$

(d) $\arctan\left(\dfrac{x}{\sqrt{1 - x^2}}\right) = \arcsin x$

(e) $\arcsin\left(\frac{x-1}{x+1}\right) = 2\arctan(\sqrt{x}) - \pi/2$

19. (a) Show that $2\arcsin x = \arccos(1 - 2x^2)$ when $0 \leq x \leq 1$.

(b) Find an identity similar to the one in part (a) that is valid when $-1 \leq x \leq 0$.

20. Let $f(x) = \arctan\left(\dfrac{1 + x}{1 - x}\right)$.

(a) Assume that $x < 1$. Show that there is a constant C such that $f(x) = C + \arctan x$.

(b) Use the result in part (a) to evaluate $\lim\limits_{x \to 1^-} f(x)$.

(c) Evaluate $\lim\limits_{x \to 1^+} f(x)$.

21. (a) Show that $\tan(\operatorname{arcsec} x) = \sqrt{x^2 - 1}$ when $x \geq 1$.

(b) Find an algebraic expression for $\tan(\operatorname{arcsec} x)$ when $x \leq -1$.

22. (a) Derive the identity $\arctan x + \arctan y = \arctan\left(\dfrac{x+y}{1-xy}\right)$ from the identity $\tan(x+y) = \dfrac{\tan x + \tan y}{1 - \tan x \tan y}$.

    (b) What conditions must be satisfied by x and y for the identity derived in part (a) to hold?

23. (a) Use the identity in part (a) of the previous problem to show that $\pi/4 = \arctan(1/2) + \arctan(1/3)$.

    (b) Show that $\pi/4 = 2\arctan(1/4) + \arctan(7/23)$. [HINT: Use the identity in part (a) of the previous problem twice.]

Chapter 4

Applications of the Derivative

4.1   Differential equations and their solutions

Such as Pythagoras's formula,
$a^2 + b^2 = c^2$.

Equations that involve derivatives.

Algebraic equations◄ describe relations among varying quantities. *Differential equations*◄ go one step—a giant step—further: they describe relations among changing quantities and the *rates* at which they change. Only by solving differential equations can many real-life phenomena be usefully described. As someone put it:

> *Nature's voice is mathematics; its language is differential equations.*

The idea of a differential equation first appeared, informally, in Section 2.1. Here and in the next section we take a closer look.

Differential equations and initial value problems: basic ideas

Observe the various derivative notations; all are used in practice.

Or more than one.

A **differential equation** (**DE**, for short) is any equation that contains one or more derivatives. Each of the following is a DE:◄

$$f'(x) = 6x + 5; \qquad y' = y; \qquad \frac{dy}{dx} = 5; \qquad g'(x) = kg(x).$$

To **solve** a differential equation means to find a *function*◄ that satisfies the DE. Any such function is called a **solution** to the DE. This definition bears repeating:

> *A solution to a DE is a* function *that satisfies the DE. A solution is* not *a number.*

You'll get no arguments here.   DE's are often stated "without arguments." For example, all of the DE's

$$y' = y, \quad y'(x) = y(x), \quad \text{and} \quad y'(t) = y(t)$$

say the same thing: *the desired function y is its own derivative.* The input variable name, or **argument**—either x or t in the examples above—can safely be "understood," so it's often omitted. Once gotten used to, omitting arguments causes little confusion.

Omitting inessential function arguments sometimes unclutters and simplifies a DE's appearance. For example, both

$$\frac{d^2 y}{dt^2} + 2t\frac{dy}{dt} + t = y(t) \quad \text{and} \quad y'' + 2ty' + t = y$$

say the same thing, but the latter form is more economical and (depending on the method) easier to handle.

356

Finding solutions vs. checking solutions.   Finding solutions to DE's, starting from scratch, can be very difficult. Many thick books treat the subject. *Checking* whether a given function satisfies a DE, by contrast, is usually much easier.➤

Solving DE's is, in this sense, similar to antidifferentiation: finding antiderivatives can be hard, but checking answers is easy.

The next examples show what it means to solve a DE, and that a single DE has many solutions.

■ **Example 1.**   Solve the DE $f'(x) = 6x + 5$. How many solutions are there? How are they related?

Solution: The DE asks for an *antiderivative* f of the function $6x + 5$. It's easy to check➤ that for any constant C, the function $f(x) = 3x^2 + 5x + C$ does the job. Thus this DE has infinitely many solutions; they differ by *additive* constants.   □

But do so now!

■ **Example 2.**   Solve the DE $y' = y$. How many solutions are there? How are they related?

Solution: The function $y(t) = e^t$ solves the DE.➤ Why we chose *this* function might seem mysterious, but it's easy to check that our guess works:

$$y(t) = e^t \implies y'(t) = e^t = y(t).$$

We wrote y as a function of t, but any other input variable name (x, z, u, etc.) would have done as well.

Thus $y' = y$, so $y(t) = e^t$ is indeed a solution of the DE.

How *did* we guess that $y(t) = e^t$? The DE itself tipped us off—the equation $y' = y$ says, in words:

The function y is its own derivative.

As we know, the natural exponential function e^t has that property.

Are there other solutions? Given previous work, we might try $y(t) = e^t + C$ for a nonzero constant C. Good guess, but no cigar:

$$y(t) = e^t + C \implies y'(t) = e^t \neq y(t).$$

Other solutions *do* exist. For any real constant C, $y(t) = Ce^t$ works:➤

$$y(t) = Ce^t \implies y'(t) = Ce^t = y(t).$$

Is C = 0 OK? Why or why not?

Again we have infinitely many solutions; this time they differ from each other by *multiplicative* constants.   □

Initial value problems

The examples above are typical. DE's often have *families* of solutions; family members differ from each other by additive or multiplicative constants. If we specify the value of a solution for some particular input—often called an **initial value condition**, we expect *only one solution*.➤ Such a combination of differential equation and initial value condition is called an **initial value problem—IVP** for short.

A "unique" solution, in math-talk.

■ **Example 3.**   Solve the IVP

$$f'(x) = 6x + 5; \quad f(0) = 42.$$

Solution: Any function➤ of the form $f(x) = 3x^2 + 5x + C$ is a solution of the DE. With the initial value condition we can find C:

As we saw above.

$$f(0) = 3 \cdot 0^2 + 5 \cdot 0 + C = 42 \implies C = 42.$$

Thus $f(x) = 3x^2 + 5x + 42$ is the *unique* solution to the IVP.   □

■ **Example 4.** Solve the initial value problem

$$y' = y; \quad y(0) = 42.$$

As we saw above.

Solution: Any function of the form $y(t) = Ce^t$ solves *this* DE.◀ Here, again, the initial value requirement forces C to be 42; $y(t) = 42e^t$ is the *unique* solution to the IVP.   □

Solving DE's and IVP's: guessing, checking, and parameters

Checking a guess

Given a DE and a *guess* at a solution, it's straightforward to *check* whether the guess works. The next example shows what we mean.

■ **Example 5.** Consider the DE $y' = k(y - T)$, where k and T are constants. Is $y(t) = T + Ae^{kt}$ a solution? Does the value of A matter?

Solution: Deciding whether $y(t) = T + Ae^{kt}$ solves the DE is a straightforward calculation—we'll calculate both sides (LHS and RHS) of the DE using this y and compare:

$$\text{LHS} = y'(t) = k \cdot Ae^{kt}; \quad \text{RHS} = k(y - T) = k(T + Ae^{kt} - T) = k \cdot Ae^{kt}.$$

The two sides *are* equal, so y *is* a solution.

See the next example.

So far the value of A hasn't mattered—y solves the DE for *any* constant A. Had we specified an initial condition, a specific value of A would have been needed.◀ □

From one solution, many: choosing values of parameters

The DE $y' = k(y - T)$ involves two constant parameters, k and T. The general solution, $y(t) = T + Ae^{kt}$, involves still another parameter, A.

Such parameters offer an important "leverage" advantage. Assigning them specific values as needed permits us, in effect, to solve infinitely many differential equations for the price of one. We illustrate by example.

Cooling is, in reality, more complicated than this simple model suggests.

■ **Example 6.** Freshly-poured coffee has temperature 190° F. As it cools, the coffee obeys **Newton's law of cooling:**◀

> *The rate at which an object cools is proportional to the temperature* difference *between the object and its environment.*

Assume that room temperature is a constant 65° F, and that after 2 minutes, the coffee has cooled to 160° F. Find a formula for temperature at time t.

We'll evaluate k below.

Solution: Under the verbiage lies the DE of the previous example. If $y(t)$ represents temperature at time t, then Newton's law says in this situation that for some constant k,◀

$$y' = k(y - 65), \quad y(0) = 190, \quad \text{and} \quad y(2) = 160.$$

As we saw in Example 5, a solution of this DE has the general form

$$y(t) = 65 + Ae^{kt},$$

for some constants A and k. All that remains is to give them numerical values.

From the initial condition $y(0) = 190$ it follows immediately that

$$190 = y(0) = 65 + Ae^{k \cdot 0} = 65 + A \implies A = 125.$$

Similarly, $y(2) = 160$ means➤ that

Check the steps.

$$160 = y(2) = 65 + 125e^{2k} \implies e^{2k} = \frac{95}{125}.$$

Solving the last equation➤ gives $k = (\ln 95 - \ln 125)/2 \approx -0.13722$. Putting it all together gives (approximately)

Take logs of both sides.

$$y(t) = 65 + 125e^{-0.13722t}.$$

Is the answer plausible? A graph of y over a 50-minute period suggests so:

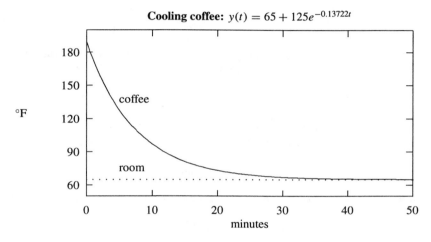

Cooling coffee: $y(t) = 65 + 125e^{-0.13722t}$

The moral: For best results, drink the coffee quickly. □

Finding solutions: antiderivatives and the chain rule

Checking candidate solutions to DE's is easy enough, but how can we *find* them?

There's no single (or simple) answer. Solving DE's in general is a nontrivial problem; techniques depend strongly on the algebraic form of the DE itself.➤

We'll see several such techniques later.

In the best possible circumstances, solving a DE boils down to finding an antiderivative, and then interpreting it properly. Below we illustrate the process in several relatively simple cases. Notice carefully how antiderivatives and the chain rule are involved.

■ **Example 7.** Solve the DE $g'(x) = 2x \cos\left(x^2\right)$.

Solution: A solution to this DE is a function g for which $g'(x) = 2x \cdot \cos\left(x^2\right)$. The general form of g suggests the result of applying the *chain rule* to some composite function. So it is: if $g(u) = \sin u$ and $u(x) = x^2$, then $g(x) = \sin\left(x^2\right)$, so by the chain rule, $g'(x) = \cos\left(x^2\right) \cdot 2x$, just as we wanted. Thus $y = \sin\left(x^2\right)$ is one solution. So is *any* function of the form

$$g(x) = \sin\left(x^2\right) + C,$$

with C any constant. □

■ **Example 8.** Solve the IVP

$$y' = 0.05y; \quad y(0) = 29.$$

Solution: A solution function $y(t)$ must satisfy both requirements above. We'll *find* such a function y by manipulating the DE.

First we rewrite the DE in the form

$$\frac{y'(t)}{y(t)} = 0.05.$$

The next step is to *antidifferentiate* both sides of the new equation. The right side is easy: For any constant C_1, $0.05t + C_1$ is an antiderivative. For the left side, the chain rule is again the key. It says that

$$\frac{d}{dt}\big(\ln(y(t))\big) = \frac{1}{y(t)}y'(t) = \frac{y'(t)}{y(t)}.$$

Therefore, for any constant C_2, $\ln(y(t)) + C_2$ is an antiderivative of the left side.

Putting these pieces together gives

$$\ln(y(t)) + C_2 = 0.05t + C_1 \quad\Longrightarrow\quad \ln(y(t)) = 0.05t + C_1 - C_2.$$

One arbitrary constant is enough. Because both C_1 and C_2 are arbitrary, so is $C = C_1 - C_2$. Thus we have, simply,

$$\ln(y(t)) = 0.05t + C.$$

Now we can solve for $y(t)$: exponentiating both sides strips away the logarithm, to give

$$y(t) = e^{0.05t+C} = e^{0.05t} \cdot e^C = A \cdot e^{0.05t},$$

where $A = e^C$ is *still* an undetermined constant.

Do check it.

And easy to check!

We're almost done. We've found the formula $y(t) = Ae^{0.05t}$; it's easy to check◄ that this y does indeed solve our DE. The initial condition $y(0) = 29$, finally, imposes a specific value for A. Because $y(0) = Ae^0 = A$, it's clear at last◄ that

$$y(t) = 29e^{0.05t}$$

solves our original IVP.   □

■ **Example 9.** We saw above, by a direct check, that for any constants T, A, and k, $y = T + Ae^{kt}$ solves the DE $y' = k(y - T)$. How did we find this y?

Solution: We found it from the chain rule. Here's how.

First we rewrote the DE, bringing all y's to the left:

$$y'(t) = k\,(y(t) - T) \quad\Longrightarrow\quad \frac{y'(t)}{y(t) - T} = k.$$

Check this—differentiate $\ln(y(t) - T)$ *for yourself.*

Then we recognized the left side as the *derivative*—via the chain rule—of $\ln(y(t) - T)$.◄ Therefore, reasoning as in the previous example, we have

$$\ln(y(t) - T) = kt + C \quad\Longrightarrow\quad y(t) - T = e^{kt+C} \quad\Longrightarrow\quad y(t) = T + Ae^{kt},$$

where (as before) $A = e^C$, and C is an arbitrary constant.   □

> **Hard work, but worth doing.** Solving DE's can be much more complicated than our simple examples suggest. A gigantic literature exists on the theory and methods of solving many sorts of differential equations. This fact attests to the depth and subtlety of the subject—but even more to its practical importance.

Exercises

1. Decide whether each of the following functions is a solution of the differential equation given with it.

   (a) $y(t) = t^2/2$; $y' = t$

   (b) $y(t) = t^2/2$; $y' = y$

   (c) $y(t) = e^t$; $y' = y$

   (d) $y(t) = 42e^t$; $y' = y$

   (e) $y(t) = -e^t$; $y' = -y$

   (f) $y(t) = e^{-t}$; $y' = -y$

   (g) $y(t) = 42e^{-t}$; $y' = -y$

   (h) $y(t) = Ce^{kt} + A$; $y' = k(y - A)$

   (i) $y(t) = e^t + \frac{1}{2}t^2$; $y' = y + t$

   (j) $y(t) = \frac{1}{2}e^{t^2}$; $y' = ty$

   (k) $y(t) = e^{t^2/2}$; $y' = ty$

2. Show that $y(x) = x^{-1} - 1$ is a solution of the differential equation $x^3 y'' + x^2 y' - xy = x$.

3. Is $y(t) = \frac{1}{2}t^4 + \frac{3}{2}t^2 + \frac{1}{4}$ a solution of the differential equation $y' - 2y/t = t^3$? Justify your answer.

4. (a) Show that $y(t) = 65 + 125\left(\sqrt{95/125}\right)^t = 65 + 125\,(95/125)^{t/2}$.

   (b) How long does the coffee in Example 6 remain above $100°$ F?

   (c) In a foam cup, coffee cools only to $180°$ F in two minutes. Let $z(t)$ be the temperature of the coffee in such a cup. Find a formula for $z(t)$. How long does coffee in a foam cup stay above $100°$ F?

   (d) Plot $y(t)$ and $z(t)$ on the same axes.

5. Verify by direct calculation that the function

$$P(t) = \frac{C}{1 + d\,e^{-k \cdot C \cdot t}}$$

   is a solution of the **logistic DE** $P' = kP(C - P)$ for any constants k, C, and d. (We'll see the logistic DE in later applications.)

6. Find real numbers a and b so that $y = e^x$ and $y = e^{2x}$ are both solutions of the differential equation $ay'' + by' + y = 0$.

7. Find functions $f(x)$ and $g(x)$ so that $y = x$ and $y = x^2$ are both solutions of the differential equation $f(x)y'' + g(x)y' + y = 0$.

8. Consider the function f shown below.

Graph of f

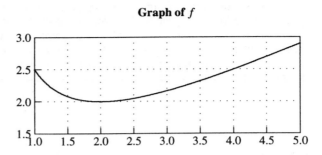

$y = f(x)$ is a solution of one of the following differential equations. Decide which one. Then explain why each of the others is not a solution.

(a) $y' = \dfrac{y - x}{x}$ (b) $y' = \dfrac{x - y}{x}$ (c) $y' = \dfrac{x^2 - y}{x}$

9. When the valve at the bottom of a cylindrical tank is opened, the rate at which the level of the liquid in the tank drops is proportional to the square root of the depth of the liquid. Thus, if $y(t)$ is the liquid's depth at time t minutes after the valve is opened, $y' = -k\sqrt{y}$ for some positive constant k.

(a) Does the water level drop faster when the tank is full or when it is half full? Justify your answer.

(b) Verify that the function $y(t) = (C - kt)^2 / 4$ is a solution of the differential equation $y' = -k\sqrt{y}$ for any values of the constants C and k.

(c) Suppose that $y(0) = 9$ feet and $y(20) = 4$ feet. Find an equation for $y(t)$.

(d) At what time is the water level dropping at a rate of 0.1 feet per minute?

10. Consider the following model for the growth of a city: The shape of the city always remains roughly circular so that the maximum travel time between two locations in the city is proportional to the diameter of the city. The population of the city is proportional to the area of the city. The rate of increase of the city's population is inversely proportional to the maximum travel time.

(a) This model predicts that the population of the city $P(t)$ satisfies the differential equation

$$\frac{dP}{dt} = \frac{K}{\sqrt{P(t)}}$$

where K is a constant. Explain why.

(b) Verify (by differentiation) that $P(t) = \left(\frac{3}{2}Kt + C\right)^{2/3}$ is a solution of the differential equation given in part (a).

(c) If the population of the city was 5000 in 1900 and 20,000 in 1950, what is the predicted population of the city in the year 2000?

11. Psychologists in learning theory study learning curves, graphs of "performance functions." A performance function $P(t)$ describes someone's skill at a task as a function of the training time t. It has been noted that learning is at first rapid, then tapers off (the rate of learning decreases) as the value of $P(t)$ approaches M, the maximal level of performance.

(a) Explain how the differential equation $P' = k(M - P)$, where k is a positive constant, describes this situation.

(b) Suppose that for a certain learning activity it is known that $P(0) = 0.1M$ and $k = 0.05$ when t is measured in hours. How long does it take to reach 90% of the maximum level of performance?

12. Assume that influenza spreads through a university community at a rate proportional to the product of the number of those already infected and the number of those not yet infected. If the total number of students at the institution is 3000 and $P(t)$ represents the number of students infected after t days, express the preceding statement as a differential equation.

13. It's a winter day in Northfield, Minnesota—10 degrees below zero Celsius. Ole and Lena stop at a convenience store for hot (initially, 90° C) coffee to warm them on the cold, windy walk back to Manitou Heights. Two types of cups are available: environmentally destructive foam and politically correct cardboard.

Ole wants foam. "What do I care about the greenhouse effect?" he asks. "Everything in Edina is air-conditioned anyway. Besides, remember the proportionality constant k in Newton's Law of Cooling:

$$T'(t) = -k \cdot (T(t) - C),$$

where C is the environmental temperature, $T(t)$ is the Celsius temperature of the coffee at time t, and t is measured in minutes? Well, for foam, $k = 0.05$. For cardboard, k is a pathetic 0.08."

"Do what you want, Attila," says Lena. "I'd rather save the world. I'm having cardboard. And make mine Nicaraguan decaf."

(a) How long does Ole's coffee stay above 70 degrees? How about Lena's? How hot is each cup after 5 minutes?

(b) Answer the questions posed in part (a) assuming that Ole and Lena drink their coffee in the overheated (25° C) convenience store.

(c) What value of k is needed to assure that coffee, starting at 90 degrees, is still at least 70 degrees after 5 minutes, outdoors. How about indoors? [HINT: Use the solution function to set up an appropriate equation; then solve for k.]

14. A crucible is removed from an oven whose temperature is $500°C$ and placed in a room in which the air temperature is kept at $20°C$. If the temperature of the crucible has fallen to $260°C$ after one-half hour, how long will it take before the temperature of the crucible reaches $80°C$? (Assume that Newton's Law applies.)

15. By Newton's Law, the rate of cooling of an object in air is proportional to the difference between the object and the temperature of the air. If the temperature of the air is $20°C$ and boiling water cools in 20 minutes to $60°C$, how long will it take for the water to drop in temperature to $30°C$?

16. An indoor thermometer, reading 60°, is placed outdoors. Ten minutes later it reads 70°. After another ten minutes it reads 76°. What is the outdoor temperature?

17. Experience suggests that the rate at which people contribute in a charity drive is proportional to the difference between the current total and the announced goal. A fund drive is announced with a goal of $90,000 and an initial contribution of $10,000. One month later, $30,000 has been contributed. What does the model predict at the end of two months?

18. A certain classroom has volume 6000 cubic feet. When the concentration of carbon dioxide in the air reaches 0.16%, the ventilation system is activated. The ventilation system brings in air containing 0.04% carbon dioxide at a rate of 1000 cubic feet per minute and drives out the mixture of stale air and fresh air at the same rate. Assuming complete and instantaneous mixing, what is the percentage of carbon dioxide in the air in the room after 5 minutes of ventilation?

19. A man eats a diet of 2500 cal/day; 1200 of them go to basal metabolism (i.e., get used up automatically). He spends approximately 16 cal/kg times his body weight in weight-proportional exercise each day.

    (a) Determine how the man's weight varies with time.

    (b) Does the man reach an equilibrium weight?

    (Assume that the storage of calories as fat is 100% efficient and that 1 kg fat contains 10,000 cal.)

4.2   More differential equations: modeling growth

Exponential growth

Exponential functions have an important property, best expressed in terms of rates:

> *Exponential functions grow at a rate proportional to their size, i.e., "with interest."*

Thanks to this property,➤ exponential functions model many real phenomena which, too, grow "with interest."

We discussed this property in Section 2.9.

Let's interpret this property in the language of DE's and IVP's. For any constants A and k, let $y(t) = Ae^{kt}$. Two key properties➤ of y matter here:

Check these claims carefully!

$$y'(t) = Ake^{kt} = ky(t) \quad \text{and} \quad f(0) = Ae^0 = A.$$

In other words:

> **Theorem 1.** For any constants A and k, the exponential function $y = Ae^{kt}$ solves the initial value problem
>
> $$y' = ky; \quad y(0) = A.$$

This theorem offers considerable practical horsepower. Just by choosing k and A judiciously, we can solve a surprising number of useful and important DE's.

Modeling things that grow "with interest"

Many real-world quantities grow or "decay" like exponential functions, at rates proportional to their size. The Consumer Price Index, the value of a financial deposit, college fees,➤ radioactive decay, biological populations—all can be modeled, more or less accurately, by the IVP of Theorem 1.

Recall Uncle Eric, in Section 1.6.

The federal deficit: keeping rates straight

Differential equations relate varying quantities to their derivatives, i.e., their *instantaneous rates* of change. Practical problems, on the other hand, often mention other sorts of rates: average rates, interest rates, percentages, etc. Teasing out the derivatives from all this loose talk may take some care.

■ **Example 1.**   At time $t = 0$ years, the federal deficit was \$300 billion and growing fast. "*How* fast?" worries Senator Smith. "What will the deficit be 6 years hence,➤ when I face re-election?" In Senate hearings Economist A predicts:

Politicians say "hence."

> *The federal deficit will grow at an instantaneous rate of 15% per year.*

"Balderdash," says Economist B,

> *The federal deficit will grow 15% each year.*

Do Economists A and B disagree? How seriously? Whom would Senator Smith rather believe?

Solution: Economists A and B agree on one thing: that the deficit grows at a rate proportional to itself. In DE terms, if $y(t)$ represents the deficit in billions, t years out, A and B agree that for some constant k,

$$y'(t) = ky(t).$$

They *disagree* on the value of k.

Economist B's statement refers to an *average* rate: over a one-year *interval*, the deficit rises 15%. Economist A's statement means something else: at any *instant*, the deficit's rate of increase, if it remained unchanged for a year, would raise the deficit by 15%.

To understand the difference, and see its effects, we'll recast the predictions as IVP's. Combined with the initial value ($300 billion) of the deficit, A's claim comes to this:

$$y'(t) = 0.15 \cdot y(t); \qquad y(0) = 300.$$

B says something else:

$$y'(t) = k \cdot y(t); \qquad y(t + 1) = 1.15 \cdot y(t); \qquad y(0) = 300.$$

We'll solve both IVP's and compare the results. A's claim fits perfectly into Theorem 1's template:

Since $y'(t) = 0.15 \cdot y(t)$ and $y(0) = 300$, $y(t) = 300e^{0.15t}$.

Again.

Solving B's IVP takes one more step. By Theorem 1,◂

Since $y'(t) = k \cdot y(t)$ and $y(0) = 300$, $y(t) = 300e^{kt}$.

Check the algebra carefully.

Now we'll find a value for k. Our new formula $y(t) = 300e^{kt}$ gives $y(t + 1) = 300e^{k(t+1)}$. Therefore◂

$$
\begin{aligned}
y(t + 1) = 1.15 \cdot y(t) &\implies 300e^{k(t+1)} = 1.15 \cdot 300e^{kt} \\
&\implies e^k = 1.15 \\
&\implies k = \ln 1.15 \approx 0.13976.
\end{aligned}
$$

Economist B models the deficit with the function $y(t) = 300e^{0.13976t}$.

What does each scenario predict over the next 6 years? Graphs answer best:

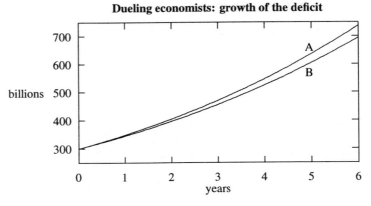

Dueling economists: growth of the deficit

A's and B's predictions diverge as time goes on. At $t = 6$, the difference is

$$300e^{0.15 \times 6} - 300e^{0.13976 \times 6} \approx 43.97 \text{ billion dollars.}$$

Which view will Senator Smith prefer? It depends, perhaps, on Senator Smith. A Washington insider might prefer B's "rosier" view; an outsider, A's "gloomy" view.□

Money in the bank

Bank interest offers the most familiar example of "growth with interest." Its crucial feature—shared by exponential growth in *all* forms—is that *the rate at which the account's value grows is proportional to the value itself.* The rich don't only get richer—they also get richer *faster.*

Bank interest may even be *too* familiar: some care is needed in translating banking language into mathematical terms. The next example illustrates this.

■ **Example 2.**   A savings account pays a **nominal annual interest rate** of 6%, compounded daily. Model the situation in DE language. How much does the account increase over one year?

Solution: If $b(t)$ represents the balance (in dollars) at time t (in years), the terms above mean[➤] that each day, the balance is multiplied by the same small factor: *In a 365-day year.*

(4.2.1) $$b\left(t + \frac{1}{365}\right) = b(t) \cdot \left(1 + \frac{0.06}{365}\right).$$

With initial balance $b(0)$, for instance, here's what happens for the first two days:

$$b\left(\frac{1}{365}\right) = b(0) \cdot \left(1 + \frac{0.06}{365}\right); \qquad b\left(\frac{2}{365}\right) = b(0) \cdot \left(1 + \frac{0.06}{365}\right)^2.$$

Over a full year, therefore, the balance increases by the factor

$$\left(1 + \frac{0.06}{365}\right)^{365} \approx 1.0618.$$

The **effective annual interest rate**, therefore, is 6.18%—over a one-year interval, the balance rises by this percentage.

How do interest rates, nominal or effective, translate into DE language?

- In reality, the balance $b(t)$ varies **discretely,**[➤] at regular intervals and in jumps   *Not continuously.* of at least one cent. Any DE model, being continuous, must commit *some* error. In practice such errors are usually small enough to ignore.

- On a time-scale measured in years, a single day is effectively an "instant." The *nominal* interest rate, therefore, can serve as "k" in the DE of Theorem 1. In other words, the balance $b(t)$ can be taken[➤] to satisfy the DE   *With very little error.*

$$b'(t) = 0.06 \cdot b(t).$$

Theorem 1 *solves* the DE above. If the initial balance is b_0, then

$$b'(t) = 0.06 \cdot b(t) \quad \text{and} \quad b(0) = b_0 \implies b(t) = b_0 \cdot e^{0.06t}.$$

After one year the balance is

$$b(1) = b_0 \cdot e^{0.06} \approx b_0 \cdot 1.0618.$$

As it should, the balance rises by 6.18%, the *effective* interest rate.   □

Interest rates and derivatives: continuous compounding.
The DE $b'(t) = 0.06 \cdot b(t)$ does a good—but not perfect—job of modeling 6% annual interest, compounded *daily*. The model would do better if interest were compounded *hourly*, or even every *minute*. The DE represents **continuous compounding**, i.e., the limiting case as the number of compounding periods tends to infinity.

Why does the DE $b'(t) = 0.06 \cdot b(t)$ approximate daily compounding as well as it does? Equation 4.2.1 and the following computation (for those who wish to follow it) explains. The result shows, too, how the frequency of compounding affects the "goodness" of approximation.

$$
\begin{aligned}
b'(t) &\approx \frac{b(t + 1/365) - b(t)}{1/365} \\
&= \left(b(t) \cdot \left(1 + \frac{0.06}{365} \right) - b(t) \right) \cdot 365 \\
&= b(t) \cdot \left(1 + \frac{0.06}{365} - 1 \right) \cdot 365 \\
&= 0.06 \cdot b(t).
\end{aligned}
$$

Radioactive decay

A sample of carbon-14 (a radioactive isotope of ordinary carbon) decays at time t (in years) at a rate proportional (with $k = -0.000121$) to its weight. In DE language,

$$
W'(t) = -0.000121 W(t).
$$

(The constant k is *negative* because W *shrinks*. In banking terms, carbon-14 pays negative interest.)

■ **Example 3.** If a sample of carbon-14 weighed W_0 pounds in 1991, how much will it weigh in 2991? What did it weigh in 991? How long does it take the sample to lose *half* its weight? (This period is called the sample's **half-life**.)

So $t = 0$ in 1991.

Solution: If $W(t)$ represents the sample's weight t years after 1991,[◄] then $W(t)$ solves the IVP
$$
W'(t) = -0.000121 W(t); \quad W(0) = W_0.
$$

According to Theorem 1, $W(t) = W_0 e^{-0.000121t}$. Therefore

$$
\begin{aligned}
\text{weight in 2991} &= W(1000) = W_0 e^{-0.121} \approx 0.88603 W_0; \\
\text{weight in 991} &= W(-1000) = W_0 e^{0.121} \approx 1.12862 W_0.
\end{aligned}
$$

The sample loses a bit more than 12% of its weight over 1000 years.

The sample's half-life is the value of t for which $W(t) = W_0/2$. To find it, we solve for t:

$$
W(t) = \frac{W_0}{2} \iff W_0 e^{-0.000121t} = 0.5 W_0 \iff -0.000121t = \ln(0.5)
$$
$$
\iff t \approx 5728 \text{ years.} \qquad \square
$$

Some biological populations grow exponentially

Certain biological populations grow exponentially—at least for a while.[*] A fruit fly population in laboratory conditions, for example, might reasonably grow at an *instantaneous* rate of 5% per day, i.e., at a rate proportional (with $k = 0.05$) to the population itself. In DE language, if the population $P(t)$ starts at P_0, then P satisfies the IVP

$$P'(t) = 0.05 P(t); \quad P(0) = P_0.$$

We'll see below why such growth can't last forever.

Notice the analogy with ordinary interest: each day's crop of new fruit flies can safely (if unpalatably) be thought of as "interest" added to the previous day's "balance."

■ **Example 4.**   A population of fruit flies grows at an instantaneous rate of 5% per day. How long does it take for the population to double?

Solution: Given an initial population[*] of P_0 flies, Theorem 1, page 365, gives the formula

I.e., at $t = 0$.

$$P(t) = P_0\, e^{0.05t}.$$

A little algebra shows that the population doubles in less than 14 days:

$$P(t) = P_0\, e^{0.05t} = 2\, P_0 \iff 0.05t = \ln 2 \iff t \approx 13.8629. \qquad \square$$

Not all populations grow exponentially: logistic growth

Real biological populations can't grow exponentially forever. An "initial deposit" of 100 fruit flies, growing as above, would number over 8 billion in a year. In a decade, the flies would outnumber Earth's molecules.[*]

Any exponentially-growing population sooner or later exceeds the limits (physical or biological) of its environment. Just this possibility, applied to human population, worried the 19th century clergyman Thomas Malthus, who predicted an early and unpleasant end to the human race.[*]

Whether Malthus himself was right or wrong, exponentially-growing biological populations *do* eventually outgrow their environments. How real populations cope with this mathematical certainty[*] varies: some (e.g., lemmings and Dungeness crabs) have boom-bust cycles; others eat up their environments and die off completely.

A third (happier) possibility is **logistic growth**. As a population approaches an upper limit C (the environment's **carrying capacity**), growth slows. For our flies, logistic growth might lead to the lower graph shown below:

A logical—not just biological—impossibility.

Was Malthus right? It's too early to tell. Some evidence may even suggest that Malthus was too optimistic.

And with other vagaries of real life.

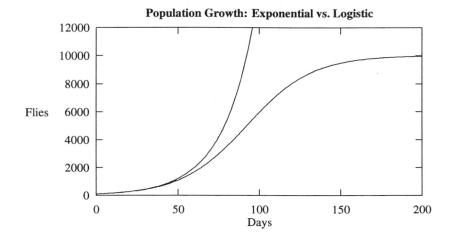

Population Growth: Exponential vs. Logistic

Notice that the two graphs are, toward the beginning, almost identical. This reflects the fact that when populations are small, the effects of "crowding" are scarcely felt, so logistic growth is essentially exponential.

Like exponential growth.

Logistic growth[*] is characterized by a rate property, and therefore by a DE (called the **logistic DE**):

> *The population's growth rate is proportional both to the population itself and to the difference between the carrying capacity and the population.*

In symbols:

$$P' = kP(C - P),$$

P and P' vary with time; k and C are parameters.

where P represents the population, P' the population's growth rate, C the carrying capacity of the environment, and k a constant of proportionality.[*]

Wise flies: a case study in logistic growth

A population of prudent fruit flies, eager to avoid crowding, choose to reproduce logistically. This is their story.

Fly specs

On a certain day, the fly population is 1000, and growing at the (net) instantaneous rate of 50 flies per day. The environment can support 10,000 flies.

Questions.   Given what we know of growth *rates*, it's natural to wonder about population *amounts*:

- What will the population be in 10 days? 100 days? 200 days?

- When is the population 5000? 9000? 9900?

- When does the population grow fastest? What's the population then?

Solving the logistic DE

Modeling logistic population growth boils down to solving an IVP. If a population has P_0 members at $t = 0$ and grows logistically, then the population function $P(t)$ is a solution of the IVP

$$P' = kP(C - P); \quad P(0) = P_0.$$

It will get easier with techniques we'll study later.

Searching systematically for a solution, given our present tools, is a slightly messy process.[*] It turns out, suffice it to say, that for any constants k, C, and d, the function

$$P(t) = \frac{C}{1 + d\,e^{-k \cdot C \cdot t}}$$

Try checking this: it's messy but not really hard.

solves the logistic DE $P' = kP(C - P)$.[*] (The constants k and C in the solution are the same as those in the DE; d depends, as we'll see, on the initial population.)

What solutions mean for our flies

Measuring population in thousands removes pesky zeros from later computations.

To put things in context, let's define:[*]

$$
\begin{aligned}
t &= \text{time, in days, since the original measurement;} \\
P(t) &= \text{population of flies, in thousands, at time } t; \\
P'(t) &= \text{growth rate, in thousands of flies per day, at time } t.
\end{aligned}
$$

Evaluating the parameters. We know three things: $P(0) = 1$, $P'(0) = 0.05$, and $C = 10$. We'll use what we know to evaluate k and d. Since

$$P'(0) = 0.05 = k \cdot P(0) \cdot (10 - P(0)) = k \cdot 1 \cdot 9,$$

$P'(0) = 0.05$ corresponds to 50 flies/day.

it follows that $k = 1/180 \approx 0.00556$.

Now we can evaluate d. By assumption,

$$P(0) = 1 = \frac{10}{1 + d\,e^{-k \cdot C \cdot 0}} = \frac{10}{1 + d},$$

so $d = 9$. Putting everything together gives, at last, an explicit formula for $P(t)$:

$$P(t) = \frac{C}{1 + d\,e^{-k \cdot C \cdot t}} = \frac{10}{1 + 9\,e^{-0.0556t}}.$$

Here's a graph of P:

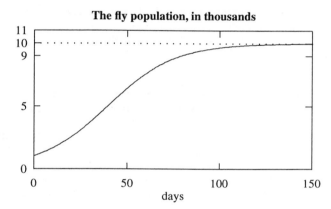

The fly population, in thousands

Our questions answered: What the graph and formula say. With the formula and the graph we can answer all the questions posed above.

Approximately, at least.

- The graph shows that on day 100, the population was about 9700; room is quickly running out. Day 200 isn't shown, but the graph's asymptotic behavior shows that at $t = 200$, the population should *still* be just short of 10,000. The formula agrees:

$$P(200) = \frac{10}{1 + 9\,e^{-0.0556 \cdot 200}} \approx 9.99867.$$

Thus the population is *at* full capacity—rounding to the nearest whole fly.

The only reasonable way to round flies.

- Both graph and formula show that the population reaches 5000 (half the upper limit) in about 40 days. Around day 80, the population reaches 9000—90% capacity. Around day 130, 99% capacity is reached.

All of these numbers can be checked with the formula.

- Intuition says that under logistic conditions a population increases slowly at first (when there are few breeding members), faster as the population rises, and then slowly again as the upper limit nears.

The graph agrees. Population grows fastest when our graph points most steeply upward, i.e., around day 40. At this time the population is about 5000—exactly *half* the carrying capacity. This isn't a coincidence. Under logistic conditions, it's a disadvantage for a population to be either too large *or* too small. Thus, intuitively, we'd expect growth conditions to be optimal at the *halfway* point.

Take a close look; do you agree?

Afterword: discrete vs. continuous growth

All mathematical models simplify, more or less, the phenomena they describe. Using DE's and IVP's to model phenomena of growth is no exception to this rule.

One sense in which DE's and IVP's may describe reality imperfectly concerns the difference between **discrete** and **continuous** phenomena of change. Human population, for instance, grows "discretely"—in jumps of at least one unit. Hot coffee in a cool room, by contrast, cools "continuously"—its temperature takes *all* intermediate values on the way down.

DE's and IVP's are *continuous* models of growth. Solutions to DE's and IVP's are *continuous* functions, so they can't, by definition, describe *discrete* phenomena exactly. (Neither, for the same reason, can discrete models perfectly describe continuous phenomena.)

Such philosophical problems, though undeniable, need not be fatal. In practice, differences between discrete and continuous growth models are often unimportant—especially when "jumps" are small. It matters little, for instance, whether coffee temperature falls continuously or in tiny increments of 0.0001 degrees. Perfection is impossible, but DE's and IVP's *can* be⁴⁴ used to model growth—even discrete growth—easily, effectively, and accurately.

And are.

Discrete models of continuous phenomena. DE's and IVP are *continuous* entities; they're often used to describe phenomena that vary *discretely*. Sometimes things go the other way: continuously-changing phenomena can be modeled discretely. Movies, digital watches, and electronic computing all depend, in various ways, on discrete approximations to continuous reality.

Exercises

1. Use Theorem 1, page 365, to solve each of the following IVP's. Check answers by differentiation.

   (a) $y' = 0.1y$; $y(0) = 100$

   (b) $y' = -0.0001y$; $y(0) = 1$

   (c) $y' = (\ln 2) \cdot y$; $y(2) = 4$

   (d) $y' = ky$; $y(10) = 2y(0)$
   [HINT: Solve for k using the initial condition.]

2. Suppose that the "wise flies" described in this section had begun by breeding at the net rate of 100 (rather than 50) flies per day. How would things be different? More specifically, answer these questions, and compare the results with those given on page 371.

   (a) What will the population be in 10 days? 100 days?

   (b) When is the population 5000? 9000? 9900?

   (c) When does the population grow fastest? What's the population then?

3. A certain menacing biological culture (aka The Blob) grows at a rate proportional to its size. When it arrived unnoticed one Wednesday noon in Chicago's Loop, it weighed just 1 gram. By the 4:00 p.m. rush hour it weighed 4 grams.

   The Blob has its "eye" on the Sears Tower, a tasty morsel weighing around 3000000000000 (i.e., 3×10^{12}) grams. The Blob intends to *eat* the Sears Tower as soon as it weighs 1000 times as much (i.e., 3×10^{15} grams). By what time must the Blob be stopped? Will Friday's rush-hour commuters be delayed?

4. A mold grows at a rate that is proportional to the amount present. Initially, its weight is 2 grams; after two days, it weighs 5 grams. How much does it weigh after eight days?

5. A bacterial culture is placed in a large glass bottle. Suppose that the volume of the culture doubles every hour, and the bottle is full after one day.

   (a) If the culture was placed in the bottle at time $t = 0$ hours, when was the bottle half full?

   (b) Assume that the bottle is "almost empty" when the culture occupies less than 1% of its volume. How long is the bottle "almost empty"?

6. The police guard gave Sara a cold look, but his voice was polite as he directed her to the room she sought. "Don't touch anything, please, Ms. Abrams. The Chief said I had to let you in, but he said to tell you to mind your fingers." "Thank you," Sara replied cooly. "The Chief knows he can trust me." The guard opened his mouth as if to speak, but he merely shook his head and withdrew.

   Sara was standing in what appeared to be a combination bedroom and laboratory. A relative had found Dr. Howell's body on the floor of this room that morning. By 9:00 a.m., the coroner had completed his examination; he stated that death was due to a severe blow to the head, and that Dr. Howell had been dead between 36 and 40 hours. It seemed critical to Sara to know exactly when Dr. Howell had died so that she could eliminate certain suspects. But how could she possibly discover exactly when he was killed? Puzzled, she wandered around the small, cluttered room, being careful not to touch anything. The old doctor apparently was conducting an experiment when he was killed. Sara absent-mindedly read from the notebook which was lying open on the bench.

   > The fungus grows at a rate proportional to its current weight.

   "Great," she thought, "I'm here to investigate a murder and instead I'm getting a biology lesson." At a loss for what else to do, she continued reading.

   > To exemplify this biological truth, I place the fungus on a scale and record its weight at various times:

10 g	5:30 p.m.
12 g	6:15 p.m.
13	

   "Hmm," Sara mused, "the poor guy didn't even get to finish the last entry." Sara suddenly frowned in concentration. She searched her pockets until she found a pencil stub and an old receipt. When the guard entered the room a few minutes later, Sara had just finished scribbling on the receipt. She smiled as she shoved the receipt and the pencil stub back into her pocket.

"Don't worry," Sara said cheerfully, "I'm leaving. I now know exactly when Dr. Howell was killed." The guard looked sourly at Sara's back as she left the room.

When was Dr. Howell killed? How did Sara know?

7. Oil is pumped continuously from a well at a rate proportional to the amount of oil left in the well. Initially there were 1,000,000 barrels of oil in the well; 6 years later 500,000 barrels remain.

   (a) At what rate was the amount of oil in the well decreasing when there were 600,000 barrels of oil remaining?

   (b) It will no longer be profitable to pump oil from the well when there are fewer than 50,000 barrels remaining. When should pumping stop?

8. Human skeletal fragments showing ancient Neanderthal characteristics were found in a Palestinian cave and brought to a laboratory for carbon dating. Analysis showed that the proportion of C^{14} to C^{12} is only 6.25% of the value in living tissue. How long ago did this person die?

9. A tank initially contains 100 gallons of water and 10 pounds of salt, thoroughly mixed. Pure water is added at the rate of 5 gallons per minute, and the mixture is drained off at the same rate. (Assume complete and instantaneous mixing.)

   (a) Explain why $S(t)$, the amount of salt in the tank at time t, is the solution of the IVP
   $$S' = -\frac{5}{100}S; \quad S(0) = 10.$$

   (b) Using Theorem 1, page 365, find a solution of this IVP.

   (c) How much salt is left in the tank after 1 hour?

4.3   Linear and quadratic approximation; Taylor polynomials

In this section we consider how to find and use polynomial functions that approximate other (non-polynomial) functions. What does it *mean* for a function f to approximate another function g? How can "good" polynomial approximations be chosen? How are derivatives involved?

We'll start with the simplest cases: linear and quadratic polynomials. Then we'll describe how to find polynomials of *any* degree, called **Taylor polynomials**, that approximate a given function.

Why bother?   Often in mathematics and applied mathematics one wants to approximate one function, say f, with another, say g. Why bother? For several possible reasons: if f is complicated, imperfectly understood, or otherwise inconvenient, it's useful to replace f with a simpler, better behaved, more tractable, or better understood g.

Polynomial functions are simple, convenient, well understood, and easy to use, so it's natural to use them to approximate more complicated functions.

Tangent lines and linear approximation

The line tangent to the graph of a function f at $x = x_0$ can be thought of in several ways. Geometrically, the tangent line is the straight line that best "fits" the graph of f at $x = x_0$. Analytically, the tangent line represents the linear *function* that best "approximates" f near $x = x_0$.

■ **Example 1.**   If $f(x) = \sqrt{x}$, then $f(64) = 8$. How closely does the tangent line function approximate values of $f(x)$ for x *near* 64?

Solution: Since $f'(x) = 1/(2\sqrt{x})$, $f'(64) = 1/16$. Therefore the line tangent to the f-graph at $(64, 8)$ has equation➡   *Verify this.*

$$\ell(x) = f(64) + f'(64)(x - 64) = 8 + \frac{x - 64}{16} = 4 + \frac{x}{16}.$$

Here are graphs of both f and ℓ:

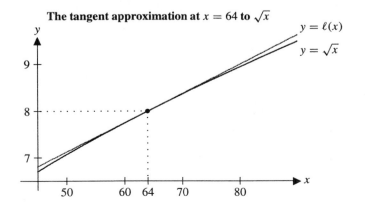

The tangent approximation at $x = 64$ to $\sqrt{x}$

Observe:

- The tangent line hugs the graph of f closely; ℓ should therefore approximate f well for x near $x = 64$—the nearer the better.

But only slightly!

- The graph of f is concave *down*, so it lies *below* its tangent line. Therefore, the tangent line function ℓ *overestimates* f[*] for each $x \neq 64$.

Constant, linear, and quadratic approximations

How might we approximate *any* function f near *any* convenient "base point" $x = x_0$? The previous example showed that the tangent line at $x = x_0$ defines a good **linear approximation** to f at x_0. **Constant** and **quadratic approximations** are also possible.

◼ **Example 2.** Which *constant* function best approximates $f(x) = \sqrt{x}$ near $x = 64$? Which *quadratic* function?

Solution: Since $f(64) = 8$, the constant function $c(x) = 8$ is the only sensible choice; c has the "right" value at $x = 64$.

How should we choose a good *quadratic* approximation q? For the *linear* approximation ℓ we required that

$$\ell(64) = f(64) = 8 \quad \text{and} \quad \ell'(64) = f'(64) = \frac{1}{16}.$$

Why is $f''(64) = -1/2048$? Check for yourself.

It's reasonable, then, to ask that a good *quadratic* approximation q go one step further:[*]

$$(4.3.1) \qquad q(64) = 8; \quad q'(64) = \frac{1}{16}; \quad q''(64) = f''(64) = -\frac{1}{2048}.$$

In other words, our desired quadratic function q should "agree" with f at $x = 64$, both in its value and in its first *two* derivatives.

Writing q in powers of $(x - 64)$, not x, greatly simplifies some of the algebra.

The easiest way to construct such a q is to write it in the form[*]

$$q(x) = a + b(x - 64) + c(x - 64)^2,$$

and then choose the coefficients a, b, and c appropriately. Since we'll need them below, we'll first calculate several values and derivatives of q:[*]

Watch carefully—here's where the form of q pays off.

$$\begin{aligned} q(x) &= a + b(x - 64) + c(x - 64)^2 \implies q(64) = a; \\ q'(x) &= b + 2c(x - 64) \implies q'(64) = b; \\ q''(x) &= 2c \implies q''(64) = 2c. \end{aligned}$$

Choosing a, b, and c is now easy:

Choosing a. Equation 4.3.1 requires that $q(64) = 8$. But $q(64) = a$, so $a = 8$.

Choosing b. Equation 4.3.1 requires that $q'(64) = 1/16$. But $q'(64) = b$, so $b = 1/16$.

Choosing c. Equation 4.3.1 requires that $q''(64) = -1/2048$. But $q''(64) = 2c$, so $c = -1/4096$.[*]

Where did the factor of 2 come from?

Putting everything together gives

$$q(x) = 8 + \frac{x - 64}{16} - \frac{(x - 64)^2}{4096}$$

as the "best" quadratic approximation to f at $x = 64$.

Graphs of c, ℓ, q, and f show what's happening:

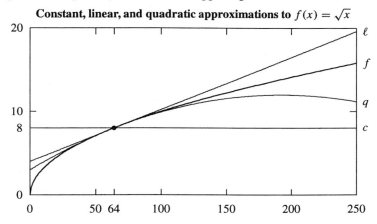

Observe:

- All three approximating functions (c, ℓ, and q) have the "right" value at $x = 64$.

- The graphs of ℓ and q (but not of c!) are *tangent* to the f-graph➤ at $x = 64$. *I.e., have the same slope.*

- The function q "fits" f *better* than ℓ➤ near $x = 64$. The fact that $q''(64) = f''(64)$ means that q and f have the same *concavity* at $x = 64$. *And much better than c.*

How well, numerically, do c, ℓ, and q approximate f? A table gives some idea:

Three approximations to $f(x) = \sqrt{x}$							
x	50	63	63.9	64	64.1	65	80
$c(x)$	8.00000	8.00000	8.00000	8	8.00000	8.00000	8.00000
$\ell(x)$	7.12500	7.93750	7.99375	8	8.00625	8.06250	9.00000
$q(x)$	7.07715	7.93726	7.99375	8	8.00625	8.06226	8.93750
$f(x)$	7.07107	7.93725	7.99375	8	8.00625	8.06226	8.94427

The numbers agree with the graphs: q approximates f best, especially near $x = 64$. □

Definitions

Linear and quadratic approximations can be chosen for *any* well-behaved function f, at *any* base point x_0. The formal definition summarizes, in general language, what we did above for $f(x) = \sqrt{x}$ and $x_0 = 64$.

Definition: Let f be any function for which $f'(x_0)$ and $f''(x_0)$ exist. The **linear approximation to f, based at x_0**, is the linear function

$$\ell(x) = f(x_0) + f'(x_0)(x - x_0).$$

The **quadratic approximation to f, based at x_0**, is the quadratic function

$$q(x) = f(x_0) + f'(x_0)(x - x_0) + \frac{f''(x_0)}{2}(x - x_0)^2.$$

■ **Example 3.** Let $f(x) = 10^x$. Find ℓ and q, the linear and quadratic approximations to f, based at $x = 3$. Use each to estimate $f(3.1)$.

Check our work.

Solution: We'll need two derivatives and a value for f, all at $x = 3$. Calculation gives $f(x) = 10^x$, $f'(x) = 10^x \ln 10$, and $f''(x) = 10^x (\ln 10)^2$, so ◀

$$f(3) = 10^3 = 1000; \quad f'(3) = 1000 \ln 10 \approx 2302.5851;$$
$$f''(3) = 1000(\ln 10)^2 \approx 5309.8981.$$

The definitions say:

$$\ell(x) = f(3) + f'(3) \cdot (x - 3) \approx 1000 + 2302.5851(x - 3);$$
$$q(x) = \ell(x) + \frac{f''(3)}{2} \cdot (x - 3)^2 \approx 1000 + 2302.5851(x - 3) + 2650.9491(x - 3)^2.$$

By calculator!

Numerical calculation ◀ shows:

$$\ell(3.1) \approx 1230.2585; \quad q(3.1) \approx 1256.7680; \quad f(3.1) \approx 1258.9254.$$

The graphs below, like the numbers above, show how much better q does than ℓ in "fitting" f:

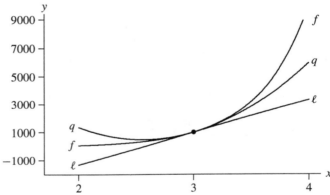

Linear and quadratic approximations to $f(x) = 10^x$

The picture shows why ℓ fits f so poorly: f bends sharply away from ℓ, even quite near $x = 3$. The second derivative, f'', explains why. Any *linear* function ℓ has *zero* second derivative. Our target function f, by contrast, has a *huge* second derivative: $f''(3) \approx 5000$. Because f is so drastically non-linear, it is badly approximated by any straight line—even the best-fitting one.   □

Taylor polynomials

Linear and quadratic approximations ℓ and q are built to "match" f as well as possible at x_0: more precisely, ℓ and f have the same value and *first* derivative at x_0; q and f have the same value and first *two* derivatives at x_0.

An important example.

Still better "agreement" is possible if polynomials of degree 3, 4, and higher are used to approximate f. An example ◀ will illustrate the idea.

Made to order. The constant, linear, and quadratic approximations are sometimes called **zeroth-order**, **first-order**, and **second-order** approximations to f near x_0. In each case **order** refers to the number of derivatives of f that a particular approximation "agrees" with at x_0.

■ **Example 4.**   Find a fifth-degree➤ polynomial p_5 that agrees with $f(x) = e^x$ to   *A "quintic."*
order 5 at $x_0 = 0$, i.e., such that

$$p_5(0) = f(0) \qquad p_5'(0) = f'(0) \qquad p_5''(0) = f''(0)$$
$$p_5^{(3)}(0) = f^{(3)}(0) \quad p_5^{(4)}(0) = f^{(4)}(0) \quad p_5^{(5)}(0) = f^{(5)}(0)$$

Solution: We'll need values for several derivatives of f, all at $x = 0$. That
$f(x) = e^x$ makes calculating derivatives very, very easy:➤   *That's why we chose this f!*

$$f(x) = e^x \implies f(0) = f'(0) = f''(0) = f^{(3)}(0) = f^{(4)}(0) = f^{(5)}(0) = 1.$$

We want a polynomial➤ p_5 with the *same* derivatives at $x = 0$. To begin, we write   *Recall that e^x itself is not a*
$p_5(x)$ in standard form: *polynomial.*

$$p_5(x) = a_0 + a_1 x + a_2 x^2 + a_3 x^3 + a_4 x^4 + a_5 x^5;$$

the coefficients a_i are still to be found.➤   *For each i, a_i is the coefficient of the*
                                                                                              ith power x^i.
    Finding a_0 through a_5 is surprisingly easy. The key idea is that the a_i are closely
related to *derivatives* of p_5 at $x = 0$. Specifically:➤   *Check these easy but important*
                                                                                              calculations.

$$
\begin{aligned}
p_5(x) &= a_0 + a_1 x + a_2 x^2 + a_3 x^3 + a_4 x^4 + a_5 x^5 &\implies& \quad p_5(0) = a_0 \\
p_5'(x) &= a_1 + 2a_2 x + 3a_3 x^2 + 4a_4 x^3 + 5a_5 x^4 &\implies& \quad p_5'(0) = a_1 \\
p_5''(x) &= 2a_2 + 3\cdot 2a_3 x + 4\cdot 3a_4 x^2 + 5\cdot 4a_5 x^3 &\implies& \quad p_5''(0) = 2a_2 \\
p_5'''(x) &= 3\cdot 2a_3 + 4\cdot 3\cdot 2a_4 x + 5\cdot 4\cdot 3a_5 x^2 &\implies& \quad p_5'''(0) = 6a_3 \\
p_5^{(4)}(x) &= 4\cdot 3\cdot 2a_4 + 5\cdot 4\cdot 3\cdot 2a_5 x &\implies& \quad p_5^{(4)}(0) = 24a_4 \\
p_5^{(5)}(x) &= 5\cdot 4\cdot 3\cdot 2a_5 &\implies& \quad p_5^{(5)}(0) = 120a_5.
\end{aligned}
$$

The pattern of coefficients should now be clear:➤ *For each i, $p_5^{(i)}(0) = i!\, a_i$.* In   *$i!$ denotes the **factorial** of i. See*
other words: *below for more on factorials.*

$$a_0 = p_5(0); \quad a_1 = p_5'(0); \quad a_2 = \frac{p_5''(0)}{2!}; \quad a_3 = \frac{p_5'''(0)}{3!}; \quad \dots \quad a_5 = \frac{p_5^{(5)}(0)}{5!}.$$

Now we can find p_5. By design,

$$p_5(0) = f(0) = 1; \quad p_5'(0) = f'(0) = 1; \quad p_5''(0) = f''(0) = 1; \quad \dots \quad p_5^{(5)}(0) = f^{(5)}(0) = 1.$$

Thus

$$a_0 = 1; \quad a_1 = 1; \quad a_2 = \frac{1}{2!}; \quad a_3 = \frac{1}{3!}; \quad a_4 = \frac{1}{4!}; \quad a_5 = \frac{1}{5!}.$$

Our search is over:

$$p_5(x) = \frac{1}{0!} + \frac{x}{1!} + \frac{x^2}{2!} + \frac{x^3}{3!} + \frac{x^4}{4!} + \frac{x^5}{5!} = 1 + x + \frac{x^2}{2} + \frac{x^3}{6} + \frac{x^4}{24} + \frac{x^5}{120}.$$

How well does p_5 approximate f near $x = 0$? Very well indeed:

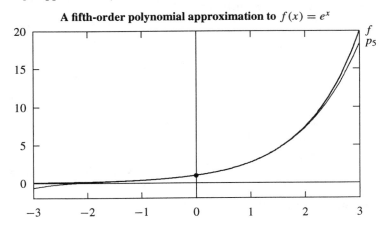

A fifth-order polynomial approximation to $f(x) = e^x$

A calculator helps.

The graphs of f and p_5 are almost identical near $x = 0$. Compare, for instance, $f(1)$ and $p_5(1)$. Easy calculations show that

$$p_5(1) = 1 + 1 + \frac{1}{2} + \frac{1}{6} + \frac{1}{24} + \frac{1}{120} = \frac{163}{60} \approx 2.717; \quad f(1) = e \approx 2.718. \quad \square$$

Factorials

The repeated product $i \cdot (i - 1) \cdot (i - 2) \cdots 2 \cdot 1$, denoted $i\,!$, is called the **factorial** of i. (By convention, $0! = 1$.) Factorials appear often in mathematics, especially when (as above) an operation is performed repeatedly. Factorials are important enough in practice that many calculators have a factorial key.

Does yours? If so, try it. If not, calculate some factorials "from scratch."

Definitions

The polynomial p_5 constructed above is the **fifth-order Taylor polynomial for** $f(x) = e^x$ **based at $x = 0$.** Here is the general definition:

> **Definition:** (**Taylor polynomials.**) Let f be any function whose first n derivatives exist at $x = x_0$. The **Taylor polynomial of order n, based at x_0,** is defined by
>
> $$\begin{aligned} p_n(x) \;=\;& f(x_0) + f'(x_0)(x - x_0) + \frac{f''(x_0)}{2!}(x - x_0)^2 + \\ & \frac{f^{(3)}(x_0)}{3!}(x - x_0)^3 + \cdots + \frac{f^{(n)}(x_0)}{n!}(x - x_0)^n. \end{aligned}$$

Taylor's idea? The idea of Taylor polynomials predates Brook Taylor (1685–1731), a contemporary of Isaac Newton, for whom they are named. According to the author George Simmons, the underlying idea of Taylor polynomials appears in work of John Bernoulli, published in 1694, when Taylor was only 9. The same versatile Bernoulli is also credited with having discovered—behind the scenes—the famous technical lemma known as l'Hôpital's Rule. (We'll see it later in this book.)

■ **Example 5.** Let $f(x) = \sqrt{x}$. Find the first three Taylor polynomials p_1, p_2, and p_3, all based at $x_0 = 64$.

Solution: We'll need the value and first three derivatives of f at $x = 64$:➤

Check the last one; we found the others above.

$$f(64) = 8; \quad f'(64) = \frac{1}{16}; \quad f''(64) = -\frac{1}{2048}; \quad f'''(64) = \frac{3}{262144}.$$

Plugging these values into the definition gives our results:➤

The first two are familiar.

$$\begin{aligned}
p_1(x) &= 8 + \frac{x - 64}{16}; \\
p_2(x) &= 8 + \frac{x - 64}{16} - \frac{(x - 64)^2}{2! \cdot 2048} = 8 + \frac{x - 64}{16} - \frac{(x - 64)^2}{4096}; \\
p_3(x) &= 8 + \frac{x - 64}{16} - \frac{(x - 64)^2}{4096} + \frac{3}{3! \cdot 262144}(x - 64)^3 \\
&= 8 + \frac{x - 64}{16} - \frac{(x - 64)^2}{4096} + \frac{(x - 64)^3}{524288}. \qquad \square
\end{aligned}$$

More on the definition

The Taylor polynomial definition deserves a closer look:

Why powers of $(x - x_0)$? Polynomials usually involve powers of x.➤ Here, though, powers of $(x - x_0)$➤ are more convenient. Above, for instance, using powers of $(x - 64)$ simplified our calculations, most of which involved $x = 64$.

x, x^2, x^3, etc.

$(x - x_0), (x - x_0)^2, (x - x_0)^3$, etc.

Maclaurin polynomials. Taylor polynomials that happen to based at $x = 0$ are known as **Maclaurin polynomials**. Maclaurin polynomials look a little simpler:➤

They're simpler only in appearance.

$$P_n(x) = f(0) + f'(0)x + \frac{f''(0)}{2}x^2 + \cdots + \frac{f^{(n)}(0)}{n!}x^n.$$

In Example 4 we found the **fifth-order Maclaurin polynomial for $f(x) = e^x$.**

Order, degree. A Taylor polynomial P_n of **order** n is chosen so that its value and first n derivatives at $x = x_0$ agree with those of f. The definition shows that P_n also has **degree** n, i.e., it involves powers of $(x - x_0)$ up through the nth. (If, by chance, $a_n = 0$, then P_n has smaller degree.)

Many symbols, just one variable. The definition involves many symbols: a, x, i, x_0, f, and n. It helps to remember, while in this thicket, that x *is the only variable.* In specific cases➤ all the other symbols take *numerical* values.

As in Example 5.

From P_n to P_{n+1}. Let f be a function and P_n the Taylor polynomial of order n based at x_0. The *next* Taylor polynomial, P_{n+1}, has just one more term:

$$P_{n+1}(x) = P_n(x) + \frac{f^{(n+1)}(x_0)}{(n + 1)!}(x - x_0)^{n+1}.$$

Approximating functions with Taylor polynomials

An important use of Taylor polynomials is in approximating other, more complicated functions. For example, the sine function—which has no algebraic formula—is a good candidate for approximation by polynomials.

■ **Example 6.** Find P_1, P_3, P_5, and P_7, the Taylor (aka Maclaurin) polynomials for $f(x) = \sin x$ based at $x = 0$. Plot everything on one set of axes. How closely does each polynomial approximate $\sin 1$?

Check this easy calculation.

Solution: To find P_7 we'll need the value and the first through seventh derivatives of $f(x) = \sin x$ at $x = 0$. Here they are, in order: ◄ 0, 1, 0, −1, 0, 1, 0, −1. It follows, therefore, that

$$P_7(x) = 0 + 1x + 0x^2 - \frac{1}{3!}x^3 + 0x^4 + \frac{1}{5!}x^5 + 0x^6 - \frac{1}{7!}x^7,$$

and hence that

$$P_1(x) = x; \quad P_3(x) = x - \frac{x^3}{6}; \quad P_5(x) = x - \frac{x^3}{6} + \frac{x^5}{120};$$

$$P_7(x) = x - \frac{x^3}{6} + \frac{x^5}{120} - \frac{x^7}{5040}.$$

Here are all five graphs:

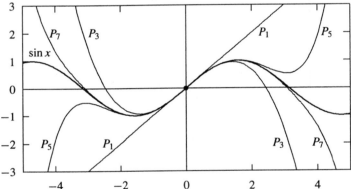

Several Taylor polynomial approximations to $f(x) = \sin x$

The picture shows:

In either direction.

Snug fits near $x = 0$. All four Taylor polynomials "fit" the sine curve well *near* the base point $x = 0$. Differences appear as we move *away* ◄ from $x = 0$.

Better and better. Higher-order Taylor polynomials approximate the sine function *better* than lower-order ones. P_7, for instance, "follows" the sine curve much farther than P_1 or P_3. than

Trace P_3 carefully to see this.

Wiggly is good when approximating a sine curve!

Standard shapes. The graphs of P_1 and P_3 have "standard" shapes for linear and cubic functions. ◄ The wigglier appearance ◄ of P_5 and P_7 reflects their higher degree.

Intuition suggests that P_1, P_3, P_5, and P_7 should give successively closer and closer estimates for $\sin 1 \approx 0.84147$. So they do:

$$P_1(1) = 1; \quad P_3(1) = \frac{5}{6} \approx 0.83333;$$

$$P_5(1) = \frac{101}{120} \approx 0.84167; \quad P_7(1) = \frac{4241}{5040} \approx 0.84147. \qquad \square$$

Bounding the error of linear approximation *(optional)*

An error bound formula

How large, numerically, are the errors committed when Taylor polynomials approximate a function f? We'll address the question only for *linear* approximations; for higher-order approximations the ideas are similar but the calculations are more complicated.

Comparing results from Example 2 and Example 3 shows that the error of a linear approximations depends on two factors:

1. the distance from x to x_0;

2. the size of the second derivative f'' near x_0. (Large values of $|f''|$ means that the f-graph bends *sharply* away from its tangent line; small⬝ values of $|f''|$ mean that the f-graph is nearly straight near x_0, and therefore closely approximated by its tangent line.)

I.e., near 0.

The following theorem restates these ideas quantitatively. Throughout, f is a differentiable function, ℓ is the linear approximation based at $x = x_0$, and I is an interval containing x_0:

> **Theorem 2. (An error bound for linear approximation.)** Suppose that the inequality
>
> $$|f''(x)| \le K$$
>
> holds for all x in I. Then for all x in I,
>
> $$|f(x) - \ell(x)| \le \frac{K}{2}(x - x_0)^2.$$

Note:

The error function. The function $f(x) - \ell(x)$ on the left above is called an **error function**: we'll call it $e(x)$. For a given x, $e(x)$ is the **error** committed by ℓ in approximating f. Ideally, $e(x) \approx 0$.

An error bound formula. The inequality

$$|f(x) - \ell(x)| \le \frac{K}{2}(x - x_0)^2.$$

is (appropriately) called an **error bound formula**; it guarantees that the error function is no larger than the *computable* quantity on the right.

It's OK to overestimate K. *Any* upper bound for $|f''(x)|$ on I can serve as K in the inequality. Choosing K as small as possible "improves" the error bound, but the inequality holds even for larger values of K.⬝

It's safe to overestimate K, but not to underestimate.

Worst-case scenario. The function $f(x) = K(x - x_0)^2/2$ represents the "worst case" that the hypothesis allows. For this f,

$$f(x_0) = 0, \quad f'(x_0) = 0; \quad \text{and} \quad f''(x) = K.$$

In this case the linear approximation is simply $\ell(x) = 0$, and error ℓ commits is as large as the theorem permits:

$$|f(x) - \ell(x)| = f(x) = \frac{K}{2}(x - x_0)^2.$$

Before proving the error bound theorem, we'll show it in action.

Satisfaction guaranteed

The practical value of an error bound formula is in *guaranteeing* the accuracy of an approximation.

■ **Example 7.** We showed above that $\ell(x) = 4 + x/16$ is the linear approximation to $f(x) = \sqrt{x}$ at $x = 64$. What accuracy does the error bound formula guarantee if ℓ approximates f on the interval $[50, 80]$? What does the formula guarantee for the interval $[63, 65]$?

Check ours.

Solution: First we'll need a suitable K. Calculation[◀◀] shows that

$$f''(x) = \frac{d^2}{dx^2}\left(\sqrt{x}\right) = -\frac{1}{4x^{3/2}} \implies |f''(x)| = \frac{1}{4x^{3/2}}.$$

Or a look at the graph of $|f''|$.

$|f''(x)|$ is largest at the left endpoint.

Watch all inequalities carefully.

A moment's thought[◀◀] shows that $|f''(x)|$ *decreases* as x increases. For the interval $[50, 80]$, therefore, we may use $K = |f''(50)| \leq 0.00071$.[◀◀] On $[63, 65]$, $K = |f''(63)| \approx 0.0005$ works.

All ingredients are ready. The theorem says:[◀◀]

- For x in $[50, 80]$ the largest possible error is no more than

$$|\sqrt{x} - \ell(x)| \leq \frac{0.0007}{2}(x - 64)^2 \leq \frac{0.0007}{2}(80 - 64)^2 \approx 0.09.$$

(We used $x = 80$, not $x = 50$, to make $(x - 64)^2$ as large as possible.) In other words: On $[50, 80]$, ℓ approximates f to about *one decimal place accuracy*.

- For x in $[63, 65]$ the largest possible error is even less:

$$|\sqrt{x} - \ell(x)| \leq \frac{0.0005}{2}(x - 64)^2 \leq \frac{0.0005}{2}(65 - 64)^2 \approx 0.00025.$$

In other words: On $[63, 65]$, ℓ approximates f to about *three decimal place accuracy*.

The table on page 377 shows that these error estimates are conservative: ℓ and q behave a little "better" than the theorem predicts. □

A proof of the error bound formula

We'll take f, ℓ, and K to be as in the theorem. From these data we'll construct two new functions—an **error function** e and and a **bounding function** b, defined as follows:

$$e(x) = f(x) - \ell(x); \qquad b(x) = K\frac{(x - x_0)^2}{2}.$$

The same proof, with minor variations, shows that the inequality holds for $x < x_0$ and in absolute value.

The theorem claims—and we need to show—that for x in I, $|e(x)| \leq b(x)$. To simplify things slightly we'll show only that $e(x) \leq b(x)$ for $x \geq x_0$.[◀◀]

To our aid come three key properties of e and b:

$(4.3.2)\quad e(x_0) = 0 = b(x_0); \quad e'(x_0) = 0 = b'(x_0); \quad |e''(x)| \leq K = b''(x).$

(All three properties of e follow from the "construction" of ℓ: by design, $\ell(x_0) = f(x_0)$, $\ell'(x_0) = f'(x_0)$, and $e''(x) = f''(x) - l''(x) = f''(x)$. All three properties of b come from direct calculations.)[◀◀]

Make them!

We met it first in Section 2.1.

To finish the proof we'll use the "racetrack principle."[◀◀] Recall what it says about *any* functions g and h:

If $g(x_0) = h(x_0)$ and $g'(x) \leq h'(x)$ for $x \geq x_0$, then $g(x) \leq h(x)$ for $x \geq x_0$.

Let's apply the racetrack principle to e and b. As we saw above, $e(x_0) = 0 = b(x_0)$. We'll be done, says the racetrack principle, if we can somehow show that $e'(x) \leq b'(x)$ for $x \geq x_0$.

To do so we apply the racetrack principle *again*, but this time to e' and b', and *their* derivatives e'' and b''. Equations 4.3.2 contain everything we need: $e'(0) = b'(0)$ and $e''(x) \leq b''(x)$. Thus the racetrack principle says that $e'(x) \leq b'(x)$ for $x \geq x_0$—exactly what we need to finish the proof. $\square$

Exercises

1. In each part below, first find the indicated Taylor polynomial P_n of order n for the function f with base point x_0. Then plot both f and P_n on the same axes. Choose your plotting window appropriately to show clearly the relationship between f and P_n.

   (a) $f(x) = \dfrac{1}{1-x}, n = 3, x_0 = 0.$

   (b) $\sin x + \cos x, n = 4, x_0 = 0.$

   (c) $f(x) = \ln x, n = 3, x_0 = 1,$

   (d) $f(x) = \tan x, n = 2, x_0 = 0,$

   (e) $\sqrt{x}, n = 3, x_0 = 4.$

2. This problem refers to Example 6, p. 382—especially the graphs. Throughout this problem, $f(x) = \sin x$, and $P_n(x)$ is the Taylor (Maclaurin) polynomial of order n based at $x = 0$.

   (a) Every function mentioned in Example 6—f, P_1, P_3, P_5, and P_7—is odd. How is this fact "reflected" in their graphs?

   (b) Find the *even-order* Maclaurin polynomials P_2, P_4, P_6, and P_8 for $f(x) = \sin x$. How are they related to P_1, P_3, P_5, and P_7?

3. Throughout this problem, $g(x) = \cos x$ and $P_n(x)$ is the Taylor (Maclaurin) polynomial of order n based at $x = 0$. For use below, here are graphs of g, P_0, P_2, P_4, and P_6.

Several Taylor polynomial approximations to $f(x) = \cos x$

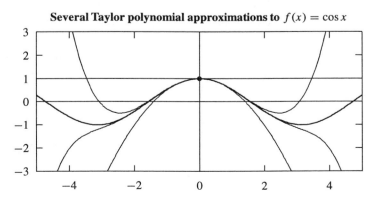

(a) Find formulas for P_0 through P_6, the Maclaurin polynomials through order 6 for g based at $x = 0$. Use your results to label the graphs shown above.

(b) Is g odd, even, or neither? What about the Maclaurin polynomials for g found in part (a)? How do the graphs "reflect" the situation?

(c) In Example 6 we found the Maclaurin polynomials P_1, P_3, P_5, and P_7 for $f(x) = \sin x$. Find the *derivatives* $P_1'(x)$, $P_3'(x)$, $P_5'(x)$, $P_7'(x)$, and $f'(x)$ of these functions. How are the results related to the rest of this problem?

4. In Example 2 we worked with the quadratic function $q(x) = a + b(x - 64) + c(x - 64)^2$. Here are more details:

(a) We claimed that $q'(x) = b + 2c(x - 64)$ and that $q''(x) = 2c$. Verify these claims by differentiation.

(b) Rewrite q in powers of x, i.e., in the form $q(x) = A + Bx + Cx^2$.

(c) Rewrite $q'(x) = b + 2c(x - 64)$ in powers of x. Does your answer agree as it should with the previous part?

5. Estimate the value of each of the following expressions using a linear approximation. Then compute the difference between your estimate and the value given by a scientific calculator.

(a) $\sqrt{103}$ (c) $\tan 31°$

(b) $\sqrt[3]{29}$ (d) 0.8^{10}

6. For each part of the previous exercise, compare the "actual" approximation error with the theoretical error bound.

7. Repeat the previous exercise using a quadratic approximation.

8. The line tangent to the curve $y = g(x)$ at the point $(3, 5)$ intersects the y-axis at the point $(0, 10)$.

(a) What is $g'(3)$?

(b) Estimate $g(2.95)$ using a linear approximation.

9. Suppose that f is a function such that $f'(x) = \sin\left(x^2\right)$ and $f(1) = 0$. [NOTE: No explicit formula for f is given.]

(a) Estimate the value of $f(0.5)$ using a linear approximation.

(b) Is the estimate you computed in part (a) greater than or less than the exact value of $f(0.5)$? Explain.

(c) Estimate the value of $f(0.5)$ using a quadratic approximation.

10. The linear approximation $(1 + x)^k \approx 1 + kx$ is often useful for "back of the envelope" computations. Find the error made in using this expression to approximate each of the following expressions.

(a) $(1.03)^2$ (c) 1.001^{-1}

(b) $\sqrt{1.0404}$

11. Find the linear and quadratic approximations of each of the following functions at $x_0 = 0$.

(a) $f(x) = \sin x$

(b) $f(x) = \cos x$

(c) $f(x) = \tan x$

(d) $f(x) = e^x$

(e) $f(x) = \tan^{-1} x$

(f) $f(x) = \sin^{-1} x$

12. (a) For each function in the previous problem, find an interval over which the linear approximation makes an error no greater than 0.01.

(b) For each function in the previous problem, find an interval over which the quadratic approximation makes an error no greater than 0.01.

13. Let g be a well-behaved function defined on the interval $[0, 10]$ such that the graph of g passes through the point $(5, 2)$ and the derivative of g is the function sketched below.

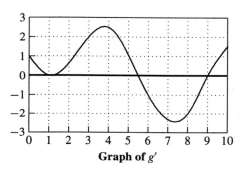

Graph of g'

(a) Estimate $g(0)$ using a linear approximation.

(b) Estimate $g(8)$ using a quadratic approximation.

14. Suppose that h is a well-behaved function such that $h(2) = 3$, $h'(2) = -2$, and $-2 \le h''(x) \le 1$ for all $x \in (0, 4)$. Show that $0 \le h(3) \le 2$.

4.8   Parametric equations, parametric curves

A point P wanders about the xy-plane, tracing its path as it goes. As P travels, its coordinates $x = f(t)$ and $y = g(t)$ are *functions of time*. In this situation t is called a **parameter**,◄ f and g are **coordinate functions**, and the figure traced out by P is a **parametric curve**.◄

Here the "parameter" is a variable—in other situations "parameters" are often constants.

Our first example illustrates the idea and some specialized vocabulary.

Most plotting programs and graphing calculators handle parametric curves. Try yours.

■ **Example 1.**   At any time t with $0 \le t \le 10$, the coordinates of P are given by the **parametric equations**

$$x = t - 2\sin t; \qquad y = 2 - 2\cos t.$$

What curve does P trace out? Where is P at $t = 1$? In which direction is P moving?

But most tedious.

I.e., the $y = f(x)$ case.

Check several entries.

Solution: The simplest◄ way to draw a curve is to calculate many points (x, y), plot each one, and "connect the dots." The first step works as usual:◄ for many inputs t, calculate the corresponding x and y. Here's a sampler of results,◄ rounded to two decimals:

Parametric plot points for $x = t - 2\sin t$, $y = 2 - 2\cos t$												
t	0	0.1	0.2	0.3	0.7	0.8	0.9	1.0	...	9.8	9.9	10
x	0	−0.10	−0.20	−0.29	−0.59	−0.63	−0.67	−0.68	...	10.53	10.82	11.09
y	0	0.01	0.04	0.09	0.47	0.61	0.76	0.92	...	3.86	3.78	3.68

Here's the result, a curve C:

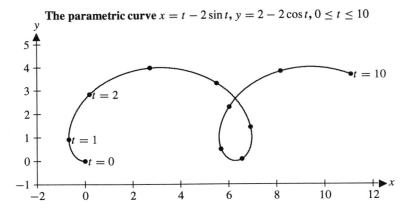

The parametric curve $x = t - 2\sin t$, $y = 2 - 2\cos t$, $0 \le t \le 10$

Notice:

• Points corresponding to *integer* values of t are shown bulleted. At $t = 1$, P has coordinates $(-0.68, 0.92)$; at this instant, P is heading almost due north.

$x = 6$, for instance.

• The full curve C is *not the graph of a function* $y = f(x)$—some x-values correspond to more than one y-value.◄ Certain *pieces* of C, however, *do* define such functions. (The part of C from $t = 2$ to $t = 5$ does so.)

• The picture shows the x- and y-axes, *but no t- axis*: t-values are indicated only by the bulleted points. In most parametric curves t-values don't appear at all.◄

Most graphing calculators don't show t-values graphically, but some calculate t-values numerically.

- The bullets on the graph appear at equal *time*➤ intervals, but not at equal *distances* from each other, because P speeds up and slows down as it moves.➤ We'll see below how to *calculate* the speed of a parametric curve at a point.

In this example, t represents time.

When is P moving fastest? Slowest?

- If t measures time we can visualize C *dynamically*, as a curve traced by a moving point. Curves defined by ordinary equations in x and y are *static* objects, lying passively on the page.

A sampler of parametric curves

Parametric curves come in mind-boggling variety. *Any* choice of two equations $x = f(t)$ and $y = g(t)$ and a t-interval produces one. Surprisingly often the result is beautiful, useful, or interesting.➤ The next several examples hint at some of the possibilities, and at connections between parametric curves and ordinary function graphs.

Or all three.

■ **Example 2.   Every ordinary function graph can be written in parametric form.** To produce a sine curve, for example, we can use

$$x = t; \qquad y = \sin t; \qquad -2\pi \le t \le 2\pi.$$

Here's the graph; integer multiples of $\pi/2$ are bulleted:

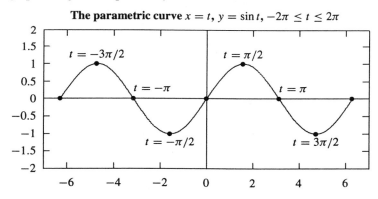

The parametric curve $x = t$, $y = \sin t$, $-2\pi \le t \le 2\pi$

■ **Example 3.   Parametric curves may have loops, cusps, vertical tangents, and other peculiar features.** Consider the curve

$$x = 2\cos(t) + 2\cos(4t); \qquad y = \sin t + \sin(4t); \qquad 0 \le t \le 5.$$

Here it is; bullets show the starting and ending points.

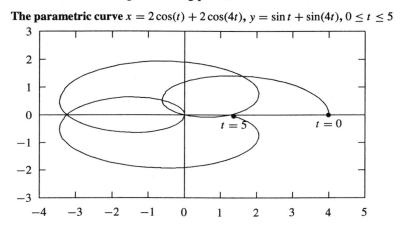

The parametric curve $x = 2\cos(t) + 2\cos(4t)$, $y = \sin t + \sin(4t)$, $0 \le t \le 5$

The previous example has this property.

■ **Example 4.   The unit circle.** If a parametric curve's coordinate functions are periodic (i.e., repeat themselves), then the curve's shape reflects this fact.➤ The simplest and most important such curve is the **unit circle**, often written in xy-form as $x^2 + y^2 = 1$. The simplest *parametric* description of the unit circle is this:

$$x = \cos t; \qquad y = \sin t; \qquad 0 \le t \le 2\pi.$$

Here's the picture; integer multiples of $\pi/2$ are bulleted:

The parametric curve $x = \cos t, y = \sin t, 0 \le t \le 2\pi$

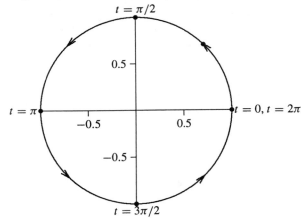

Notice:

- The parameter t *can* be thought of, as usual, as time. Alternatively, t can be thought of as the *radian measure* of the angle determined by the x-axis and the line from the origin to P. Then for any *angle* t, $(x, y) = (\cos t, \sin t)$ is the point on the unit circle lying t radians counterclockwise from $(1, 0)$.

- Because $x = \cos t$ and $y = \sin t$,

$$x^2 + y^2 = (\cos t)^2 + (\sin t)^2 = 1.$$

  This shows that every point on our parametric curve satisfies the Cartesian equation $x^2 + y^2 = 1$, and so—as we intended—lies on the unit circle.

  "Reducing" two parametric equations to xy-form, as we just did, is called **eliminating the parameter**.   (Doing so is sometimes possible, sometimes not.)

- The circle is traced *once* as t runs from $t = 0$ to $t = 2\pi$. With a larger t-interval the same circle would be traced *repeatedly*; with a smaller t-interval, the circle would be traced only partially.

■ **Example 5.   Other circles.** The idea in the previous example extends to *all* circles. If (a, b) is any point in the plane, and $r > 0$ any radius, then the parametric equations

$$x = a + r \cos t; \qquad y = b + r \sin t; \qquad 0 \le t \le 2\pi$$

produce the circle with center (a, b) and radius r. (See the exercises for more details.)

■ **Example 6. Other curves with periodic coordinate functions.** Many curves defined by periodic➤ coordinate functions have striking, beautiful shapes. A typical **Lissajou curve**, for instance, is defined by

I.e., repeating.

$$x = \sin(5t); \qquad y = \sin(6t); \qquad 0 \le t \le 2\pi.$$

(Replacing 5 and 6 above with other integers produces other Lissajou figures.) Here's ours:

The Lissajou curve $x = \sin(5t)$, $y = \sin(6t)$, $0 \le t \le 2\pi$

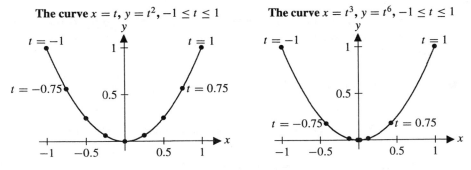

■ **Example 7. Same curve, different parametric equations.** Different pairs of parametric equations may produce exactly the same curve in the xy-plane. In such cases, labeling t-values can make differences appear. For example, setting

$$x = t; \qquad y = t^2; \qquad -1 \le t \le 1$$

produces a parabolic arc. So does

$$x = t^3; \qquad y = t^6; \qquad -1 \le t \le 1.$$

Here are both; the bullets mark 0.25-second time intervals:

The curves are geometrically identical; they differ only in how they are traced out.

Calculus with parametric curves: speed and slope

Suppose that a parametric curve C has differentiable coordinate functions $x = f(t)$ and $y = g(t)$. What do the *derivatives* f' and g' say about the situation?

Speed

As the bullets on some graphs above show, speed varies with time.

If t tells time, then $(f(t), g(t))$ gives P's *position* in the xy-plane at time t, and it makes sense to consider the **speed** of P at time t. Since the derivatives f' and g' tell, respectively, how fast x and y increase, their appearance below is no surprise:

> **Definition:** Suppose that the position of a point P at time t, $a \le t \le b$, is given by differentiable coordinate functions $x = f(t)$ and $y = g(t)$. Then
>
> $$\text{speed of } P \text{ at time } t = \sqrt{f'(t)^2 + g'(t)^2}.$$

Making sense. Is the definition "right"? Definitions aren't subject to proof, but they *should* appeal to common sense. Does this one?

Speed, "properly" defined, should tell how fast the distance traveled by P increases with respect to time. Showing carefully that the definition above actually does so involves a rather subtle definition of **arclength**, i.e., distance measured along a curve. We won't define arclength here, but the next example shows that the definition makes good sense for *linear* coordinate functions.

■ **Example 8.** At time t seconds, a point P has coordinates $x = at + b$, $y = ct + d$; a, b, c, and d are constants. How fast is P moving at time t_0?

Solution: The definition says that the speed of P at *any* time t is constant:

$$\sqrt{f'(t)^2 + g'(t)^2} = \sqrt{a^2 + c^2}.$$

Let's see why this result makes sense.

See the exercises for more on this point.

Notice first that because P has *linear* coordinate functions, P *moves along a straight line.* When $t = t_0$, P has coordinates $(x_0, y_0) = (at_0 + b, ct_0 + d)$; by time t_1, P has moved to $(at_1 + b, ct_1 + d)$—a distance of

Check our algebra.

$$\sqrt{(at_1 + b - at_0 - b)^2 + (ct_1 + d - ct_0 - d)^2} = (t_1 - t_0)\sqrt{a^2 + c^2}$$

units. Therefore from $t = t_0$ to $t = t_1$, P has

$$\text{average speed} = \frac{\text{distance}}{\text{time}} = \frac{(t_1 - t_0)\sqrt{a^2 + c^2}}{t_1 - t_0} = \sqrt{a^2 + c^2}.$$

This shows that P has the same average speed—$\sqrt{a^2 + c^2}$—over *any* time interval. Thus the *instantaneous* speed of P at any time t_0 is also $\sqrt{a^2 + c^2}$, as the definition says.   □

■ **Example 9.** Consider again the parametric curve

$$x = f(t) = t - 2\sin t; \qquad y = g(t) = 2 - 2\cos t$$

Look first at the picture; try to guess reasonable answers.

of Example 1, page 388. Find the speed at $t = 3$. When does P move *fastest*?

Solution: The definition says that at any time t,

$$\text{speed} = \sqrt{(f'(t))^2 + (g'(t))^2} = \sqrt{(1 - 2\cos t)^2 + (2\sin t)^2} = \sqrt{5 - 4\cos(t)}.$$

Check for yourself.

(The last step uses a trigonometric identity.)

The rest is easy. At $t = 3$ the speed is $\sqrt{5 - 4\cos 3} \approx 2.99$. The speed formula also shows—and the picture agrees—that the speed $\sqrt{5 - 4\cos t}$ is greatest when $\cos t = -1$, e.g., when $t = \pi$.   □

Slopes on parametric curves

On an ordinary, $y = f(x)$-style curve, $f'(x)$ gives the slope at (x, y). Not surprisingly, slopes on parametric curves also involve derivatives.

Not all functions have derivatives; not all parametric curves have slopes. To avoid needless complications we'll work only with **smooth** parametric curves, defined as follows:

Definition: The parametric curve C defined by

$$x = f(t); \quad y = g(t); \quad a \le t \le b$$

is **smooth** if f' and g' are continuous functions of t, and f' and g' are not simultaneously zero.

(The second requirement says that the moving point P *never stops*.)$^{\blacktriangleright\blacktriangleright}$ For smooth curves the next theorem tells how to find the slope at a point.

See the exercises for more on this idea.

Theorem 3.   Let the smooth parametric curve C be defined as above. If $f'(t) \ne 0$, then the slope dy/dx at the point $(x, y) = (f(t), g(t))$ is given by

$$\frac{dy}{dx} = \frac{g'(t)}{f'(t)} = \frac{dy/dt}{dx/dt}.$$

Before proving the theorem, let's use it.

■ **Example 10.**   Consider the parametric curve

$$x = f(t) = t - 2\sin t; \qquad y = g(t) = 2 - 2\cos t$$

of Example 1, page 388. Find the slope at $t = 1$. Where is the curve horizontal? Where is it vertical?$^{\blacktriangleright\blacktriangleright}$

Guess first, by looking at the picture.

Solution: The theorem says that

$$\text{slope at } t = \frac{g'(t)}{f'(t)} = \frac{2\sin t}{1 - 2\cos t},$$

unless $f'(t) = 0$.$^{\blacktriangleright\blacktriangleright}$ Setting $t = 1$ gives

Where the denominator is zero.

$$\text{slope at time } 1 = \frac{g'(1)}{f'(1)} = \frac{2\sin 1}{1 - 2\cos 1} \approx -20.88.$$

The picture agrees: at $(f(1), g(1)) \approx (-0.68, 0.92)$, the curve has large negative slope.

The general slope formula

$$\frac{dy}{dx} = \frac{2\sin t}{1 - 2\cos t}$$

shows that the curve is *horizontal* only when the numerator is zero, i.e., when t is an integer multiple of π. Again the picture agrees. At $t = \pi$, for instance, $(x, y) = (\pi, 4)$. At this point the curve appears to be horizontal.

The curve C can be *vertical* only where the denominator above is zero,$^{\blacktriangleright\blacktriangleright}$ i.e., when $\cos t = 1/2$. One such value is $t = \pi/3 \approx 1.05$; again the picture agrees.   □

Otherwise, the slope of C is finite.

Proof of the theorem

Remember—we're assuming that f is smooth.

Suppose that $f'(t_0) \neq 0$; then either $f'(t_0) > 0$ or $f'(t_0) < 0$. If $f'(t_0) > 0$, then continuity of f'★ implies that $f'(t) > 0$ for t near t_0, and therefore that $x = f(t)$ is *increasing* near $t = t_0$. If $f'(t) < 0$ then (by the same reasoning) $x = f(t)$ is *decreasing* near $t = t_0$. In either case it follows that if t is near t_0 but $t \neq t_0$, then

Denominators don't vanish unexpectedly.

$f(t) \neq f(t_0)$. This assures, in turn, that all of the following limit computations makes sense.★ Notice first that

$$\text{slope} = \left.\frac{dy}{dx}\right|_{t=t_0} = \lim_{t \to t_0} \frac{y - y_0}{x - x_0} = \lim_{t \to t_0} \frac{g(t) - g(t_0)}{f(t) - f(t_0)}.$$

A common sort of trick in proving calculus theorems.

Now for a little trick.★ We'll divide the numerator and denominator above by $(t - t_0)$, take limits of everything in sight, and see what happens. Here goes:

$$\left.\frac{dy}{dx}\right|_{t=t_0} = \lim_{t \to t_0} \frac{\frac{g(t) - g(t_0)}{t - t_0}}{\frac{f(t) - f(t_0)}{t - t_0}} = \frac{\lim_{t \to t_0} \frac{g(t) - g(t_0)}{t - t_0}}{\lim_{t \to t_0} \frac{f(t) - f(t_0)}{t - t_0}} = \frac{g'(t_0)}{f'(t_0)},$$

which is what we wanted to show. (The second last step is OK because, by assumption, the limit in the denominator is not zero.) □

Modeling with parametric equations

Parametric equations often help model physical processes that involve quantities varying in both time and space. Describing **trajectories**—paths taken by projectiles—is one such setting.★

"Projectiles" could be baseballs, BB's, ballistic missiles, etc.

■ **Example 11.** A major league fastball leaves the pitcher's hand traveling horizontally, with initial speed 150 feet per second and initial height 7 feet. Ignoring wind resistance,★ describe the ball's trajectory. When, and at what speed, does the ball cross the plate, 60.5 feet away? Is the pitch a strike? If the batter, catcher, and umpire all miss the ball, where will it land?

Major league batters don't ignore wind resistance. See the exercises for a more realistic example.

Solution: We'll use an xy-coordinate system with origin at the pitcher's feet and home plate at $(60.5, 0)$; the ball starts at $(0, 7)$.

Ignoring wind resistance means that the ball, once released, is influenced only by gravity. Thus the ball's horizontal acceleration★ is zero; its vertical acceleration, due to gravity, is -32 feet per second per second.★

Recall: acceleration is the second derivative of position.

Why is the vertical acceleration negative?

If $x = f(t)$, $y = g(t)$, and t measures time in seconds since the ball's release, then our information boils down to this:

$$f''(t) = 0; \quad f'(0) = 150; \quad f(0) = 0; \quad g''(t) = -32; \quad g'(0) = 0; \quad g(0) = 7.$$

These data lead to simple formulas for f and g. The argument for f goes like this:

$$f''(t) = 0 \implies f'(t) = a \quad \text{and} \quad f(t) = at + b,$$

where a and b are constants, as yet unknown. Finding them is easy: combining the facts that $f(t) = at + b$, $f(0) = 0$, and $f'(0) = 150$ gives $a = 150$ and $b = 0$, so $f(t) = 150t$. A similar argument applies to g—here's the result:

$$x = f(t) = 150t; \qquad y = g(t) = 7 - 16t^2.$$

These equations make sense, of course, only while the ball remains airborne, i.e., until $y = 7 - 16t^2 = 0$, or $t = \sqrt{7}/4 \approx 0.661$ seconds. Here's the parametric curve for $0 \le t \le 0.661$:

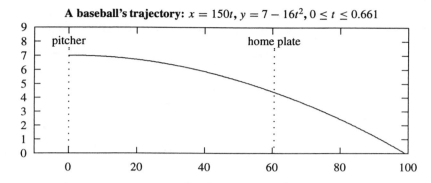

A baseball's trajectory: $x = 150t$, $y = 7 - 16t^2$, $0 \le t \le 0.661$

Notice the *parabolic* shape: in fact, *every* free-falling object either moves vertically or follows a parabolic trajectory.➤➤

See the exercises for more on parabolic trajectories.

Using the picture and the formulas we can answer our original questions. The ball crosses the plate when $x = f(t) = 150t = 60.5$, i.e., when $t = 121/300 \approx 0.4033$ seconds. At this time,➤➤

Check the arithmetic.

$$\text{speed} = \sqrt{f'(0.4033)^2 + g'(0.4033)^2} \approx 150.55$$

feet per second. At the same time, the ball's height is $y = g(0.4033) \approx 4.4$ feet—high but in the strike zone for an average batter. Left untouched, the ball hits the ground at $t = 0.661$; at this time, $x = f(t) = 150 \cdot 0.661 \approx 99.2$ feet—almost 40 feet behind home plate. □

Exercises

1. Plot each parametric curve below. (Using a machine is fine, but then draw or copy your own curve on paper.) In each case, mark the direction of travel and label the points corresponding to $t = -1$, $t = 0$, and $t = 1$.

   (a) $x = t$, $y = \sqrt{1 - t^2}$, $-1 \le t \le 1$

   (b) $x = t$, $y = -\sqrt{1 - t^2}$, $-1 \le t \le 1$

   (c) $x = \sqrt{1 - t^2}$, $y = t$, $-1 \le t \le 1$

   (d) $x = -\sqrt{1 - t^2}$, $y = t$, $-1 \le t \le 1$

   (e) $x = \sin(\pi t)$, $y = \cos(\pi t)$, $-1 \le t \le 1$

2. Each parametric "curve" below is actually a line segment. In each part, state the beginning point ($t = 0$) and ending point ($t = 1$) of the segment. Then state an equation in x and y for the line each segment determines.

   (a) $x = 2 + 3t$, $y = 1 + 2t$, $0 \le t \le 1$

   (b) $x = 2 + 3(1 - t)$, $y = 1 + 2(1 - t)$, $0 \le t \le 1$

   (c) $x = t$, $y = mt + b$, $0 \le t \le 1$

   (d) $x = a + bt$, $y = c + dt$, $0 \le t \le 1$

   (e) $x = x_0 + (x_1 - x_0)t$, $y = y_0 + (y_1 - y_0)t$, $0 \le t \le 1$

3. A parametric curve $x = f(t)$, $y = g(t)$, $a \leq t \leq b$ has **constant speed** if—what else?—its speed is constant in t. Which of the curves in Problem 1 above have constant speed?

4. Show that if a curve C has *linear* coordinate functions $x = at + b$ and $y = ct + d$, then C has constant speed. (See the previous problem.)

5. Consider the curve C shown in Example 1; suppose that t tells time in seconds.

   (a) At which bulleted points would you expect P to be moving quickly? Slowly? Why?

   (b) Use the curve to *estimate* the speed of P at $t = 3$. (One approach: Estimate how far P travels—i.e., the length of the curve—over the one-second interval from $t = 2.5$ to $t = 3.5$. If d is this distance, then d distance units per second is a reasonable speed estimate at $t = 3$.)

   (c) Use the curve to estimate the speed of P at $t = 6$.

6. Plot the parametric curve

   $$x = t^3; \qquad y = \sin t^3; \qquad -2 \leq t \leq 2.$$

   What familiar curve is produced? Why does the result happen?

7. Let (a, b) be any point in the plane, and $r > 0$ any positive number. Consider the parametric equations

   $$x = a + r \cos t; \qquad y = b + r \sin t; \qquad 0 \leq t \leq 2\pi.$$

   (a) Plot the parametric curve defined above for $(a, b) = (2, 1)$ and $r = 2$. Describe your result in words.

   (b) Show by calculation that if x and y are as above, then $(x-a)^2 + (y-b)^2 = r^2$. Conclude that the curve defined above is the circle with center (a, b) and radius r.

   (c) Write parametric equations for the circle of radius $\sqrt{13}$, centered at $(2, 3)$.

   (d) What "curve" results from the equations above if $r = 0$?

8. Let a and b be any positive numbers, and let a parametric curve C be defined by

   $$x = a \cos t; \qquad y = b \sin t; \qquad 0 \leq t \leq 2\pi.$$

   The resulting curve is called an **ellipse**.

   (a) Plot the curve defined above for $a = 2$ and $b = 1$. Describe C in words. Where is the "center" of C? Why do you think the quantities $2a$ and $2b$ are called the **major and minor axes** of C.

   (b) What curve results if $0 \leq t \leq 4\pi$? Why?

   (c) Write parametric equations for an ellipse with major axis 10 and minor axis 6.

   (d) Write parametric equations for *another* ellipse with major axis 10 and minor axis 6.

   (e) Show that for all t, $\dfrac{x^2}{a^2} + \dfrac{y^2}{b^2} = 1$.

   (f) How does the "ellipse" look if $a = b$? How does its xy equation look?

   (g) How does an ellipse look if $a = 1000$ and $b = 1$?

   (h) How does an ellipse look if $a = 1$ and $b = 1000$?

9. Consider the Lissajou curve in Example 6.

   (a) Label the points corresponding to $t = 0$, $t = 0.1$, and $t = \pi/2$. Add some arrows to the curve to indicate direction.

   (b) How often in the interval $0 \le t \le 2\pi$ does the tracing point P return to $(0, 0)$?

   (c) How would the picture be different if the t-interval $0 \le t \le 4\pi$ were used?

10. We stated in this section that if a parametric curve C has *linear* coefficient functions $x = f(t) = at + b$ and $y = g(t) = ct + b$, then C is a straight line (or part of a line). This problem explores that fact.

   (a) Plot the parametric curve $x = 2t$, $y = 3t + 4$, $0 \le t \le 1$. Where does the curve start? Where does it end? What is its shape?

   (b) Eliminate the variable t in the two equations above to find a single tion in x and y for the line of the previous part. e that the parametric curve is only part of this line.)

   (c) Find the slope at $t = t_0$ of the parametric curve $x = 2t$, $y = 3t + 4$, $0 \le t \le 1$. Why doesn't the answer depend on t?

11. This problem pursues the idea that if a parametric curve C has *linear* coefficient functions $x = f(t) = at + b$ and $y = g(t) = ct + d$, then C is a straight line (or part of a line).

   (a) Consider the parametric curve $x = at + b$, $y = ct + d$, $t_0 \le t \le t_1$. Where does the curve start? Where does it end?

   (b) Assume that $a \ne 0$. (Don't assume that $c \ne 0$.) Eliminate the variable t in the two equations above to find one equation in x and y. What line does it describe?

   (c) Assume that $c \ne 0$. (Don't assume that $a \ne 0$.) Eliminate the variable t in the two equations above to find one equation in x and y. What line does it describe?

   (d) What happens if a and c are both zero?

12. This problem explores the technical assumptions in the definition of *smooth* curves.

   (a) Consider the "curve" defined for all t by $x = 0$, $y = 0$. Plot C. Does C "deserve" to have a slope at $t = 0$? If so, what slope? If not, why not? Is C smooth in the sense of the definition?

   (b) Consider the curve defined for $-1 \le t \le 1$ by $x = t^3$ and $y = t^3$. Plot C. Does C "deserve" to have a slope at $t = 0$? What does Theorem 1 say in this case?

   (c) Consider the curve defined for $-1 \le t \le 1$ by $x = t$ and $y = t$. Plot C. Does this C "deserve" to have a slope at $t = 0$? What does Theorem 1 say about slope this time?

13. This problem concerns Example 11, p. 394.

   (a) How could the model be made more realistic? What additional information would be needed?

(b) Use the conditions $g''(t) = -32$, $g'(0) = 0$, and $g(0) = 7$ to show that $g(t) = 7 - 16t^2$, as claimed in the Example.

(c) Find a formula for $s(t)$, the ball's speed at time t. Plot $s(t)$ over an appropriate interval. How does the graph's shape reflect the physical situation?

14. Use parametric equations to describe the trajectory of a baseball thrown exactly as in Example 11, p. 394, except that this time the initial velocity is 100 feet per second. Plot the result over an appropriate interval. Is the pitch a strike? At what speed does it cross the plate? (The strike zone is roughly from 1.5 to 4.5 feet above the ground at home plate.)

15. Consider a baseball thrown horizontally, starting from height 7 feet (as in Example 11, p. 394), but with initial speed s_0.

(a) Explain briefly why the parametric equations

$$x = s_0 t, \quad y = 7 - 16t^2$$

describe the ball's trajectory (ignoring wind resistance).

(b) When does the ball reach home plate?

(c) Is the trajectory parabolic? Why?

16. (This problem is again about the situation in Example 11, p. 394, but takes wind resistance into account.) In practice, wind resistance *does* affect a baseball's trajectory. One possible model of wind resistance (we omit the details) leads to the parametric equations

$$x = \frac{\ln(150 \cdot k \cdot t + 1)}{k}; \quad y = 7 - 16t^2,$$

where the constant k can be thought of as the ball's "drag coefficient"—the smoother the ball, the lower the value of k. For a typical baseball, $k = 0.003$ is reasonable. With this value of k we get

$$x = \frac{1000 \ln(0.45t + 1)}{3}; \quad y = 7 - 16t^2.$$

The resulting trajectory is plotted below; the curve from Example 11 is also shown:

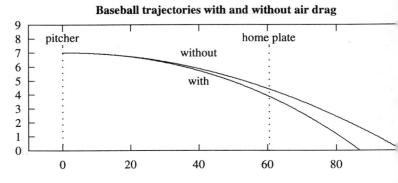

Baseball trajectories with and without air drag

In each part below, use the graphs to give approximate answers; then use appropriate formulas to improve your results.

(a) At what *time* does the air-dragged ball cross the plate? (Note: the graphs don't show time.) How much longer does it take to reach the plate than the drag-less ball?

(b) At what height does the air-dragged ball cross the plate?

(c) At what speed does the air-dragged ball cross the plate?

(d) Where does the air-dragged ball land?

17. The situation is as in the previous problem, except that this time the ball is scuffed, so that its "drag coefficient" is 0.005, not 0.003. Plot the new trajectory. Then answer the same questions as in the previous problem.

Appendix J

A graphical glossary of functions

Graphical reasoning—representing functions and their derivatives graphically—is an invaluable skill in calculus. This appendix offers a brief "atlas" of representatives of the most important classes of calculus functions. Each function is shown together with its derivative.

A linear function and its derivative

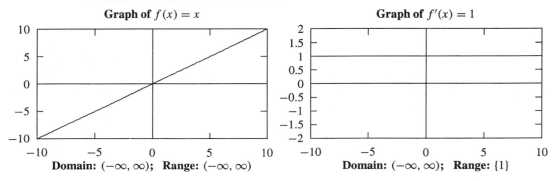

Graph of $f(x) = x$ — **Domain:** $(-\infty, \infty)$; **Range:** $(-\infty, \infty)$

Graph of $f'(x) = 1$ — **Domain:** $(-\infty, \infty)$; **Range:** $\{1\}$

Notes. The derivative of a linear function is a constant function.

A quadratic function and its derivative

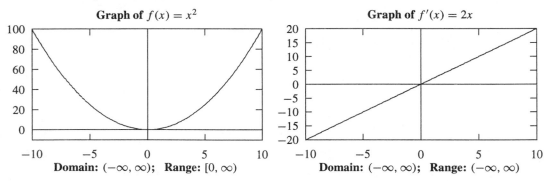

Graph of $f(x) = x^2$ — **Domain:** $(-\infty, \infty)$; **Range:** $[0, \infty)$

Graph of $f'(x) = 2x$ — **Domain:** $(-\infty, \infty)$; **Range:** $(-\infty, \infty)$

Notes. The derivative of a quadratic function is a linear function. The quadratic function's vertex corresponds to a root of the derivative.

A rational function and its derivative

Graph of $f(x) = \dfrac{x}{x^2 - 9}$

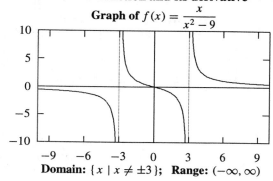

Domain: $\{x \mid x \neq \pm 3\}$; **Range:** $(-\infty, \infty)$

Graph of $f'(x) = -\dfrac{x^2 + 9}{\left(x^2 - 9\right)^2}$

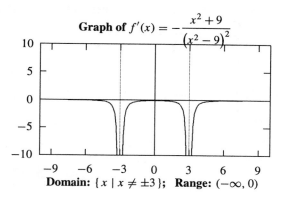

Domain: $\{x \mid x \neq \pm 3\}$; **Range:** $(-\infty, 0)$

Notes. The derivative of a rational function is another rational function. Notice the behavior of asymptotes: If f has a *vertical* asymptote at $x = a$, then so does f'. If f has a *horizontal* asymptote at $y = b$, then f' has a horizontal asymptote at $y = 0$.

The sine function and its derivative

Graph of $f(x) = \sin(x)$

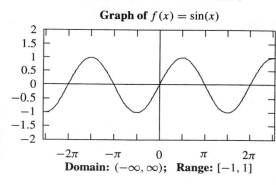

Domain: $(-\infty, \infty)$; **Range:** $[-1, 1]$

Graph of $f'(x) = \cos(x)$

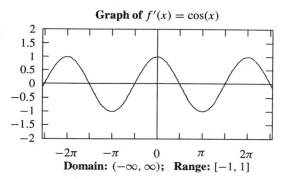

Domain: $(-\infty, \infty)$; **Range:** $[-1, 1]$

Notes. The sine function is 2π-periodic; so, therefore, is its derivative. Both the sine function and its derivative oscillate between -1 and 1. At the origin, the sine graph has slope 1.

The cosine function and its derivative

Graph of $f(x) = \cos(x)$

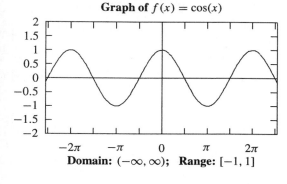

Domain: $(-\infty, \infty)$; **Range:** $[-1, 1]$

Graph of $f'(x) = -\sin(x)$

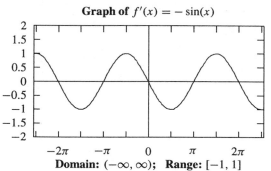

Domain: $(-\infty, \infty)$; **Range:** $[-1, 1]$

Notes. The cosine function is 2π-periodic; so, therefore, is its derivative. Both the sine function and its derivative oscillate between -1 and 1. At the origin, the cosine graph is horizontal.

The tangent function and its derivative

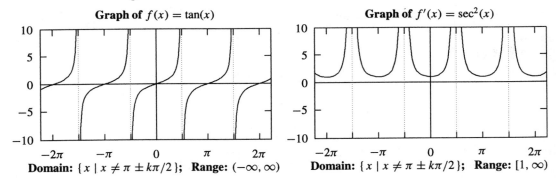

Domain: $\{x \mid x \neq \pi \pm k\pi/2\}$;   **Range:** $(-\infty, \infty)$ **Domain:** $\{x \mid x \neq \pi \pm k\pi/2\}$;   **Range:** $[1, \infty)$

Notes.   The tangent function is π-periodic; so is its derivative. Unlike the sine and cosine functions, the tangent function and its derivative have vertical asymptotes. At the origin, the tangent graph has slope 1.

Exponential functions and their derivatives

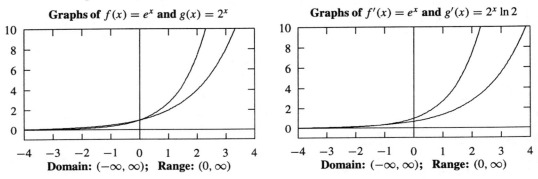

Domain: $(-\infty, \infty)$;   **Range:** $(0, \infty)$ **Domain:** $(-\infty, \infty)$;   **Range:** $(0, \infty)$

Notes.   The natural exponential function $y = e^x$ is its own derivative. Other exponential functions, such as $y = 2^x$, are *proportional* to their own derivatives. The $y = e^x$ graph has slope 1 at $x = 0$; the $y = 2^x$ graph has slope $\ln 2$ at $x = 0$. All exponential functions have horizontal asymptotes; so do their derivatives.

The natural logarithm function and its derivative

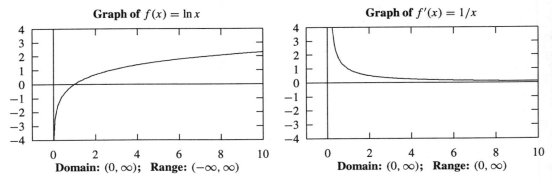

Domain: $(0, \infty)$;   **Range:** $(-\infty, \infty)$ **Domain:** $(0, \infty)$;   **Range:** $(0, \infty)$

Notes.   The natural logarithm function $y = \ln x$ is the inverse function to $y = e^x$. It is defined only for positive inputs. The graph of $y = \ln x$ has slope 1 at $x = 1$.

Index

TABLE OF INTEGRALS

Basic Forms

1. $\displaystyle\int x^n \, dx = \frac{x^{n+1}}{n+1}, \quad n \neq -1$

2. $\displaystyle\int \frac{dx}{x} = \ln|x|$

3. $\displaystyle\int e^x \, dx = e^x$

4. $\displaystyle\int b^x \, dx = \frac{1}{\ln b} b^x$

5. $\displaystyle\int \sin x \, dx = -\cos x$

6. $\displaystyle\int \cos x \, dx = \sin x$

7. $\displaystyle\int \tan x \, dx = \ln|\sec x| = -\ln|\cos x|$

8. $\displaystyle\int \cot x \, dx = \ln|\sin x| = -\ln|\csc x|$

9. $\displaystyle\int \sec x \, dx = \ln|\sec x + \tan x| = \ln\left|\tan\left(\frac{x}{2} + \frac{\pi}{4}\right)\right|$

10. $\displaystyle\int \csc x \, dx = \ln|\csc x - \cot x| = \ln\left|\tan\left(\frac{x}{2}\right)\right|$

11. $\displaystyle\int \sec^2 x \, dx = \tan x$

12. $\displaystyle\int \csc^2 x \, dx = -\cot x$

13. $\displaystyle\int \sec x \tan x \, dx = \sec x$

14. $\displaystyle\int \csc x \cot x \, dx = -\csc x$

15. $\displaystyle\int \frac{dx}{x^2 + a^2} = \frac{1}{a} \arctan\left(\frac{x}{a}\right), \quad a \neq 0$

16. $\displaystyle\int \frac{dx}{x^2 - a^2} = \frac{1}{2a} \ln\left|\frac{x-a}{x+a}\right|$

17. $\displaystyle\int \frac{dx}{\sqrt{a^2 - x^2}} = \arcsin\left(\frac{x}{a}\right), \quad a > 0$

18. $\displaystyle\int \ln x \, dx = x(\ln x - 1)$

Expressions Containing $ax + b$

19. $\displaystyle\int (ax+b)^n \, dx = \frac{(ax+b)^{n+1}}{a(n+1)}, \quad n \neq -1$

20. $\displaystyle\int \frac{dx}{ax+b} = \frac{1}{a} \ln|ax+b|$

21. $\displaystyle\int \frac{x}{ax+b} \, dx = \frac{x}{a} - \frac{b}{a^2} \ln|ax+b|$

22. $\displaystyle\int \frac{x}{(ax+b)^2} \, dx = \frac{b}{a^2(ax+b)} + \frac{1}{a^2} \ln|ax+b|$

23. $\displaystyle\int \frac{dx}{x(ax+b)} = \frac{1}{b} \ln\left|\frac{x}{ax+b}\right|$

24. $\displaystyle\int \frac{dx}{x^2(ax+b)} = -\frac{1}{bx} + \frac{a}{b^2} \ln\left|\frac{ax+b}{x}\right|$

25. $\displaystyle\int \sqrt{ax+b} \, dx = \frac{2}{3a} \sqrt{(ax+b)^3}$

26. $\displaystyle\int x\sqrt{ax+b} \, dx = \frac{2(3ax-2b)}{15a^2} \sqrt{(ax+b)^3}$

27. $\displaystyle\int \frac{dx}{\sqrt{ax+b}} = \frac{2\sqrt{ax+b}}{a}$

28. $\displaystyle\int \frac{dx}{x\sqrt{ax+b}} = \frac{1}{\sqrt{b}} \ln\left|\frac{\sqrt{ax+b} - \sqrt{b}}{\sqrt{ax+b} + \sqrt{b}}\right|, \quad b > 0$

29. $\displaystyle\int \frac{dx}{x\sqrt{ax-b}} = \frac{2}{\sqrt{b}} \arctan\sqrt{\frac{ax-b}{b}}, \quad b > 0$

30. $\displaystyle\int x^n \sqrt{ax+b} \, dx = \frac{2}{a(2n+3)} \left(x^n \sqrt{(ax+b)^3} - nb \int x^{n-1} \sqrt{ax+b} \, dx \right)$

31. $\displaystyle\int \frac{dx}{x^n \sqrt{ax+b}} = -\frac{\sqrt{ax+b}}{(n-1)bx^{n-1}} - \frac{(2n-3)a}{(2n-2)b} \int \frac{dx}{x^{n-1}\sqrt{ax+b}}$

Expressions Containing $ax^2 + c$, $x^2 \pm p^2$, and $p^2 - x^2$, $p > 0$

32. $\displaystyle\int \frac{dx}{p^2 - x^2} = \frac{1}{2p} \ln\left|\frac{p+x}{p-x}\right|$

33. $\displaystyle\int \frac{dx}{ax^2 + c} = \frac{1}{\sqrt{ac}} \arctan\left(x\sqrt{\frac{a}{c}}\right), \quad a > 0, \ c > 0$

34. $\displaystyle\int \frac{dx}{ax^2 - c} = \frac{1}{2\sqrt{ac}} \ln\left|\frac{x\sqrt{a} - \sqrt{c}}{x\sqrt{a} + \sqrt{c}}\right|, \quad a > 0, \ c > 0$

35. $\displaystyle\int \frac{dx}{(ax^2 + c)^n} = \frac{1}{2(n-1)c} \frac{x}{(ax^2 + c)^{n-1}} + \frac{2n-3}{2(n-1)c} \int \frac{dx}{(ax^2 + c)^{n-1}}, \quad n > 1$

36. $\displaystyle\int x\left(ax^2 + c\right)^n dx = \frac{1}{2a} \frac{(ax^2 + c)^{n+1}}{n+1}, \quad n \neq -1$

37. $\displaystyle\int \frac{x}{ax^2 + c} dx = \frac{1}{2a} \ln\left|ax^2 + c\right|$

38. $\displaystyle\int \sqrt{x^2 \pm p^2}\, dx = \frac{1}{2}\left(x\sqrt{x^2 \pm p^2} \pm p^2 \ln\left|x + \sqrt{x^2 \pm p^2}\right|\right)$

39. $\displaystyle\int \sqrt{p^2 - x^2}\, dx = \frac{1}{2}\left(x\sqrt{p^2 - x^2} + p^2 \arcsin\left(\frac{x}{p}\right)\right), \quad p > 0$

40. $\displaystyle\int \frac{dx}{\sqrt{x^2 \pm p^2}} = \ln\left|x + \sqrt{x^2 \pm p^2}\right|$

Expressions Containing Trigonometric Functions

41. $\displaystyle\int \sin^2(ax)\, dx = \frac{x}{2} - \frac{\sin(2ax)}{4a}$

42. $\displaystyle\int \sin^3(ax)\, dx = -\frac{1}{a}\cos(ax) + \frac{1}{3a}\cos^3(ax)$

43. $\displaystyle\int \sin^n(ax)\, dx = -\frac{\sin^{n-1}(ax)\cos(ax)}{na} + \frac{n-1}{n}\int \sin^{n-2}(ax)\, dx, \quad n > 0$

44. $\displaystyle\int \cos^2(ax)\, dx = \frac{x}{2} + \frac{\sin(2ax)}{4a}$

45. $\displaystyle\int \cos^3(ax)\, dx = \frac{1}{a}\sin(ax) - \frac{1}{3a}\sin^3(ax)$

46. $\displaystyle\int \cos^n(ax)\, dx = \frac{\cos^{n-1}(ax)\sin(ax)}{na} + \frac{n-1}{n}\int \cos^{n-2}(ax)\, dx$

47. $\displaystyle\int \sin(ax)\cos(bx)\, dx = -\frac{\cos((a-b)x)}{2(a-b)} - \frac{\cos((a+b)x)}{2(a+b)}, \quad a^2 \neq b^2$

48. $\displaystyle\int x\sin(ax)\, dx = \frac{1}{a^2}\sin(ax) - \frac{x}{a}\cos(ax)$

49. $\displaystyle\int x\cos(ax)\, dx = \frac{1}{a^2}\cos(ax) + \frac{x}{a}\sin(ax)$

50. $\displaystyle\int x^n \sin(ax)\, dx = -\frac{x^n}{a}\cos(ax) + \frac{n}{a}\int x^{n-1}\cos(ax)\, dx, \quad n > 0$

51. $\displaystyle\int x^n \cos(ax)\, dx = \frac{x^n}{a}\sin(ax) - \frac{n}{a}\int x^{n-1}\sin(ax)\, dx, \quad n > 0$

52. $\displaystyle\int \tan^n(ax)\, dx = \frac{\tan^{n-1}(ax)}{a(n-1)} - \int \tan^{n-2}(ax)\, dx, \quad n \neq 1$

53. $\displaystyle\int \sec^n(ax)\, dx = \frac{\sec^{n-2}(ax)\tan(ax)}{a(n-1)} + \frac{n-2}{n-1}\int \sec^{n-2}(ax)\, dx, \quad n \neq 1$

Expressions Containing Exponential and Logarithm Functions

54. $\displaystyle\int xe^{ax}\, dx = \frac{e^{ax}}{a^2}(ax - 1)$

55. $\displaystyle\int x^n e^{ax} = \frac{1}{a}x^n e^{ax} - \frac{n}{a}\int x^{n-1}e^{ax}\, dx, \quad n > 0$

56. $\displaystyle\int e^{ax}\sin(bx)\, dx = \frac{e^{ax}}{a^2 + b^2}\left(a\sin(bx) - b\cos(bx)\right)$

57. $\displaystyle\int e^{ax}\cos(bx)\, dx = \frac{e^{ax}}{a^2 + b^2}\left(a\cos(bx) + b\sin(bx)\right)$

58. $\displaystyle\int x^n \ln(ax)\, dx = x^{n+1}\left(\frac{\ln(ax)}{n+1} - \frac{1}{(n+1)^2}\right), \quad n \neq -1$

59. $\displaystyle\int (\ln x)^n\, dx = x(\ln x)^n - n\int (\ln x)^{n-1}\, dx$

60. $\displaystyle\int \frac{dx}{a + be^{px}} = \frac{x}{a} - \frac{1}{ap}\ln\left|a + be^{px}\right|$

Expressions Containing Inverse Trigonometric Functions

61. $\displaystyle\int \arcsin(ax)\, dx = x\arcsin(ax) + \frac{1}{a}\sqrt{1 - a^2x^2}$

62. $\displaystyle\int \arccos(ax)\, dx = x\arccos(ax) - \frac{1}{a}\sqrt{1 - a^2x^2}$

63. $\displaystyle\int \operatorname{arccsc}(ax)\, dx = x\operatorname{arccsc}(ax) + \frac{1}{a}\ln\left|ax + \sqrt{a^2x^2 - 1}\right|$

64. $\displaystyle\int \operatorname{arcsec}(ax)\, dx = x\operatorname{arcsec}(ax) - \frac{1}{a}\ln\left|ax + \sqrt{a^2x^2 - 1}\right|$

65. $\displaystyle\int \arctan(ax)\, dx = x\arctan(ax) - \frac{1}{2a}\ln(1 + a^2x^2)$

66. $\displaystyle\int \operatorname{arccot}(ax)\, dx = x\operatorname{arccot}(ax) + \frac{1}{2a}\ln(1 + a^2x^2)$